# STATISTICAL DYNAMICS

## A Stochastic Approach to Nonequilibrium Thermodynamics

# STATISTICAL DYNAMICS

## A Stochastic Approach to Nonequilibrium Thermodynamics

### Second Edition

### R. F. Streater
King's College London, UK

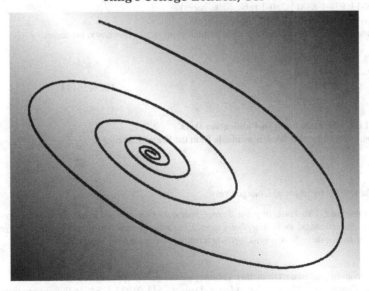

Imperial College Press

*Published by*

Imperial College Press
57 Shelton Street
Covent Garden
London WC2H 9HE

*Distributed by*

World Scientific Publishing Co. Pte. Ltd.

5 Toh Tuck Link, Singapore 596224

*USA office:* 27 Warren Street, Suite 401-402, Hackensack, NJ 07601

*UK office:* 57 Shelton Street, Covent Garden, London WC2H 9HE

**Library of Congress Cataloging-in-Publication Data**
Streater, R. F.
    Statistical dynamics : a stochastic approach to nonequilibrium thermodynamics / by R.F.
Streater. -- 2nd ed.
        p. cm.
    Includes bibliographical references.
    ISBN-13: 978-1-84816-244-0 (hardcover : alk. paper)
    ISBN-10: 1-84816-244-8 (hardcover : alk. paper)
    ISBN-13: 978-1-84816-250-1 (pbk. : alk. paper)
    ISBN-10: 1-84816-250-2 (pbk. : alk. paper)
    1. Nonequilibrium thermodynamics--Statistical methods. 2. Stochastic processes.
I. Title.
    QC318.I7S768 2009
    536'.7--dc22

                                                                    2008050134

**British Library Cataloguing-in-Publication Data**
A catalogue record for this book is available from the British Library.

Printed in Singapore.

# Preface

This book grew out of lectures at Virginia Tech. in 1989-1991, and continued at King's College London. The first half is suitable as a text for a course of one semester, being part of a programme for students of mathematics or physics pursuing a degree at the level of the M. Sc., M. Sci., M. Phys. or M. Math. The second half, which is undoubtedly harder, might be used as a course-book for a more challenging course or as a source for project work at the M. Sc. level, or as a guide to research work leading to the M. Phil.

The author is indebted to many students and friends for help: E. B. Davies, A. Uhlmann, B. Crell, B. Luffman, L. Rondoni, S. Koseki, F. Behmardi and C. Connaughton have been particularly generous. He owes the idea of 13.3 to V. Jaksic. The author was introduced to cotransport by Prof. R. J. Naftalin, of the Department of Physiology, King's College, who also kindly drew the pictures using the Stella II 2.2.1 programme for differential equations, produced by High Performance Systems.

King's College London                                                R. F. S. May 1995

# Preface to the Second Edition

This edition is larger and, it is hoped, more accurate than the first edition. The notation has been simplified. The author thanks R. Balian for his remark that information can be lost to inaccessible parts of a large system, thus leading to entropy increase of the reduced dynamics. He thanks F. Leppington for the remark that any theory of fluid dynamics without a velocity field 'has not got off the ground'. He thanks M. R. Grasselli for a critical reading of the first edition, and comments. Thanks are also due to

H. Brenner for sending him his work on diffusion in the continuity equation. Thus, the new chapter on fluid dynamics has been added. Another new chapter reinterprets the work of W. De Roeck, T. Jacobs, C. Maes and K. Netočný, on their quantum Kac model, as the proof that the von Neumann entropy increases along the approximate dynamics. The extra pages on information dynamics were made possible by visits to the Science University of Tokyo, CNRS Luminy, the University of Madeira, the Erdos Institute Budapest, the University of Sao Paulo, Torino Politecnico, and MaPhySto Aarhus. The author is endebted to M. Ohya, S. Watanabe, H. Hasegawa, F. Combe, H. Nencka, D. Petz, W. Wrezsinski, G. Pistone, P. Gibilisco, and O. Barndorff-Nielson for help and discussions. The extra work on chemistry was done while the author was at the Polytechnic, Helsinki; he thanks Prof. Nieminen for the invitation, and for correcting his version of the Arrhenius law. Any remaining errors are the responsibility of the author.

King's College London                                    R. F. S. Dec 2008

# Contents

# PART 1
# Classical Statistical Dynamics

# Chapter 1

# Introduction

Probability theory has been used in physics since the work of Maxwell, Boltzmann and Gibbs, if not before. However, many-body theory was not very successful until Planck introduced the quantum of action. Then many puzzles about black-body radiation, specific heats, and the like, were resolved. Looking today at Planck's argument we see that he used classical probability theory, even though his energy was quantised, to get the famous Planck black-body law. The same can be said of Einstein's seminal paper on stimulated emission. Even Bose's paper on the derivation of Planck's law from radiation theory is a correct use of classical probability, if we admit that we are discussing the statistics of field configurations, rather than configurations of particles. It was Einstein who noticed that what Bose had done was to use a new counting rule for identical particles (assuming that photons were particles). At first, the new quantum mechanics was regarded as a modification of *mechanics*, not probability theory. Indeed, it was not until 1933 that Kolmogorov formally defined what a probability theory was. The upshot of Bell's notable work is that the Copenhagen interpretation of quantum theory, formulated between 1925 and 1930, is not actually a consistent set of rules within classical probability, but needs a generalization, which we call *quantum probability*. This was formalised by von Neumann in his book (von Neumann, 1932) before Kolmogorov [Neveu (1965)] set up the foundations of the classical theory.

There were two reasons for the lack of success of the statistical mechanics of classical particles; the first is the continuous nature of the phase space, with the result that the statistical entropy of a distribution is infinite. In statistical mechanics, we hope to identify the statistical entropy with the experimental entropy, which is finite if the system has finite volume. We may introduce a coarse-graining by replacing phase-space by a discrete set of points, such as the set of labels for phase-cells; but then the entropy depends on the choice of coarse-graining, and increases as we make the divisions finer and finer. The second trouble with the statistical mechanics of Newtonian particles comes from their distinguishability. Thus, if there are two particles in three-dimensional space, phase-space is $\Gamma = \mathbf{R}^{12}$; the configuration $(\boldsymbol{x}, \boldsymbol{p}; \boldsymbol{y}, \boldsymbol{q})$ is taken to be a different point in $\Gamma$ from $(\boldsymbol{y}, \boldsymbol{q}; \boldsymbol{x}, \boldsymbol{p})$ (unless the points $\boldsymbol{x}$ and $\boldsymbol{y}$ coincide, as well as the momenta $\boldsymbol{p}$ and $\boldsymbol{q}$). A consequence of this was noticed by Gibbs in the early days, and is known as the Gibbs paradox. The problem arises when we try to relate the physical entropy of a gas (say) to the statistical entropy. To define the latter, we need a sample space $\Omega$ with sample points $\omega \in \Omega$, and a distribution $p(\omega)$, that is, a probability measure on $\Omega$. The statistical entropy, for a discrete space, is

$$S(p) = - \sum_\omega p(\omega) \log p(\omega).$$

The entropy is measured in bits if the logarithm is taken to base two, and could be infinite; it is always non-negative. If the sample space has a finite number of elements, $N$, then the probability giving rise to the largest entropy is the uniform distribution

$$p(\omega) = 1/N.$$

Then the entropy is $\log N$, which should be compared with Boltzmann's formula for the physical entropy of a gas in equilibrium,

$$S = k_B \log N$$

where here $N$ is the number of configurations of a given number of the molecules with a given total energy, and $k_B$ is Boltzmann's constant. For a system in the usual continuum $\mathbf{R}^3$ the number of configurations is infinite, but if we divide space into cells (and do the same in momentum space), and restrict to a finite volume, then the entropy can be calculated. Gibbs's paradox arises when we compare the number of configurations of $n$ particles in $m$ cells in phase space with the number of configurations of $2n$ particles in $2m$ cells, corresponding to twice the volume in $x$-space and the same volume in momentum space. According to thermodynamics, which is confirmed by experiment, the physical entropy of the second system is twice that of the first; we say that entropy is an *extensive* variable. Looking at Boltzmann's formula, we see that for this to be true, the number of configurations in the second system must be $N^2$, the product of the numbers in the two halves. This is just not true for Newtonian particles. Not only can each of the $N$ configurations in the first half be accompanied by each of the $N$ configurations in the second half (already giving us the $N^2$ configurations), but a particle in the first half of space can be in any cell in the second half, and vice versa; there are far more than $N^2$ configurations, and the entropy of twice the volume is more than twice that of its parts (other things being equal). Gibbs called this extra entropy *the entropy of mixing*. He noted that its presence is not observed experimentally when two gases of identical molecules are mixed. Gibbs's paradox has been clearly discussed by [Collie (1982)] and by [Amit and Verbin (1999)]. It is not so much a paradox within probability theory as a demonstration that molecules do not obey classical mechanical laws, given that we have faith in the use of probability theory. Gibbs's proposed solution to his paradox is to divide the number of configurations by suitable factorials. In modern parlance, this amounts to altering the configuration space of $n$ particles, by collecting together the $n!$ configurations that differ only by a permutation of the particles into an equivalence class; one then takes the whole equivalence class as a single point in a new sample space. The Gibbsian sample space has $N/n!$ points; it is the quotient of the Newtonian phase space on $n$ particles by the permutation group $S(n)$, which acts on it. It was not appreciated at the time, but this is tantamount to introducing Bose statistics for the particles, and using the field description of the configuration. For liquids, $n$ is about $10^{23}$ for each cubic centimetre, so the states in Newtonian physics are over-counted by a factor of $(10^{23})!$, one of the biggest errors in any theory still taught to students. No wonder that classical statistical mechanics was not able to explain even the main features of microscopic many-body systems. Statistical methods applied to Newtonian particles do give good answers for the behaviour of large numbers of ball bearings or tossed coins, which are consequently said to obey Boltzmann statistics. This is true even if the samples are so accurately similar that no difference between them can be discerned. The failure of molecules to obey Boltzmann statistics is not related to the difficulty, in practice, of following such a small individual along its path, but is due to the need for a new formulation, which we call the 'field' point of view. We describe this in the next section, and contrast it with the 'particle' point of view of Newton. This will resolve Gibbs's paradox; but the first difficulty, the lack of a natural unit of size for a cell in phase space, will not be resolved without quantum theory.

The first half of the book is devoted to giving the theory of classical statistical dynamics, based on a discrete sample space and using the field point of view. The objective is to provide a theory of nonequilibrium thermodynamics which obeys the first and second laws. The formalism is constructed in a way that makes clear that the classical theory is a special case, rather than a limiting case, of the quantum version, which is the subject of Part II. Both classical and quantum estimation theory are treated in the final chapter.

We start with Chapter 2, with some remarks on probability theory. It is emphasised that the sample space must be clearly defined; one's choice of sample space, denoted $\Omega$, is just as much part of the physics as is the choice of dynamics, and indeed must come first. The number of points in $\Omega$, for example, determines the entropy of the uniform state. We also mention the concept of *local structure*. There should be a set $\Lambda$, which can be finite or infinite, a point of which represents a point or small region in physical space; to each $x \in \Lambda$ is assigned the local sample space $\Omega_x$. A configuration of the system, a point $\omega \in \Omega$, is then a *field*, that is, a map $\omega$ from $\Lambda$ such that $\omega(x) \in \Omega_x$. This formalism avoids Gibbs's paradox, since the entropy is extensive. To make the models tractable, we choose a discrete, but possibly infinite, sample space at each $x \in \Lambda$. In this context, marginal probabilities and conditional expectations are entirely elementary, and can be introduced gently. The idea of a random variable, often disguised as a mystery in books on Physics, is simply a real function on $\Omega$. To get a model, we must specify a random variable, $\mathcal{E}$, the energy; this divides $\Omega$ into energy-shells $\Omega_E$; this is the set of all sample points with energy $E$. We shall always assume that $\Omega_E$ is finite; this is needed for thermodynamic stability, in that the microcanonical state, known to physicists as an 'ensemble', then exists. The assumption of finiteness also enables us to avoid measure theory and the many niceties of functional analysis, until the final two chapters of the book.

The bounded random variables form an algebra, denoted by $\mathcal{A}$; its dual, the set of linear functionals on this algebra, contains the positive cone. The normalised part of this is the set of possible states of the system. We shall limit our attention to the set $\Sigma$ of *normal* states, each given by a (countably additive) probability measure. In the finite case we can forget this refinement, and the duality between states and algebras is mutual. The characteristic of Kolmogorov's probability is that the algebra of observables is *abelian*, that is, commutative. For this reason we call the first half of this book the *classical* theory, even though there are not necessarily Newtonian particles present. The discrete sample space leads to a discrete energy-spectrum, so we are able to incorporate this aspect of quantum theory without which a classical theory would give very poor results. The main mathematical tool is convexity, which plays a large role in the entropy-estimates, the paradigm of which is Kullback's inequality.

In Chapter 3 we define reversible dynamics as a group $\{\tau_t\}$ of invertible maps on $\Omega$, labelled by discrete time $t \in \mathbf{Z}$. Here $\mathbf{Z}$ denotes the group of all integers. We prefer the term reversible to 'conservative dynamics', since all our dynamics, including dissipitative ones, will obey the first law, conservation of energy. This includes the more interesting cases called dissipative, which are non-reversible. Our tenet is that it is *information* that is dissipated, not energy. We shall see that reversible dynamics is given by a linear operator $\tau_t$ on $\mathcal{A}$; it is not necessarily given by linear equations, as in a noninteracting system; indeed, it gives rise to a group of automorphisms of $\mathcal{A}$. The dynamics will be limited to energy-conserving maps. Thus $\tau_t$ maps $\Omega_E$ bijectively onto itself. This allows us to extend the duality theory (valid if $|\Omega| < \infty$) to all cases of interest to us, even if $|\Omega| = \infty$. Here we get our first success: we can construct models in which the mean energy is independent of time (the first law). We introduce random dynamics by forming the convex sum of reversible dynamics. This is a generalisation of the construction of ergodic means and leads to the class of bistochastic Markov chains with discrete time. These have the Markov property: the state at time $t + 1$ depends on the state at time $t$, but not on the history, that is, the state at all times before $t$. Not only does the discrete time lead to a readily computable theory and avoids some hard questions on the existence and uniqueness of solutions to non-linear equations, but the positive size of the time-step mitigates the Markov assumption. It may be a poor approximation to real systems to assume the Markov property if time is continuous, but if the time-step is chosen to be large on the time-scale of the fast variables, it should give a good account of the dynamics of the slow variables. This division into fast and slow variables is well explained in [Balian (2006a,b)]. The present theory is not designed to describe the phenomenon of hysteresis, an effect that depends on the history of the state. If hysteresis occurs experimentally, it could be argued that the description of the state contains too few slow variables; a Markovian description might well be possible if more detail of the state is retained. The dynamics given by a mixture of reversible maps is entropy-increasing, and so we get models obeying the second law. Here we meet the first 'ensemble', the microcanonical state, and show that it is the limit of any ergodic bistochastic process if the initial state has a sharp energy. This uses Lyapunov's direct method, which is explained simply. We illustrate with models of discrete diffusion and of reacting particles.

In this chapter, we note that random variables form the algebra, $\mathcal{A}$, being functions of the sample point; probabilities are also functions of the sample point, but are limited to positive values. In this sense, state-space is the normalised positive cone inside $\mathcal{A}$; the time-evolution of the states is the inverse of the time-evolution of the observables. More generally, if inverses do not exist, the time-evolution of the states is the transpose of the action on the observables. The same is true of any other symmetry group. This can be summarised as follows: the algebra $\mathcal{A}$ has two actions by the symmetry group, a left-action and a right-action. The left-action is the way observables transform, and the right-action is the way the states transform. Thus, the algebra is a *bimodule* for the group of symmetries. This structure is also true for quantum mechanics, where there the algebra is non-abelian.

In Chapter 4 we introduce the Boltzmann map. This is an abstract version of the Stosszahlansatz of Boltzmann. Given any algebra $\mathcal{A} = \mathcal{A}_1 \bigotimes \mathcal{A}_2$ (a tensor product), we can replace any state $p$ on the algebra $\mathcal{A}$ by the product of its marginal states, relative to the algebras $\mathcal{A}_1$ and $\mathcal{A}_2$. This is the state of greatest entropy having the same marginal distributions as $p$. It follows that the map increases the entropy (or leaves it the same if it was already a product). This was the step in Boltzmann's derivation of his equation that caused most controversy. We do not attempt to prove the validity, or approximate validity, of the step in this chapter; rather, the step is added to the dynamics to model randomness which increases entropy but does not alter the mean energy, and which is not accounted for by the random reversible dynamics. The increase in entropy is caused by the fast variables' having correlations between the factors $\mathcal{A}_1$ and $\mathcal{A}_2$ involved in the map. The resulting dynamics is non-linear on the space of states, which causes worry to some people. In particular, the powerful and popular theory of Markov chains, briefly described in Chapter 3, does not apply. But, as with Boltzmann's original equation, the non-linearity is necessary if the isolated system is to converge to equilibrium at its own temperature. It is the Boltzmann map that mixes up the different energy-shells, and leads to the canonical state at large times. A further simplifying map, called the *LTE*-map, standing for *local thermodynamic equilibrium* is introduced. This increases the entropy still further, by replacing the state after one time-step by the product of states which, on each local algebra, are in a canonical state relative to the local energy. After the *LTE*-map the state is independent at each site $x$. This map conserves energy only if there is no energy of interaction between sites. In that case it leads to very tractable equations of motion, in which the state is parametrised by a field of intensive variables, such as the beta and chemical potential, far simpler than the detail needed to specify a general state. The non-linear equations then converge to equilibrium. This is illustrated with simple models of chemical kinetics, and a free gas in a gravitational field. This last model allows us to give a natural definition of the pressure.

Then we introduce an idea possibly even more controversial than non-linearity: the heat-particle. This is a convenient way to incorporate the first law of thermodynamics, which requires that the total energy, including heat, must be conserved by the dynamics. In statistical mechanics, heat is most often identified with random kinetic energy. More precisely, it is any form of energy that has been randomised. By introducing a new degree of freedom explicitly to carry the heat, we mitigate the omission of all velocity variables, such as would occur in the particle picture. Then we can formulate a stochastic law that conserves total energy. It is one of the tenets of statistical dynamics that in irreversible dynamics it is *information* that is dissipated as time progresses, not energy; energy simply moves between the different forms in which it can appear. In the simplest models there is just one heat-particle, which represents the shared, thermalised kinetic energy of the fields in the system. In more detailed models, the heat-particle is located on the bond between two molecules, and represents the energy of vibration of the bond, and the shared electromagnetic energy contained in the photons moving between them, and the thermalised phonons. The chapter ends with a general theory of chemical reactions. Chemical kinetics, started in 1850 in [Wilhelmy (1850)], has been an active field since 1864, when the law of mass action was formulated by [Guldberg and Waage (1864)] in the form we learn at school: the rate of a reaction is proportional to the product of the concentrations of the reactants. Later, it was recognised that this could not be completely correct; the problem is that if the forward reaction rate is in balance with the backward rate, so that we are in equilibrium, then we get an equation between products of concentrations, and this equation is not in general satisfied by the canonical or grand canonical state. So the law of mass action was rewritten, so that the rate was proportional to the product of the *activities* of the reactants, rather than their concentrations. This fitted the data even better than the original form of the law, and, in addition, the grand canonical states are among its fixed points. The original law became known as the 'density-led' form, and the new form was known as the 'activity-led' version. Part of the Chapter is used to show that the activity-led theory can be fitted into statistical dynamics by a suitable choice of sample space, energy-function, and bistochastic map. It is clear from this construction that not all kinetic equations must be activity-led. A simple but quite general set of models for chemical reactions, based on the existence of molecules with hard cores of the same size, is shown to lead to activity-led equations if we make a particular assumption about the reaction rates. The good properties of the dynamics (energy-conservation, increasing entropy, convergence to equilibrium) then follow from our general theory.

In Chapter 5 we introduce the concept of isothermal dynamics. This arises from the isolated dynamics by adjusting the state after each time-step so that the heat-particle is in thermal equilibrium at a specified temperature. The motion then reduces to a linear dynamics through the states of the system, and so the theory can again be described by a linear stochastic process. The adjustment to the state after each time-step models the heat-flows that must occur to keep the temperature constant. We show that this is related to the Legendre transform. The theory is applied to chemical reactions, and we quickly obtain the famous results of chemical kinetics: the principle of detailed balance, which says that for closed systems of chemical reactions at equilibrium each chemical reaction is separately in equilibrium. Moreover the *'law' of Arrhenius* (which says that the rate of an isothermal reaction depends on the temperature according to the Boltzmann factor $e^{-\beta E}$, where $E$ is the energy of activation, and $\beta = 1/(k_B \Theta)$, where $\Theta$ is the temperature and $k_B$ is Boltzmann's constant) is a consequence of the simplest possible dynamics; however, this is not a property of the general theory, and so the law of Arrhenius is not a law of chemistry. Nevertheless it *can* be shown that the ratio of the forward and backward rates in isothermal chemistry is proportional to $e^{\beta(E_2 - E_1)}$ as $\beta$ varies; here, $E_1$ and $E_2$ are the activation energies of the forward and backward reactions. The main theorem is that the free energy decreases along the orbit, and so is a Lyapunov function. This contrasts with the isolated dynamics, in which the entropy increases, and is the Lyapunov function in that case. The difference between these is the heat that must be put in or taken out to maintain constant beta, and indeed we derive the axiom of thermodynamics, that entropy changes by $\geq \beta dQ$ when heat $dQ$ is added at beta $\beta$. We use the free-energy theorem to prove the 'fundamental theorem of chemical kinetics', which states that isothermal chemical reactions converge to thermal and chemical equilibrium at large times.

We end Chapter 5 with a convergence theorem and a structure theorem. The convergence of systems to equilibrium is shown to occur in the very strong sense defined by a large norm. We also state the law of 'equivalence of ensembles'. This says that the state of a subsystem of fixed size is close to a canonical state if the full system is large and is in a microcanonical state. In our version, the 'closeness' is measured by the norm. This should not be construed as saying that the isothermal and isolated dynamics are nearly the same; as every chemist knows, isothermal dynamics needs carefully controlled heat-flows through the boundary. The distinction between the microcanonical and the canonical state disappears in the thermodynamic limit: it is one of the casualties of the idealisation implicit in the infinite-volume theory. This makes the boundary conditions difficult to specify, and some ambiguity in the dynamics seems inevitable. Ingarden [Ingarden (1992)] says 'At present, many physicists think that this assumption [that the volume is infinite] is the only possible one for thermodynamical and statistical theories. Such an extreme point of view seems to be, however, absurd.'

The structure theorem says that every Markov chain with positive fixed point, and small time step, can be regarded as the isothermal dynamics of the system coupled to its heat-particle. This shows that the theory, and in particular the introduction of the heat-particle, leads by passing to its isothermal version to the most general Markov chain with the canonical state as its fixed point.

In Chapter 6 we consider driven systems; these are open in the sense of chemistry, in that particles can flow in and out, in order to keep the chemical potentials of certain chemicals fixed. These are called the driving chemicals. In driven systems the fundamental theorem of chemical kinetics does not apply. This is illustrated by heat diffusion through a lattice gas. There is a brief introduction to linear stability theory and the Lyapunov indices. It is applied to a simple version of the Brussellator of Prigogine, both with and without noise. There is also a numerical study of this model. The noise in the model is *thermal* and we show how to include it in the equations. It is shown by example that noise can destroy a limit cycle.

Chapter 7 is entirely new. We generalise the methods of statistical dynamics to include a transition probability $T$ which itself depends on the state $p$, following [Alicki and Messer (1983)]. We then set up a lattice model of hydrodynamics, in which the kinetic energy of the particles is included; that is, it is not replaced by heat particles. We show that entropy increases with time (or stays the same) for the lattice model. By taking the limit as the size of the lattice goes to zero, we get the Euler equations. We show that the Euler equations on the lattice do not lead to any production of entropy, provided that the pressure is taken to be the equilibrium pressure at each point of the lattice, equal to that predicted by the model, rather than that given by Boyle's law, $R\Theta/V$.

There follows a treatment of dissipative fluid dynamics; we suggest that a more useful system than the usual Navier-Stokes equations is obtained by modifying the continuity equation by adding a dissipative term such as $\lambda\nabla^2\rho$ or $\lambda\left(\nabla \cdot \rho^{-1}\nabla\left(\Theta^{\frac{1}{2}}\rho\right)\right)$, where $\lambda > 0$; this arises if the system is not exactly in a Maxwell-Boltzmann Gaussian state. Suitable changes to the heat equation and the Navier-Stokes equation are then necessary. We show that these modified equations are invariant under Galilean transformations; this is in contrast to the view that Galilean invariance leads to the usual (dispersion-free) continuity equation. We argue for the view that for any smooth initial conditions of compact support our new system has smooth strong solutions for all time. These solutions are not expected to remain finite for all times in the limit as $\lambda \to 0$. Our new theory is in contrast to the usual Navier-Stokes equations; for them, while there exist smooth strong solutions with small-enough initial conditions, it is not known whether or not there exist strong solutions, satisfying general smooth initial conditions, that remain finite for all time.

The classical methods of Part I have some success, and are the guide for a similar treatment of a theory based on quantum mechanics, which is treated in Part II.

# Chapter 2

# Probability Theory

## 2.1 Sample Spaces and States

The sample space of the system will be taken to be a countable space $\Omega$, a point of which will represent a configuration. A point in $\Omega$ will not represent a point in phase space of a Newtonian particle, because of the difficulty with Gibbs's paradox, explained in the Introduction. In the simplest models, the velocity or momentum of the particle will not be explicitly mentioned; the kinetic energy will be represented in an average way by the energy of the heat-particle, or by a label attached to the site where the particle is. Only in Chapter 7, on fluid dynamics, will we discuss a model with enough detail to show the momentum of the system.

Space will be approximated by a finite space $\Lambda$, and might have a regular structure such as the set of vertices of a periodic crystal. A subset $\Lambda$ of the cubic lattice in three dimensions, $\mathbf{Z}^3$ will be our usual choice. We want the theory to have a local structure, which means that at each point of space, $x \in \Lambda$, we are given the sample space $\Omega_x$. This is the set of values that the field or fields at $x$ can take. Then we require that

$$\Omega(\Lambda) = \prod_{x \in \Lambda} \Omega_x. \tag{2.1}$$

This notation means that an element $\omega$ of $\Omega$, that is, a configuration of the field, is given by a labelled set $\{\omega_x\}$ as $x$ runs over $\Lambda$, with each $\omega_x \in \Omega_x$. If all the sample spaces $\Omega_x$ are the same, say $\Omega_x = \mathbf{N} = \{0, 1, 2, \ldots\}$, then the configuration $\omega$ is a map $\omega : \Lambda \to \mathbf{N}$. In this example, if $\Lambda$ have $N$ points, then $\omega$ is a finite sequence of natural numbers $\{\omega_1, \omega_2, \ldots, \omega_N\}$. The non-negative integer $\omega_x \in \Omega_x$ could represent the number of molecules of a chemical that sit at $x$, or, lie in the volume labelled by $x$. We see that $\omega$ is indeed a map from $\Lambda$ to $\mathbf{N}$. In this case $\omega$ is a configuration of the number

13

field. This specification of the way particles are distributed is called the *field point of view*. It should be contrasted with the *particle point of view*; in the same example, with $|\Lambda| = N$, a configuration of $n$ particles is given by, in turn, considering the first particle, and then specifying its location, at $x_1 \in \Lambda$, then considering the second particle, and specifying its location, at $x_2 \in \Lambda, \ldots$. Thus, from the particle point of view, a configuration is an ordered sequence of $n$ elements of $\Lambda$. This is a different space from that given by the field point of view; see Exercise 2.1. In particular, the space $\Omega(\Lambda)$ of (2.1) obeys the product law: if $\Lambda_1$ and $\Lambda_2$ are disjoint subsets of $\Lambda$, then

$$\Omega(\Lambda_1 \cup \Lambda_2) = \Omega(\Lambda_1) \times \Omega(\Lambda_2). \tag{2.2}$$

In Exercise 2.40 you could prove this when $\Omega$ is finite, and show that it is not true in the particle point of view.

The product structure of the sample space is a feature of several very successful models of equilibrium statistical mechanics, such as the Ising model, interpreted as a gas of indistinguishable particles on a lattice [Simon (2005)]. The question then arises how to describe the dynamics of such a model, so that it approaches equilibrium at large times. This question was first considered by Glauber [Glauber (1963)]. His proposal, now called the Glauber dynamics, does not involve any momentum for the particles. Our sample space is constructed in the same spirit, and we shall be able to describe diffusion in such a model. In Chapter 7, we include velocities in the description of the local sample point, and obtain a model showing convection as well as diffusion.

In a probabilistic theory, in general we do not know the configuration of the system precisely at a given time; the best we can do is to describe the possible configurations by a probability. Thus, we describe the *state* of the system at a given time by a probability measure on $\Omega$, typically denoted $p$:

**Definition 2.1.** A probability measure on $\Omega$ is a function $\omega \mapsto p(\omega)$ obeying

(1) $p(\omega) \geq 0$
(2) $\sum_\omega p(\omega) = 1$.

If $n$ is the number of points in $\Omega$, thus $n = |\Omega|$, then it is common to write $p$ as a row-vector with $n$ components, and we shall do this:

$$p = (p(\omega_1), p(\omega_2), \ldots, p(\omega_n)) = (p_1, \ldots, p_n).$$

The set of all probability measures on $\Omega$ forms the set $\Sigma(\Omega)$, called the *state space* of the model. We shall use the word state to mean a probability measure. If we have complete information about the system, for example that the configuration is $\omega_0$, then the state is the point measure $p(\omega) = \delta(\omega, \omega_0)$. At the other extreme, if we know nothing about the configuration, and $\Omega$ is finite, then it is sometimes reasonable to postulate that the state is the uniform measure $p(\omega) = 1/|\Omega|$; here, $|\Omega|$ denotes the number of points in $\Omega$.

The use of probability theory in physics can lead to long and heated debates about philosophy. Historically, many people were worried by Boltzmann's work. This debate should not be surprising, since the theory of probability itself was not formalised until much later. To explain probability, Gibbs asked his readers to imagine all possible configurations of the system laid out on a table; together, these form the 'ensemble'; the actual system will be one of these, picked out blindly at random. He had in mind that each configuration was equally likely. We recognise the ensemble as the sample space, provided with the uniform probability. Other states, with non-uniform probability on the sample space, are not so easy to explain using the ensemble. Gibbs was able to formulate mixing (ensembles of ensembles) and conditioning (subensembles), but there remained some confusion between the ensemble of atoms in a gas (an actual large collection of atoms) and the ensemble of configurations of the gas as a whole. In statistical dynamics, it is imperative to choose the sample space firmly at the start, as part of the specification of the model. We shall see that there is no sample space in quantum probability, and so the Gibbsian point of view can become even more confusing. The 'many-worlds' interpretation of quantum mechanics is a modern relic, a brave (or foolhardy) attempt to rescue the notion of an ensemble in quantum mechanics. Balian [Balian (2006b)] uses the word ensemble to mean an indefinite collection of independent, identically prepared physical samples of the system, and thus, all in the same state. In this book, we replace 'ensemble' with the mathematical notion, 'state'.

There are three philosophies concerning what choice of probability, $p \in \Sigma(\Omega)$, we should make to describe the state at hand. The first, used in some books on statistics, is the *frequency* definition. The second uses the notion of a *prior* distribution, and has gained ground with the popularity of information theory. The third considers the state $p$ to be part of the model, along with the sample space and the choice of dynamics.

In the frequency definition, we do not commit ourselves as to $p$ until we have made several tests on the system. We take it that the system is reproducible in the same state as often as necessary. We find $p(\omega_1)$ for example by counting the number of times the configuration $\omega_1$ occurs in a long run, and dividing it by the total number of trials. The method is rather deficient if we want to use it as the *definition* of $p$. It is difficult to think of one theorem that can be proved using this idea, which depends on an infinite sequence of experiments on a physical system.

In the definition using prior knowledge, we take it that in the absence of any knowledge as to what the configuration is, the 'correct' probability distribution is the uniform one. Bernoulli and then Laplace used this in their work. It only makes sense when the sample space is finite; for example if we try to use the uniform distribution when $\Omega$ is continuous, the paradoxes of Bertrand are very puzzling [Woyszynski and Saichev (1996)]. Suppose we are in the finite case, and we start with a known sample space but do not know the state. We therefore postulate the uniform distribution as the best (or rather, least bad) guess. Suppose we now get some information about the system. We must distinguish between two ways in which this can happen. In the first type, the sample point has already been chosen, but we do not know which one it was. Imagine a punter with the problem of laying a bet on a throw of a die, which has already occurred, but in another room. The chance is $1/6$ that a particular number $\omega$ has occurred. If he finds out that the number was in fact even, then taking into account this information, the new improved probability is zero for odd $\omega$, and $1/3$ for even $\omega$. More generally, the rules for conditioning allow us to modify the uniform distribution, or any other prior distribution, in a systematic manner, to take account of new information. This leads to the school of statistics known as Bayesian, whose methods are used in operations research, decision theory and planning. This type of information, in which some hints are given as to which sample point has occurred, should not be confused with a second type of knowledge, in which we are not dealing with one sample, but the whole sample space. Suppose that we do not know what the probability is, but that we have measured the mean of one or more random variables by examining a large number of samples. Information theory then tells us how to best choose the prior distribution, given these mean values. If no variables have been measured, then the method gives the uniform distribution as its best prediction. If the mean of at least one random variable is known, the method leads to analogues of the canonical states occurring in statistical physics. These states are not conditioned in the true sense. They can be modified by conditioning if information about the sample point is acquired, as in Bayes's theorem, and nowadays the whole theory is known as the Bayesian method. It has been introduced into physics and chemistry by Jaynes [Jaynes (1957)]. One can counter the criticism, that the hypothesis of equal probabilities (in the absence of information) is purely arbitrary, with the question 'can you suggest anything better?'. All we can say is that people with no information should not be taking decisions. We shall see that the uniform distribution is the one with the largest entropy (and least information, which is negative entropy); this is why information theory starts with this as the prior distribution. Some people find this a completely convincing justification; in Chapter 15 we shall justify Jaynes's method by the modern method of estimation theory.

A finite sample space $\Omega$ is often obtained by setting up cells in a continuous space; this coarse graining is then quite arbitrary, and the corresponding discrete uniform distribution is no more special than another based on a different division into cells. The Bayesian method can be quite misleading for the application of probability in physics. Consider for example the problem of setting up a theory of the interior of a distant star, about which we have no information except that it is composed of protons and electrons. Suppose that we set up a large but finite sample space $\Omega$ where a configuration expressed, for each point $x$ in the star (where $x$ runs over a finite set $\Lambda$) whether the site were occupied or not. Suppose that we formulate some dynamics for the system. We are then faced with the problem, what is the initial state of the system? In the absence of information, a Bayesian would say: take $p$ to be the uniform distribution on $\Omega$. Now, such a theory might well be able to predict that the star will collapse, or will explode, depending on the initial state. Say, that if the initial state is uniform, it will collapse. Thus a Bayesian with no further information would predict that it will collapse: this would be his best estimate. But since the initial state could have been anything, this conclusion is not justified. We could seek more information about the star, such as its brightness. This would allow the Bayesian to condition his probability, and he might now predict that it will not collapse, but will explode. We cannot believe that the puny human's knowledge, obtained by measuring the brightness, makes any change whatsoever to the state of the star, since it is large and far away. So we conclude that the uniform state was not a good description of the situation. A physical theory cannot do better than allow us to work out which $p$ lead to a collapse, and which $p$ lead to explosion. Physics is not in the gambling business, and is not obliged to hazard a guess as to what the initial state was. Another example of the odd conclusions that can be got from the uniform distribution is that all systems must be in equilibrium. Under this assumption, the equilibrium distribution is not only the most probable one, but the chance of any discernable deviation from it is astronomically small; Grandy [Grandy (1992)] says that the probability of being in equilibrium is 'overwhelmingly' large. If it is so overwhelming, how come systems still manage to exist out of equilibrium for minutes, hours, years at a time? The answer is that the overwhelming probability calculated here is based on the assumption that all states are equally likely, which is not true if we are given an initial state far from equilibrium. The idea that equilibrium states are the only possible ones has become part of the folklore of statistical physics, but it is a non-starter in a theory such as statistical dynamics, which is trying to describe non-equilibrium phenomena.

The third philosophy about probability is more recent than the Bayesian, but does use ideas from it, as well as from the frequency definition. It is to regard the choice of $p$ as part of the model, along with the choice of $\Omega$ and the dynamics. We are led to this point of view by comparison with quantum theory, where the state of the system is not regarded as something one can predict *a priori*, by pure reasoning. Thus any $p \in \Sigma(\Omega)$ is a possible state of the system (at the time in question), and we do not attempt to give preference to some over others without recourse to experiment. In particular, we do not use the idea that there is an *a priori* probability that the probability is $p$. The procedure is as follows: we decide on $p$ as the *hypothesis H*, and then test the theory by the method of significance tests; we reject the model if it fails the test. We might use the first method, frequency, to help us with the hypothesis. This comes under the legitimate subject of estimation theory. Bayes's rule can also be used if more information is available; it is the correct way to incorporate information. It should be kept in mind that the conditioned probability is still part of the model under the hypothesis H. This is the philosophy we shall adopt in this book.

We are led to define a probabilistic model as a pair $(\Omega, p)$, where $\Omega$ is a countable space and $p$ is a probability on $\Omega$.

**Definition 2.2.** Let $\Omega$ be a countable space. We say that a function $p : \Omega \to \mathbf{R}$ is a probability if

(1) $p \geq 0$
(2) $\sum_{\omega \in \Omega} p(\omega) = 1$

This is less elaborate than Kolmogorov's definition of a probability space, which involves a $\sigma$-ring consisting of measurable sets and a countably additive measure $\mu$ [Neveu (1965)]. However, we do not need this much generality. Suffice it to say that each subset $\Omega_0$ of $\Omega$ is measurable, and represents an *event*. We say that the event $\Omega_0$ has occurred if the sample point $\omega$ chosen lies in $\Omega_0$. The probability that the event $\Omega_0$ occur is given by the measure of the set $\Omega_0 \subseteq \Omega$, which is

$$\text{Prob}\{\Omega_0\} = \sum_{\omega \in \Omega_0} p(\omega).$$

You are asked to prove that Prob is countably additive in the exercises. Conversely, every countably additive measure Prob obeying $\text{Prob}\{\Omega\} = 1$ defines a probability by

$$p(\omega) = \text{Prob}\{\omega\}$$

where $\{\omega\}$ is the set containing the single point $\omega$.

We now give some important examples of states:

- Point measure at $\omega_0$:
$$p(\omega) = \delta(\omega, \omega_0) := \begin{cases} 1 & \text{if } \omega = \omega_0 \\ 0 & \text{if } \omega \neq \omega_0 \end{cases}$$

- Uniform state: here $|\Omega| = n < \infty$, and for all $\omega$, $p(\omega) = 1/n$.

- Microcanonical state: here $|\Omega|$ can be infinite; we are given an energy function $\mathcal{E}$ on $\Omega$ such that the *level sets* $\Omega_E$ of $\mathcal{E}$ are finite. $\Omega_E$ is the *energy-shell*, that is the set of points $\omega$ where $\mathcal{E}(\omega)$ takes the value $E$. Let $|\Omega_E| = n(E)$. Then the microcanonical state with energy $E$ is
$$p(\omega) = \begin{cases} 0 & \text{if } \omega \notin \Omega_E \\ 1/n(E) & \text{if } \omega \in \Omega_E \end{cases}. \qquad (2.3)$$

- The canonical states: here the energy function $\mathcal{E}$ is taken to be bounded below. Let $\beta$ be a real number, and suppose that $Z_\beta = \sum_\omega \exp(-\beta\mathcal{E}(\omega))$ is finite. In this case, $Z_\beta$ is called the *canonical partition function*. The canonical state is then
$$p_\beta(\omega) = Z_\beta^{-1} \exp\left(-\beta\mathcal{E}(\omega)\right). \qquad (2.4)$$

We note that if $\Omega$ is infinite, then $\beta$ is necessarily positive.

- The grand canonical states: here we are given an energy function $\mathcal{E}$, and another function $\mathcal{N}(\omega)$ which takes the values 0, 1, .... Let $\beta$ and $\mu$ be real numbers such that
$$Z_{\beta,\mu} = \sum_\omega \exp\left(-\beta\{\mathcal{E}(\omega) - \mu\mathcal{N}(\omega)\}\right) < \infty.$$

In this case, $Z_{\beta,\mu}$ is called the *grand partition function* and $\mu$ is the *chemical potential*. The grand canonical state is then
$$p(\omega) = Z_{\beta,\mu}^{-1} \exp\left(-\beta\{\mathcal{E}(\omega) - \mu\mathcal{N}(\omega)\}\right). \qquad (2.5)$$

Please look at Exercise 2.39 at this point. The parameter $\beta$ is related to the inverse temperature: $\beta = 1/(k_B\Theta)$, where $k_B$ is Boltzmann's constant. The parameter $\mu$ is called the chemical potential.

There are other interesting states, the great grand canonical states with several numbers $\mathcal{N}$, that we shall meet later. One of our problems is to show why these arise as equilibrium states of systems, and why many systems settle down to one of these at large times, and what determines which one.

The set of states, $\Sigma$, is a convex set. It can be regarded as a subset of the set of all real functions on $\Omega$; this set is a real vector space. The concepts of linear sums, and of linear independence are well defined for functions, and so have a meaning for states. If $p$ and $q$ are states and $\lambda \in [0,1]$, then $\lambda p + (1 - \lambda)q$ is a function of $\omega$ which obeys (2.1), and so lies in $\Sigma$. Geometrically, this means that if $p \in \Sigma$ and $q \in \Sigma$ then the whole segment of the line joining $p$ and $q$ lies in $\Sigma$. This is the definition of a convex set. Please prove that $\Sigma$ is convex, and do Exercise 2.45. For any $n$, if $p_1, p_2, \ldots, p_n$ are linearly independent states, then the set of linear combinations with positive weights

$$p = \sum_j \lambda_j p_j, \qquad \text{where } \lambda_j \in [0,1] \text{ and } \sum_j \lambda_j = 1, \qquad (2.6)$$

form a closed convex set, which is a simplex. The set $\Sigma(\Omega)$ is also a simplex. The meaning of simplex is the following. Given any convex subset $K$ of a vector space, we say that a point $p \in K$ is an *extreme point* of $K$ if it cannot be written as $p = \lambda p_1 + (1 - \lambda)p_2$ for any pair of different elements $p_1$, $p_2$ of $K$, where $0 < \lambda < 1$. We say that a convex set $K$ is generated by its extreme points if $K$ is the smallest closed convex set containing the extreme points of $K$. When $K = \Sigma(\Omega)$ with $\Omega$ a countable set, we can prove this, which is a special case of the Krein-Milman theorem. For, a probability $p$ can obviously be written as

$$p(\omega) = \sum_{j \in \Omega} p_j \delta(\omega, j).$$

It follows that $p$ is extremal if and only if it is a point measure; also, this decomposition of $p$ into point measures is unique (the coefficients $p_j$ are unique). Non-extremal points of $\Sigma$ are called *mixed states*. We say that a convex set $K$ is a simplex if the decomposition of any point in it into extreme points is unique. Thus, we have proved that $\Sigma(\Omega)$ is a simplex. Note that when $K$ is not the whole of $\Sigma$, the extreme points of $K$ might not be point measures, since extreme points are limited to belong to $K$. An extreme point of $K$ might not be an extreme point of a larger convex set containing $K$, as can be seen by choosing $K$ to be a line, and extending one end to form a larger convex set. Please look at Exercise 2.47.

The interpretation of the act of forming the mixture of the states $p$ and $q$ with weights $\lambda, 1 - \lambda$ is given by considering the states of the system got by physically mixing the states $p$ and $q$ with probabilities $\lambda$ and $1 - \lambda$. One way to describe the statistics of the mixture is as follows: we start with two systems, one in the state $p$ and the other in the state $q$. We put each in a black box, and juggle them up, and select $p$ with probability $\lambda$ and $q$ with probability $1 - \lambda$. Whereas this does give the same statistics as the state $\lambda p + (1 - \lambda)q$ on the space $\Omega$, there is no real mixing of the physical systems, which remain separated. The physical development in time of the two black boxes will be different from that of the state $\lambda p + (1 - \lambda)q$, regarded as the state of a single system in one container. For example, if the two black boxes are Thermos flasks, each system will approach equilibrium in time at its own temperature, but if they are in a single container, the limit state will have a single, common temperature. We are led to conclude that not all physical information is contained in the state; we need to specify the dynamics, just as in a reversible theory the Hamiltonian must be given. The local dynamics is determined physically by what type of molecule we have, and this should be the same whether we put the molecules in one container or two; but the walls of the container should be specified (whether permeable to particles, or heat conducting, or not). These properties are usually modelled by imposing boundary conditions on the dynamics, and this will be how it is done in statistical dynamics. The boundary effects are casualties of the idealisation involved in the thermodynamic limit; by taking $\Lambda$ to become infinitely extended, we take it that there is only one container. The mixing obtained by considering two unrelated black boxes is an every-day sort of mixing, coped with by ordinary probability theory. This is to be contrasted with the more intimate mixing of two states $p$ and $q$, obtained by considering $\lambda p + (1 - \lambda)q$ as a state of the system regarded as being in the one and the same container as the original system. This is the interpretation we shall put on this mixture, but how it is achieved physically is not going to be discussed.

The real vector space, called $\ell^1(\Omega)$, spanned by the states on $\Omega$ can be easily shown to consist of functions of the form $f = \alpha p - \beta q$, where $p$ and $q$ are states and $\alpha$ and $\beta$ are non-negative. It is exactly the set of summable functions on $\Omega$; see Exercise 2.48. We can furnish this space with a norm by

$$\|f\|_1 = \sum_{\omega \in \Omega} |f(\omega)|. \tag{2.7}$$

It can be shown that this is a norm and that $\ell^1(\Omega)$ is complete in the topology defined by this norm; see Exercise 2.49. As usual, we can use the norm to define convergence: we say that a sequence $\{p_j\}$ of elements of $\ell^1(\Omega)$ converges to $p \in \ell^1(\Omega)$ if

$$\lim_{j \to \infty} \|p_j - p\|_1 = 0.$$

Thus the vector space $\ell^1(\Omega)$ spanned by the set of states $\Sigma$ becomes a real Banach space: that is, a complete normed real vector space [Riesz and Nagy (1975)].

The following device is sometimes useful. Let $\text{Span}\,\Omega$ denote the formal vector space spanned by $\Omega$. How can we add elements of $\Omega$, which is a set without linear structure? Easy: we just do it. This is how. We number the elements of $\Omega$, say $\omega_1, \omega_2, \ldots$, and define $\text{Span}\,\Omega$ to be the set of terminating sequences of real numbers $\{\lambda_1, \lambda_2, \ldots\}$ with the addition and multiplication by reals given elementwise. That is, $\text{Span}\,\Omega$ is the set of real functions $\lambda$ of $\omega$ having finite support, with $\lambda(\omega_j) = \lambda_j$. This is clearly a vector space. Then we write symbolically

$$\{\lambda_1, \lambda_2, \ldots\} = \lambda_1 \omega_1 + \lambda_2 \omega_2 + \ldots.$$

We identify $\omega$ with the point measure at $\omega$; $\text{Span}\,\Omega$ is then dense in $\ell^1(\Omega)$ and $\ell^1(\Omega)$ is the completion of $\text{Span}\,\Omega$ in the norm $\| \bullet \|_1$ given by (2.7). In this notation, $\Sigma$ is the closure of the convex hull of $\Omega$.

To understand the concept of *marginal probability*, consider two spaces $\Omega_1 = \{1, 2, \ldots, m\}$ and $\Omega_2 = \{1, 2, \ldots, n\}$, and take $\Omega = \Omega_1 \times \Omega_2$. A sample point is a pair of points, $\omega_1, \omega_2$, one in $\Omega_1$ and the other in $\Omega_2$. A probability on $\Omega$ is determined by an $m \times n$ matrix $p = [p_{ij}]$ of non-negative numbers, such that the sum of all the components is unity:

$$\sum_{i=1}^{m} \sum_{j=1}^{n} p_{ij} = 1. \tag{2.8}$$

This is called a bivariate distribution in statistics. The *marginal distribution on* $\Omega_1$ gives the probability of the event $\{\omega \in \Omega : \omega_1 = i\}$, and this is the sum of the entries in the $i$-th row of $p$, $\sum_j p_{ij}$. This is a probability on $\Omega_1$, since its components are non-negative and, by (2.8), they sum up to unity. We denote by $\mathcal{M}_1$ the map taking $p$ to its marginal on $\Omega_1$, and similarly, we define the marginal map on $\Omega_2$. Bearing in mind that $p$ is a row vector with $mn$ columns, we can represent the marginal map $\mathcal{M}_1$ as an $mn \times m$ matrix, such that

$$(p\mathcal{M}_1)(\omega_1) = \sum_{\omega_2} p(\omega_1, \omega_2). \tag{2.9}$$

We say that $i$ and $j$ are independent if $p$ is the product of its marginals, thus:

$$p_{ij} = (p\mathcal{M}_1)(i) \times (p\mathcal{M}_2)(j). \tag{2.10}$$

In statistical dynamics, the sample space is the product as in (2.1), from which we can show that given that $\Lambda = \Lambda_1 \cup \Lambda_2$, with $\Lambda_1$ and $\Lambda_2$ disjoint, then putting $\Omega_1 = \Omega(\Lambda_1)$, and similarly for $\Omega_2$, we have the product structure $\Omega(\Lambda) = \Omega_1 \times \Omega_2$. A sample point $\omega$ can be written as $\omega = (\omega_1, \omega_2)$ and a probability $p \in \Sigma(\Omega)$ is determined by the function $p(\omega_1, \omega_2)$. The marginal map $\mathcal{M}_1$ takes $p$ to its marginal on $\Omega_1$, and is thus a map from $\Sigma(\Omega)$ to $\Sigma(\Omega_1)$. Explicitly,

$$(p\mathcal{M}_1)(\omega_1) = \sum_{\omega_2 \in \Omega_2} p(\omega_1, \omega_2). \tag{2.11}$$

The map $\mathcal{M}_1$ can be extended to the whole of $\ell^1(\Omega)$ by linearity; it follows from the definition that $\mathcal{M}_1$ is linear on $\Sigma(\Omega)$. Similarly, $\mathcal{M}_2$ is defined. The sums involved obviously converge, and the map $\mathcal{M}_1$ is continuous; see Exercise 2.50.

Given two probability spaces $(\Omega_1, p_1)$ and $(\Omega_2, p_2)$, we can form their independent product as follows. Put $\Omega = \Omega_1 \times \Omega_2$ and $p(\omega_1, \omega_2) = p_1(\omega_1) \times p_2(\omega_2)$. We sometimes write $p = p_1 \otimes p_2$ for the independent product, since it agrees with the tensor product of elements of the Banach spaces $\ell^1(\Omega_1)$ and $\ell^1(\Omega_2)$. That is,

$$\ell^1(\Omega_1) \otimes \ell^1(\Omega_2) \equiv \ell^1(\Omega_1 \times \Omega_2). \tag{2.12}$$

If $p$ is the independent product of $p_1$ and $p_2$ as above, then the marginals of $p$ on $\Omega_1$ and $\Omega_2$ are exactly the $p_1$ and $p_2$ we started with. This, then, is the characteristic feature of independence: we say that $\Omega_1$ and $\Omega_2$ making up a probability space $(\Omega = \Omega_1 \times \Omega_2, p)$ are independent if $p$ is the product of its marginals:

$$p = p\mathcal{M}_1 \otimes p\mathcal{M}_2. \tag{2.13}$$

We say that a state $p \in \Sigma(\Omega)$ is independent over $\Lambda$, where $|\Lambda| < \infty$, if

$$p = \bigotimes_{x \in \Lambda} p\mathcal{M}_x \tag{2.14}$$

which means, if we label the points $x \in \Lambda$ by $1, 2, \ldots, n$:

$$p(\omega) = (p\mathcal{M}_1)(\omega_1) \times (p\mathcal{M}_2)(\omega_2) \times \ldots \times (p\mathcal{M}_n)(\omega_n). \qquad (2.15)$$

The marginal map $\mathcal{M}_1$ will be used to restrict the state to a subsystem; this could be a true subsystem, located in part $\Lambda_1 \subseteq \Lambda$ of the space when $\Omega = \Omega_1 \otimes \Omega_2$ and $\Omega_1$ is the sample space of $\Lambda_1$. More generally, the marginal map onto a subsystem (that are not necessarily localised in a subset of space) provides a reduced description of the state. A full account of this idea needs the concept of random variable, given in the next Section. Note that the reduced description does not involve putting the probability to zero on a region of sample space; rather, the probability $p$ is replaced by $p\mathcal{M}$ which does not depend on some of the variables needed for a full description of the state $p$.

## 2.2   Random Variables, Algebras

There is no other simple mathematical idea that is taught in such a muddled way as random variables are taught to physics students. We must distinguish the concept of random variable, and its probability distribution, from the concept of sample space and the probability $p$ on it. In our case, where $\Omega$ is countable and every subset is measurable, a random variable is a real-valued function on $\Omega$, denoted by $F$ or other capital letter. Thus, $F(\omega)$ is the value that $F$ takes when the sample point takes the value $\omega$, which it does with probability $p(\omega)$. The state $p$ itself does not enter the definition of $F$ but it does enter into the probability distribution of $F$; this is denoted $p_F$, and is defined as follows. Let $f$ denote a real number (not a function). Since $\Omega$ is countable, the set of (real) values $\{f\}$ that $F(\omega)$ takes is discrete. The probability that $F = f$ is

$$p_F(f) = \sum_{\omega : F(\omega) = f} p(\omega). \qquad (2.16)$$

For different $p$ we obviously get different $p_F$.

In gambling with dice, $\Omega$ is the set of outcomes and $F(\omega)$ might denote the reward that a particular player gets if a particular outcome arises. A random variable $F$ divides $\Omega$ into disjoint sets, on each of which $F$ takes different values. Thus, define the events,

$$\Omega_f = \{\omega \in \Omega : F(\omega) = f\}$$

for all possible values $f$ of $F$. Then $\Omega$ is the disjoint union of all possible $\Omega_f$, and $p_F(f)$ is the probability of the event $\Omega_f$. This idea, the division of

$\Omega$ into disjoint sets, will be important when we come to discuss the random variable chosen to represent the energy. In this case, the slices of $\Omega$ defined by the level surfaces of the energy are called *energy-shells*. When $\Omega$ has a local structure, as in (2.1), then a random variable is a function of all the points in $\Lambda = \{1, 2, \ldots\}$; the counterpart of writing a state as a row vector is to write a random variable as a column vector: $F(\omega) = F(\omega_1, \omega_2, \ldots)^d$. (We denote the transpose of a matrix by $^d$.)

Common notations for the mean and variance of $F$, in the state $p$, are

$$\text{mean} : p \cdot F = \mathbf{E}_p[F] = \langle F \rangle_p = \langle p, F \rangle := \sum_{\omega} p(\omega) F(\omega) \qquad (2.17)$$

$$\text{variance} : \quad (\Delta F)^2 = \sum_{\omega} p(\omega)(F(\omega) - \langle F \rangle_p)^2. \qquad (2.18)$$

In the row-column notation, the mean is simply $pF$. These definitions assume that the sums are absolutely convergent. They can be expressed in terms of the probability distribution of $F$; see Exercise 2.51. The standard deviation $\Delta$ is often called the *uncertainty* in books on quantum mechanics. The correlation between two random variables $F$ and $G$ is defined to be

$$\mathcal{C}(F, G) = \langle FG \rangle_p - \langle F \rangle_p \langle G \rangle_p \qquad (2.19)$$

where $FG$ is the product function of $F$ and $G$. The correlation cannot be expressed in terms of the distribution functions of $F$ and $G$, but needs the joint probability distribution (also called the bivariate distribution) defined as

$$p_{F,G}(f, g) = \sum_{\omega : F(\omega) = f, G(\omega) = g} p(\omega). \qquad (2.20)$$

At this point we would get into trouble if we tried to define probability theory in terms of the various $p_F$, without introducing the sample space and its state $p$, since the joint distribution would need a separate definition, and must satisfy some subtle relations in order to be derivable from a $p$. Now try Exercise 2.52.

In elementary books on probability, two random variables $F$ and $G$ are said to be independent if their joint probability distribution is the product of the two distribution functions: $p_{F,G}(f, g) = p_F(f) p_G(g)$ for all $f$ and $g$. The correlation between $F$ and $G$ is a measure of the dependence of these random variables; for example, it is zero if they are independent, but the vanishing of the correlation is not a sufficient condition for independence. In addition, to ensure independence, all the higher correlation functions involving $F$ and $G$ must be zero too. The higher moments and correlation functions are best defined using the *characteristic function*. Let us first treat the case of one random variable. So let $p$ be a probability on $\Omega$ and let $F$ be a random variable. Then the function

$$\mathcal{F}_F(x) = \mathbf{E}_p[\exp ixF] = \sum_\omega p(\omega) e^{ixF(\omega)} \qquad (2.21)$$

is called the *characteristic function* of the random variable $F$ (if $p$ is given) or of the state $p$ (if $F$ is given). It always has the properties

(1) $\mathcal{F}(0) = 1$;
(2) $\mathcal{F}(x)$ is a continuous function on the whole real line;
(3) for any $N$ complex numbers $\lambda_1, \ldots, \lambda_N$ and any real numbers $x_1, \ldots, x_N$ we have

$$\sum_{ij} \overline{\lambda}_i \lambda_j \mathcal{F}(x_j - x_i) \geq 0. \qquad (2.22)$$

A function with these properties characterises the random variable; the famous Bochner theorem [Riesz and Nagy (1975)] says that given a function $\mathcal{F}$ with the properties (1), (2) and (3), there exist a sample space $\Omega$, a probability $p \in \Sigma(\Omega)$ and a random variable $F$ such that (2.21) holds. A function satisfying (3) is said to be of *positive type*. If $\mathcal{F}$ has a power series expansion about $x = 0$ then it can be written

$$\mathcal{F}_F(x) = 1 + (ix) \sum_\omega p(\omega) F(\omega) + (ix)^2/2! \sum_\omega p(\omega) F^2(\omega) + \ldots \qquad (2.23)$$

from which the $n$-th moment can be read:

$$i^n \mathbf{E}_p[F^n] = \left. \frac{\partial^n}{\partial x^n} \mathcal{F}_F(x) \right|_{x=0}. \qquad (2.24)$$

For this reason the function $\mathcal{G}(s) = \mathcal{F}(-is)$ is called the *moment-generating function*. The $n$-th cumulant is then defined by

$$K_n(F) = \left. \frac{\partial^n}{\partial s^n} \log \mathcal{G}(s) \right|_{s=0}. \qquad (2.25)$$

We say that $\log \mathcal{G}$ is the cumulant generating function. There is a generalisation of all this to any number of random variables. Let $F_1, \ldots, F_m$ be random variables; define

$$\mathcal{F}(x_1, x_2, \ldots, x_m) = \mathbf{E}\left[ \exp(i \sum_j x_j F_j) \right]$$

$$= \sum_\omega p(\omega) \exp\{i(x_1 F_1(\omega) + \ldots + x_m F_m(\omega))\}. \qquad (2.26)$$

This always has the properties

(1) $\mathcal{F}(0, \ldots, 0) = 1$;

(2) $\mathcal{F}$ is a continuous function of the $m$ real variables $X = (x_1, \ldots, x_m)$;

(3) $\sum_{i,j=1}^{N} \bar{\lambda}_i \lambda_j \mathcal{F}(X_j - X_i) \geq 0$ for all choices of $N$ complex numbers $\lambda_i$ and $N$ real $m$-vectors $X_i$.

Parts (1) and (2) are not hard to prove; (3) follows simply from the positivity of the expectation of the random variable $|\sum \lambda_i \exp\{\sum_{j=1}^{m} x_{ij} F_j\}|^2$. If $\mathcal{F}$ is a bit more than continuous, say analytic in a neighbourhood of $X = 0$, then the multiple power series expansion converges and the coefficient of $x_1^{r_1} \ldots x_m^{r_m}$ is related by combinatorial factors to the moment

$$\mathbf{E}_p \left[ F_1^{r_1} F_2^{r_2} \ldots F_m^{r_m} \right].$$

Again, the multiple cumulants are similarly generated by $\log \mathcal{F}$. There is also a Bochner theorem, which says that any function obeying (1), (2) and (3) is the characteristic function of $m$ random variables $F_j$ on some space. If $\log \mathcal{F}$ is a quadratic form in $x_j$, then we say that the random variables are jointly Gaussian. There is the interesting result [Marcinkiewicz (1939)], which says that if all the cumulants vanish beyond some order, then they vanish beyond the second order, and the system of random variables is Gaussian.

The characteristic function is the Fourier transform of the joint distribution of the variables; see Exercise 2.53. There is no analogue of joint probability distribution in quantum mechanics, between non-compatible observables, but there are analogues of correlation functions. These are defined in terms of the state, and the algebraic properties of the observables; to keep the analogy and to allow a quantum generalisation, we develop the classical theory along these lines.

A random variable represents an *observable* in the theory. This is just another way of saying that an observable in a theory with randomness will in general be random. We shall use the terms random variable and observable interchangeably. The constant function $F(\omega) = 1$, or any other constant, is not actually random, as it has zero variance. Conversely, any random variable with zero variance is constant, except on sets of measure zero; see Exercise 2.54. Such a random variable is called a *sure* function, and is said to take its value with probability 1.

The set of random variables forms an *algebra*; an algebra is a vector space with a product, so we must specify the product in question. We choose the *point-wise* product (ordinary product of the values of the functions); the addition of functions, and real multiples of functions, is also taken point-wise. Of great technical usefulness is the set of bounded observables, denoted $\mathcal{C}(\Omega)$, which is a subalgebra of the set of all observables. A bounded observable is a map $F : \Omega \to \mathbf{R}$ such that the range of $F$, that is, its set of values, makes up a bounded set in $\mathbf{R}$. In symbols,

$$\|F\|_\infty = \sup\{|F(\omega)| : \omega \in \Omega\} < \infty. \tag{2.27}$$

Now try Exercise 2.58. Clearly any sure observable is bounded, as its range is a single point. The advantage of the bounded observables is that all means, variances and correlations are finite (the sums (2.17) etc. all converge). Moreover, the value of the supremum in (2.27) is a norm for the algebra $\mathcal{C}(\Omega)$, which is complete in the metric given by this norm. $\mathcal{C}(\Omega)$ is a Banach algebra; in particular, it is a real Banach space. In Exercise 2.59, you are asked to show that $\mathcal{C}(\Omega)$ is the topological dual of $\ell^1(\Omega)$, the span of the set of states, introduced in Sec. 2.1 and with norm given by Exercise 2.27. It is clear that each $F \in \mathcal{C}(\Omega)$ can be used to define a continuous linear functional $\phi$ on $\ell^1(\Omega)$ by the formula

$$p \mapsto \phi(p) = p \cdot F = \sum_\omega p(\omega) F(\omega). \tag{2.28}$$

This is clearly linear in the variable $p$, and it is part of the exercise to show that it is continuous in the sense that if

$$\|p_n - p\|_1 \to 0, \text{ then } p_n \cdot F \to p \cdot F. \tag{2.29}$$

The converse also holds: if $p \mapsto \phi(p)$ is a continuous linear functional on $\ell^1(\Omega)$ then there exists an element, say $F$ of $\mathcal{C}(\Omega)$, such that (2.28) holds (see Exercise 2.59). Thus, $\mathcal{C}(\Omega)$ is exactly the set of continuous linear functions from $\ell^1(\Omega)$ to $\mathbf{R}$, which is known as the topological dual. We say that $\ell^1$ is the predual of $\mathcal{C}(\Omega)$. A Banach space which is equal to the dual of its dual is said to be *reflexive*. In fact, $\ell^1$ is not reflexive, being only a tiny subspace of the set $\mathcal{C}^d$ of continuous linear functionals on $\mathcal{C}(\Omega)$. The missing states are not countably additive. We shall manage to avoid them. We write $p \cdot F$ for both the action of $F$ on $p \in \Sigma$, and for the action of $p$ on $F$. By identifying $\omega$ with the point measure at $\omega$, we can write $F(\omega) = \omega \cdot F$. By identifying $\omega$ with the random variable which is 1 at $\omega$ and zero elsewhere, we can write $p(\omega) = p \cdot \omega$. Then we get the Dirac-like identity

$$p \cdot F = \sum_{\omega \in \Omega} (p \cdot \omega)(\omega \cdot F).$$

The dot-product obeys the basic inequality

$$|p \cdot F| \leq \|p\|_1 \, \|F\|_\infty. \tag{2.30}$$

The Banach space $\mathcal{C}(\Omega)$ contains the *positive cone*, $\mathcal{C}_+$ which consists of bounded functions $F$ obeying $F(\omega) \geq 0$ for all $\omega$. The space $\ell^1(\Omega)$ contains the positive cone, $\ell^1_+$ of summable functions with non-negative values. These cones are dual to each other, in that given $F \in \mathcal{C}(\Omega)$, then $p \cdot F \geq 0$ for all $p \in \ell^1_+$ if and only if $F \in \mathcal{C}_+$; conversely, given $p \in \ell^1$, then $p \cdot F \geq 0$ for all $F \in \mathcal{C}_+$ if and only if $p \in \ell^1_+$; see Exercise 2.60. We shall frequently use the *separating* property of a dual space: $p \in \ell^1$ is determined by its values $p \cdot F$ for all $F \in \mathcal{C}$ and $F \in \mathcal{C}$ is determined by its values $p \cdot F$ for all $p \in \ell^1$.

We have seen that the pure states (the point measures) form a natural basis $\{\omega_1, \omega_2, \ldots\}$ in the completion of $\text{Span}\,\Omega$, the space $\ell^1(\Omega)$. The dual set in $\mathcal{C}(\Omega)$ is defined as the sequence of functions $F_n$ such that $\omega_m \cdot F_n = \delta_{n,m}$; clearly $F_n$ is zero except at $\omega_n$, where it is 1. The algebraic span of these consists of functions that are zero except at a finite number of points; we say such a function has *finite support*. However the completion in the norm $\| \bullet \|_\infty$ of the algebraic span of all the $F_n$ is not the whole of $\mathcal{C}$ but consists of its compact elements, $\mathcal{K}(\Omega)$. These are so called since they are compact operators when regarded as multiplication operators on the Hilbert space of square-summable sequences $\ell^2(\Omega) = \{F \in \mathcal{C} : \sum |F(\omega)|^2 < \infty\}$. For example, if $|\Omega| = \infty$, a non-zero sure function cannot be uniformly approximated by functions of finite support (Exercise 2.61). This natural basis in $\text{Span}\,\Omega$ allows us to identify the $\ell^1(\Omega)$ with a subset of $\mathcal{K}(\Omega)$, where the point measure at $\omega_n$ is identified with $F_n$. We get only a subset, (the elements of trace-class). This identification of a space with part of its dual, although natural, can lead to confusion since the action of a linear operator on a space should be distinguished from the dual action on the dual. This is illustrated in quantum mechanics, where we shall see a similar identification of density-matrices as operators; nevertheless, the time-evolution of the density matrices is the conjugate of the time-evolution of the observables [von Neumann (1932)]. We shall see in the next Chapter that this is true in classical statistical dynamics too.

If $|\Omega| < \infty$, any linear transformation $T$ on $\mathcal{C}$ can be 'thrown' onto the dual space, $\mathcal{C}^d$, which in the finite case is $\ell^1$, to provide the dual linear transformation $T^d$, the transpose of $T$, there. The transpose arises if both a space and its dual are regarded as column vectors. In our notation, no transpose occurs. In the basis $\omega_1, \ldots, \omega_n$ of $\text{Span}\,\Omega$, regarded as point measures, a linear transformation is determined by the $n \times n$ matrix $T_{ij}$, acting on the right, so that $(\omega T)_j = \sum_i \omega_i T_{ij}$, $j = 1, \ldots, n$, expressed as $\omega \mapsto \omega T$. This can be extended by linearity to act on any measure. The action on $\mathcal{C}$

of a right multiplication by $T$ on $\ell^1$ becomes left multiplication of $F$ by $T$ (not its transpose). This is because the transpose of the column vector $TF$ is the row vector $F^d T^d$. The dual relation between the transformations of row and columns can thus be expressed as the associativity of the matrix product:

$$pT \cdot F = p \cdot TF. \tag{2.31}$$

This defines the left multiplication $T : F \mapsto TF$ uniquely, since there is only one functional that, applied to $p$, gives the right-hand side. Similarly (if the sample space is finite) a left multiplication on $\mathcal{C}(\Omega)$ defines by duality a right multiplication on $\ell^1$.

More generally, the dual of a linear map from $\mathcal{V}_1$ to $\mathcal{V}_2$, where these are any vector spaces, is a map from the dual space $\mathcal{V}_2^d$ to the dual space $\mathcal{V}_1^d$. Here, $\mathcal{V}^d$ denotes the algebraic dual of $\mathcal{V}$, that is, the set of linear maps, $\mathcal{V} \to \mathbf{R}$ without any requirement of continuity.

Now consider the case where $\Omega = \Omega_1 \times \Omega_2$, and put $\mathcal{C} = \mathcal{C}(\Omega)$, $\mathcal{C}_1 = \mathcal{C}(\Omega_1)$, and $\ell^1$, $\ell_1^1$ their predual spaces. The marginal distribution (2.11) gives us a linear map from $\ell^1$ to $\ell_1^1$. The dual of this must be a map from $\mathcal{C}_1$ to $\mathcal{C}$. We claim that this is the injection of $\mathcal{C}_1$ into $\mathcal{C}$: any function of the variable $\omega_1$ can be regarded as a function of $\omega = (\omega_1, \omega_2)$ which happens not to depend on $\omega_2$. Roughly, $\mathcal{C}_1$ is a subset of $\mathcal{C}$; more exactly, there is a natural injection, $\Upsilon$ say, of $\mathcal{C}_1$ into $\mathcal{C}$, defined for all $F_1 \in \mathcal{C}_1$ by

$$\Upsilon F_1(\omega_1, \omega_2) = F_1(\omega_1). \tag{2.32}$$

We now prove our claim: for all $p \in \ell^1$ and all $F \in \mathcal{C}_1$,

$$p \cdot \Upsilon F = \sum_{\omega_1, \omega_2} p(\omega_1, \omega_2) F(\omega_1) = \sum_{\omega_1} F(\omega_1) \left( \sum_{\omega_2} p(\omega_1, \omega_2) \right)$$
$$= p\mathcal{M}_1 \cdot F$$

which shows that $\Upsilon^d = \mathcal{M}_1$ as claimed. The map $\Upsilon$ is called an *ampliation*. Because of (2.31) the notation $\Upsilon$ is redundant; it is just left multiplication by $\mathcal{M}_1$.

Now again suppose that $|\Omega| < \infty$, and that $\Omega$ is the product of $\Omega_1$ and $\Omega_2$; then corresponding algebras are related by the tensor product

$$\mathcal{C}(\Omega) = \mathcal{C}(\Omega_1) \bigotimes \mathcal{C}(\Omega_2). \tag{2.33}$$

This means that any function of two variables, $\omega_1$ and $\omega_2$, which run over a finite set can be written as a sum of product functions $F(\omega_1)G(\omega_2)$. We write $F \otimes G$ for the product function. In this notation, the ampliation is given by $F \mapsto F \otimes 1$. We can extend (2.33) to any finite number of factors. In particular, when $\Omega$ has a local structure, as in (2.1) we get

$$C(\Omega) = \bigotimes_{x \in \Lambda} C(\Omega_x). \qquad (2.34)$$

In this way, the algebra of observables acquires a local structure, since a function depending only on the $\omega_x$ for $x \in \Lambda_0$ can be regarded by ampliation as an element of $C(\Omega)$ that is localised in $\Lambda_0$. Note that this is not related to being zero outside $\Lambda_0$ in any sense. An important example is the local number, defined in the case where $\Omega_x = \{0, 1, 2, \ldots\}$ as the function $\mathcal{N}_x(\omega) = \omega_x$, which is an integer, the value of the sample point at $x$.

The formulation of independence is best done using events. We specify a state in $\Sigma(\Omega)$, which defines the probability of any subset of $\Omega$.

**Definition 2.3 (Independence of events).** *We say that two events* $\Omega_1, \Omega_2$ *(i.e., two subsets of $\Omega$) are independent if*

$$\mathrm{Prob}(\Omega_1 \cap \Omega_2) = \mathrm{Prob}(\Omega_1)\mathrm{Prob}(\Omega_2).$$

Then two observables $F$ and $G$ are independent if every level set $\Omega_f = \{\omega : F(\omega) = f\}$ is independent of every level set $\Omega_g = \{\omega : G(\omega) = g\}$. Note that there is no connection between this concept, which is *statistical* independence, and depends on the choice of $p$, and the linear independence of $F$ and $G$, although if $F$ and $G$ are functionally dependent they can be statistically independent only if they are constant, Exercise 2.62. Similarly we say that three random variables, $F, G, H$ are mutually independent if the joint probability distribution $p_{F,G,H}(f, g, h)$ is the product of its marginals. It can be shown that three random variables can be pairwise independent but not mutually independent. Mutual independence can be formulated for events, extending Def. 2.3 to more than two sets.

Another way to express independence involves the concept of a *ring of subsets* (to avoid confusion, I follow [Kunze and Segal (1978)] in not calling it an algebra; any ring can be made into an algebra over the field of two elements, $\{0, 1\}$ with addition modulo 2; we have no use for this idea).

**Definition 2.4 (Boolean Ring).** *A collection $\mathcal{B}$ of subsets of $\Omega$ is said to be a Boolean ring if $\Omega$ itself belongs to $\mathcal{B}$, the complement of any member of $\mathcal{B}$ is also a member, and if the union of two elements of $\mathcal{B}$ is also in $\mathcal{B}$.*

By working with the axioms, we can show that the empty set is a member of $\mathcal{B}$, and that the union or intersection of any finite number of members of $\mathcal{B}$ is also a member. It is a ring, with product given by intersection, and sum given by symmetric difference [Kunze and Segal (1978)]. In measure theory use is made of the concept of Boolean $\sigma$-ring, which is a Boolean-ring in which the union of countably many members is always a member. A $\sigma$-ring defines what is known as a *measurable structure* on $\Omega$, and the elements are called measurable sets. The necessity for the introduction of this idea is seen in the Banach-Tarski 'paradox', in which it is shown that a sphere can be cut up into parts, each isometric to the original, and reassembled to form a larger sphere. The sets involved are very pathological, and measure theory outlaws such sets (they are not measurable). Such troubles cannot arise if $\Omega$ is countable, but nevertheless it is worth introducing Boolean $\sigma$-rings, which provide us with a useful heuristic. Given a measurable space $(\Omega, \mathcal{B})$, we say that a function $F : \Omega \to \mathbf{R}$ is $\mathcal{B}$-measurable if the inverse image of every interval (in $\mathbf{R}$) lies in $\mathcal{B}$. Here, the inverse image of the interval $[a, b]$ is the set $\{\omega \in \Omega : a \leq F(\omega) \leq b\}$. For example the indicator function of a measurable set is measurable. There is some similarity, in the idea of a measurable function, with the idea of a continuous function, defined in terms of a collection of sets $\mathcal{S}$, called the open sets: a function $F : \Omega \to \mathbf{R}$ is continuous if the inverse image of every open interval is open. The open sets provide $\Omega$ with a topology (a topological structure), just as the measurable sets provide it with a measurable structure. The following manœuvres are very useful. Given a collection of sets, there is a unique smallest Boolean $\sigma$-ring containing every set in the collection; it is called the $\sigma$-ring generated by the sets. Similarly, given a collection of functions $\{F, G \ldots\}$ on $\Omega$, there is a unique smallest Boolean $\sigma$-ring with respect to which they are all measurable; it is the $\sigma$-ring generated by all the level sets of the functions (when $\Omega$ is countable). We denote it by $\mathcal{B}(F, G, \ldots)$. We shall use this idea when we have a local structure for $\Omega$, so that a sample point $\omega$ is written as $\omega = (\omega_1, \omega_2, \ldots)$. Here, $\omega_i \in \Omega_1$ and $\Omega = \prod_i \Omega_i$. A similar structure arises in processes, in which case $i$ labels the time. In either case, we define a *cylinder set* to be any subset of $\Omega$ defined by a condition on some finite number of components $\omega_j$, with no condition on the remaining components. For example, let $\Omega_{x_1,0}$ be a subset of $\Omega_{x_1}$ for one $x_1 \in \Lambda$. Then the condition $\omega_{x_1} \in \Omega_{x_1,0}$ defines the cylinder set of all configurations whose component at the point $x_1$ takes one of the values in $\Omega_{x_1,0}$. Where the label $x$ refers to the time, as in a process, we shall denote it by $t$; then consider the cylinder set which contains all paths which at time $t_1$ sit in $\Lambda_{t_1,0}$, which can be pictured as all paths passing through the 'gate' $\Lambda_{t_1,0}$ at time $t = t_1$. In this case, $\Lambda_{x_1,0}$ or $\Lambda_{t_1,0}$ is the base of the cylinder set. Similarly, cylinder sets with bases given by conditions on two, three or more variables $\omega_j$ can be pictured. To each region $\Lambda_0$ we may consider all cylinder sets based sets defined by conditions on variables $\omega_x : x \in \Lambda_0$. These generate a $\sigma$-ring associated to $\Lambda_0$, which we call $\mathcal{B}(\Lambda_0)$. Then a function on $\Omega$ depends only on the variables in $\Lambda_0$ if and only if it is $\mathcal{B}(\Lambda_0)$-measurable; see Exercise 2.63.

We now formulate the concept of independence using $\sigma$-rings. Given a countable probability space $(\Omega, p)$ we say that two $\sigma$-rings $\mathcal{B}_1$ and $\mathcal{B}_2$ are independent if all the events in one are independent of all the events in the other. Recall the definition (2.3) for events. Then it follows that all random variables measurable with respect to $\mathcal{B}_1$ are independent of all random variables measurable with respect to $\mathcal{B}_2$. Conversely, given two independent random variables, the $\sigma$-rings they generate are independent. This can be generalised to a family of mutually independent $\sigma$-rings. Thus, our definition of a state's being independent over $\Lambda$ can be restated in dual terms, by saying that the $\sigma$-rings $\mathcal{B}_x$ generated by the cylinder sets are mutually independent; we might also say that the algebras $\mathcal{C}_x$ are mutually independent (relative to the given $p \in \Sigma$). We can use Boolean rings to give a neat formulation of conditional probability. If $\Omega_1 \subseteq \Omega$ is an event in the probability space $(\Omega, p)$, such that $\text{Prob}(\Omega_1) \neq 0$, then the conditional probability of an event $\Omega_2$, given $\Omega_1$, is defined to be

**Definition 2.5 (Conditional Probability).**

$$\text{Prob}(\Omega_2|\Omega_1) = \frac{\text{Prob}(\Omega_1 \cap \Omega_2)}{\text{Prob}(\Omega_1)}.$$

This is the correct probability to use if, after a measurement on $(\Omega, p)$, we found that the sample point was certainly in $\Omega_1$, but that no more information was available. If $\Omega_1$ and $\Omega_2$ are independent, then conditioning relative to $\Omega_1$ makes no difference to the probability of $\Omega_2$. This is obvious, as $\text{Prob}\{\Omega_1 \cap \Omega_2\}$ is the product of the probabilities of $\Omega_1$ and $\Omega_2$. We shall denote by $p(\omega|\Omega_1)$ the function of $\omega$ got by choosing $\Omega_2$ to be the set containing the single point $\omega$. It is to be shown in Exercise 2.64 that this is indeed a probability on $\Omega$. A slight variation of this concept is the conditional probability, given the value of the random variable $F$. If we are given that $F = f$, then we can say that the event, the level set $\Omega_f$, has happened. Then the conditional probability, given $F$, is the collection of states $p(\omega|\Omega_f)$, as $f$ runs over the possible values of $F$. We might be given more information, such as the values of $F$ and $G$. Now, giving $G$ when $F$ is known might entail giving some redundant information, especially when $G$ is functionally related to $F$. Thus if $G = F^3$, conditioning given $F$ and $G$ is the same as giving just $F$. What is relevant is the Boolean ring $\mathcal{B}(F, G, \ldots)$ of joint level-sets of the functions. In our case, where $\Omega$ is countable, we can write $\Omega$ as the union of disjoint, minimal elements of $\mathcal{B}(F, G, \ldots)$, and the probability is conditioned by the information telling us which of these sets the sample point is in. This leads to the concept of $\text{Prob}(\Omega_1|\mathcal{B})$, the conditional probability given the Boolean ring $\mathcal{B}$: it is the collection of conditional probabilities of $\Omega_1$, conditioned by each of the sets in $\mathcal{B}$.

The *conditional expectation* of a random variable $G$, given $F$, naturally uses the conditional probability, and forms the mean in the usual way. It is denoted $\mathbf{E}_p[G|F]$. Thus

$$\mathbf{E}_p[G|F](f) = \mathbf{E}_p[G|\mathcal{B}(F)](f) = \sum_\omega G(\omega)p(\omega|\Omega_f). \qquad (2.35)$$

Again, if $G$ is independent of $F$, then these conditional expectations coincide with the usual expectation of $G$. The conditional expectation can be regarded as a new random variable, as it is a function on $\Omega$, constant on the level-sets of $F$. Thus, define $\mathbf{E}_p[G|F](\omega) = \mathbf{E}_p[G|F](f)$ where $f = F(\omega)$. This may be said to be the part of the function $G$ that is a function of $F$. This can be more exactly expressed as follows. Although we have avoided problems of measurability by using a discrete sample space, we find it useful to consider random variables that are measurable relative to a Boolean $\sigma$-ring $\mathcal{B}$. Then we denote by $\ell^\infty(\Omega, \mathcal{B})$ the algebra of bounded $\mathcal{B}$-measurable functions; if $\mathcal{B} = \mathcal{B}(F)$, then this is the set of bounded functions of $F$. Similarly we define the space $\ell^2(\Omega, p, \mathcal{B})$ as the set of $\mathcal{B}$-measurable functions $F(\omega)$ such that $\sum p(\omega)|F(\omega)|^2 < \infty$. This is a Hilbert space. You are asked in Exercise 2.65 to show that if $\mathcal{B} = \mathcal{B}(F)$ for some function, then the map $G \mapsto \mathbf{E}[G|F]$ is the orthogonal projection from $\ell^2(\Omega, p)$ to $\ell^2(\Omega, p, \mathcal{B}(F))$. It is then natural to define, for any $\mathcal{B}$, the conditional expectation $\mathbf{E}_p[G|\mathcal{B}]$ to be the orthogonal projection of $G$ onto $\ell^2(\Omega, p, \mathcal{B})$.

## 2.3  Entropy

The entropy of a probability $p$ on a countable space $\Omega$ is given by

**Definition 2.6.** $S(p) = -\sum_{\omega \in \Omega} p \log p$ .

In this definition, we use natural logarithms, and adopt the convention that $0 \log 0 = 0$, and that a divergent sum means that the entropy is infinite (all the terms are non-negative). In thermodynamics, this formula is multiplied by the Boltzmann constant $k_B$, while in information theory, the base 2 for the logarithm is often used; then the unit used is the *bit*.

Formula (2.6) is a special case of the quantum entropy defined in [von Neumann (1932)], but it is sometimes attributed to Shannon. Actually, Shannon's entropy is more general; given $(\Omega, p)$, it is defined for each random variable $F$, and uses (2.6) with $p$ replaced by the distribution of $F$ in the fixed state $p$. See Exercise 2.69,2.71, and [Werhl (1987, 1979)] for the interpretation of Shannon's entropy. The von Neumann entropy is additive for independent systems: if $\Omega = \Omega_1 \times \Omega_2$ and $p = p_1 \otimes p_2$, then we have:

$$S(p) = -\sum_{\omega_1, \omega_2} p(\omega) \log\left(p_1(\omega_1)p_2(\omega_2)\right)$$

$$= -\sum_{\omega_1, \omega_2} p_1(\omega_1)p_2(\omega_2)[\log p_1(\omega_1) + \log p_2(\omega_2)]$$

$$= -\sum_{\omega_2} p_2(\omega_2) \sum_{\omega_1} p_1(\omega_1) \log p_1(\omega_1)$$

$$- \sum_{\omega_1} p_1(\omega_1) \sum_{\omega_2} p_2(\omega_2) \log p_2(\omega_2)$$

$$= S(p_1) + S(p_2)$$

since $\sum_{\omega_1} p_1(\omega_1) = \sum_{\omega_2} p_2(\omega_2) = 1$. For the converse, for conditions under which $S$ is the only function with the additive property, see [Renyi (1970); Thirring (1980)]. We can now see that entropy is an extensive quantity in a model with local structure, whenever the state is independent over $\Lambda$, (Exercise 2.44). Another easy result is that the entropy is zero if and only if the state is a point measure (recall that such is an extremal point of the set $\Sigma$). The entropy is an unbounded function on $\Sigma$, but can be shown to be finite on a dense set of states, (Exercise 2.43). It cannot be defined on the whole dual space $\ell^1$, since it relies on the fact that the values of $p(\omega)$ are non-negative. But if needed it can be extended to positive but not normalised elements.

We now prove the intuitively obvious idea that the uniform distribution contains no information about the system.

**Theorem 2.7.** *Suppose that* $|\Omega| = n < \infty$. *Then the state* $p \in \Sigma(\Omega)$ *that maximises* $S(p)$ *is unique, and is given by the uniform distribution* $p(\omega) = 1/n$.

**Proof.** We must maximise $S(p)$ over the variables $p_1, p_2, \ldots, p_n$ subject to the constraint $\sum_j p_j = 1$. We use the method of Lagrange multipliers; so, maximise $L = S(p) + \lambda \sum_j p_j$ subject to no constraint, and then find $\lambda$

by fitting the solution to the constraint. We find the turning points in the region $0 < p_j < 1, \ 1 \le p_j \le n$ by

$$\frac{\partial L}{\partial p_j} = -\log p_j - p_j(1/p_j) + \lambda = 0.$$

The solution is $p_j = \exp(\lambda - 1)$ for all $j$; since they sum to 1, the unique answer is $p_j = 1/n$, and $S = \log n$. The second derivative $\partial^2 L/\partial p_j \partial p_k$ without constraint is $-\delta_{j,k}/p_j$ and this is negative. This shows that we are at a maximum, and *a fortiori* the constrained function is a maximum. The function $S(p)$ is smaller than $\log n$ on the boundary of $\Sigma$: if we are on a face of the simplex $\Sigma$, such that $m > 0$ components of $p$ are zero, the same method shows that the maximum of $S$ on the face is $\log(n - m)$, which is less.                                                                                   □

Inasmuch as information is negative entropy, it is therefore minimised by the uniform distribution.

In Exercise 2.41 it is shown by a similar method that the canonical state is the state of maximum entropy among all states with a given value of the mean energy.

The entropy is a concave function: this means that the area under the graph of $S$ is a convex set. The exact definition is

**Definition 2.8 (Concavity).** *Let $\Sigma$ be a convex set. A function $F : \Sigma \to$ **R** is said to be* concave *if for all $\lambda \in (0,1)$, we have*

$$F(\lambda p + (1 - \lambda)q) \ge \lambda F(p) + (1 - \lambda)F(q)$$

*for all $p$ and $q$ in $\Sigma$. It is said to be* strictly concave *if $\ge$ is replaced by $>$ whenever $p \ne q$.*

Note that linear functionals are concave but never strictly concave. The space $\Sigma$ must be convex (as, indeed, it is if it is $\Sigma(\Omega)$) in order for the left-hand side to make sense. We shall prove a much stronger result, which estimates the amount of concavity of $S$; in the following theorem, $\|p\|_2 :=$ $(\sum |p_j|^2)^{1/2}$ is the $\ell^2$-norm.

**Theorem 2.9 (Concavity of the Entropy).** *Let $p, q \in \Sigma(\Omega)$ and $0 \le \lambda \le 1$. Then*

$$S(\lambda p + (1 - \lambda)q) - \lambda S(p) - (1 - \lambda)S(q) \ge \frac{1}{2}\lambda(1 - \lambda)\|p - q\|_2^2. \quad (2.36)$$

**Proof.** We use Taylor's theorem with Lagrange remainder up to second order, and shall use $\xi$ and $\eta$ to denote intermediate abscissae. Let $f(x) =$

$-x \log x$; then $f' = -\log x - 1$ and $f'' = -1/x$. First let $p$ and $q$ denote numbers in $[0, 1]$. Let $x = \lambda p + (1 - \lambda)q$; then $p = x + (1 - \lambda)(p - q)$ and $q = x + \lambda(q - p)$. Then the difference in entropy is

$$
\begin{aligned}
\delta S &= f(x) - \lambda f(p) - (1 - \lambda)f(q) \\
&= f(x) - \lambda\{f(x) + (1 - \lambda)(p - q)[-\log x - 1] \\
&\quad -(1 - \lambda)^2(p - q)^2/(2\xi)\} \\
&\quad - (1 - \lambda)\{f(x) + \lambda(q - p)[-\log x - 1] - \lambda^2(p - q)^2/(2\eta)\} \\
&= \lambda(1 - \lambda)^2(p - q)^2/(2\xi) + (1 - \lambda)\lambda^2(p - q)^2/(2\eta) \\
&> \left(\lambda(1 - \lambda)^2 + (1 - \lambda)\lambda^2\right)(p - q)^2/2 \\
&= \frac{1}{2}\lambda(1 - \lambda)|p - q|^2.
\end{aligned}
$$

Now do the same for each component $p_j$ and $q_j$, and sum, to get the theorem. □

This proof is typical; the estimate involves the second derivative $-f''$ which is positive and its intermediate values are all larger than 1. We interpret the theorem as saying that if we mix two different states of a gas or liquid, then there is a gain in entropy overall. The theorem applies to states $p$, $q$ and $\lambda p + (1 - \lambda)q$ of a solid, but it is hard to see how to do the mixing experimentally.

Suppose that $\Omega$ has a product structure, so that $\Omega = \Omega_1 \times \Omega_2$, and write $\mathcal{B}_1$ for the Boolean ring consisting of cylinder sets based on subsets of $\Omega_1$; similarly, write $\mathcal{B}_2$ for the cylinder sets based on $\Omega_2$. Then a state in which $\mathcal{B}_1$ and $\mathcal{B}_2$ are independent has higher entropy than a state with the same marginals in which they are not independent. Before proving this, we need an important result. It concerns the *relative entropy* $S(p|q)$ of $p$ given $q$, defined to be

**Definition 2.10 (Relative Entropy).** *Let $p$ and $q$ be states in $\Sigma(\Omega)$. Then*

$$
S(p|q) = \sum_\omega p(\omega)(\log p(\omega) - \log q(\omega)).
$$

**Lemma 2.11 (Kullback).** $S(p|q) \geq \frac{1}{2}\|p - q\|_2^2$.

**Proof.** If $p$ and $q$ lie in the interval $(0, 1)$, then

$$
\begin{aligned}
p \log p &= q \log q + (p - q)(1 + \log q) + \frac{(p - q)^2}{2\xi} \\
&\geq p - q + p \log q + (p - q)^2/2.
\end{aligned}
$$

Do this for each $p_j$ and $q_j$, and sum, giving

$$\sum_j p_j (\log p_j - \log q_j) \geq \sum_j (p_j - q_j)^2 / 2$$

since $\sum p_j - \sum q_j = 0$. □

A version of Kullback's inequality can be proved by the same method for the generalised relative $\alpha$-entropies related to Renyi's $\alpha$-entropy:

**Definition 2.12.** Let $p$ and $q$ be probabilities. Then

$$S_\alpha(p|q) := \frac{2}{1-\alpha^2} \sum_j \left( p_j - p_j^{(1-\alpha)/2} q_j^{(1+\alpha)/2} \right). \qquad (2.37)$$

Then we have

**Theorem 2.13.** $S_\alpha(p|q) \geq \frac{1}{2} \|p - q\|_2^2$.

**Proof.**   Do Exercise 2.66. □

One can also prove a stronger version of Kullback's inequality,

$$\sum_j p_j (\log p_j - \log q_j) \geq \frac{1}{2} \left( \sum_j |p_j - q_j| \right)^2 = \frac{1}{2} \|p - q\|_1^2;$$

see Exercise 2.67.

Now let $p \in \Sigma(\Omega_1 \times \Omega_2)$, and let $p_1$, $p_2$ be the marginals onto $\Omega_1$ and $\Omega_2$, as in (2.11). Then we have

**Theorem 2.14.**

$$S(p_1 \otimes p_2) - S(p) \geq \|p - p_1 \otimes p_2\|_2^2 / 2.$$

**Proof.**   Let $q = p_1 \otimes p_2$ in Kullback's lemma; then

$$\sum p(\omega) \log p(\omega) - \sum p(\omega) \log (p_1(\omega_1) p_2(\omega_2)) \geq \frac{1}{2} \|p - q\|_2^2.$$

The second sum on the left-hand side is

$$- \sum_{\omega_1, \omega_2} p(\omega_1, \omega_2) \log p_1(\omega_1) - \sum_{\omega_1, \omega_2} p(\omega_1, \omega_2) \log p_2(\omega_2)$$

$$= S(p_1) + S(p_2). \qquad (2.38)$$

Collecting terms gives the result. □

We shall use this result a lot when we come to discuss the stoss map. It can be used to give a proof of Theorem 2.9.

## 2.4 Exercises

**Exercise 2.39.** Let $\mathcal{E}$ be a random variable such that the canonical partition function is finite. Show that any level set $\Omega_E$ of $\mathcal{E}$ obeys $|\Omega_E| < \infty$.

**Exercise 2.40.** Suppose that $\Lambda_1 \cap \Lambda_2 = \emptyset$. Show that

$$\Omega(\Lambda_1 \cup \Lambda_2) \equiv \Omega(\Lambda_1) \times \Omega(\Lambda_2)$$

and that this is not true in the particle point of view.

**Exercise 2.41.** If $\Omega = \{0, 1, 2, \ldots\}$ and $\mathcal{E}(\omega) = \kappa\omega$, show that the canonical distribution is

$$s_\beta(\omega) = (1 - \exp\{-\beta\kappa\}) \exp\{-\beta\kappa\omega\}.$$

Hence show that the mean energy is given by Planck's formula for one degree of freedom,

$$\langle \mathcal{E} \rangle = \kappa \frac{e^{-\beta\kappa}}{1 - e^{\beta\kappa}}. \tag{2.42}$$

Show that in general the state with the greatest entropy among all states with the same given mean energy is the canonical state. Show that the state having the greatest entropy among all states with the same given mean energy and the same given mean particle number is the grand canonical state.

**Exercise 2.43.** Show that if $|\Omega| < \infty$ then $S(p) < \infty$ on a dense set.

**Exercise 2.44.** Show that if $p$ is independent over $\Lambda$ then $S$ is extensive.

**Exercise 2.45.** Show that if $p_1, p_2, \ldots, p_n$ lie in a convex set $\Sigma$, then the smallest convex set containing all these points is a subset of $\Sigma$.

**Exercise 2.46.** Show that if $K \subseteq \mathbf{R}^n$ is a closed bounded convex set, then a point $k \in K$ is an extreme point if and only if $K - \{k\}$ is convex.

**Exercise 2.47.** Show that a closed triangle in the plane and a closed tetrahedron in $\mathbf{R}^3$ are simplices. Show that the state spaces of the sample spaces $\{1, 2, 3\}$ and $\{1, 2, 3, 4\}$ are simplices.

**Exercise 2.48.** Show that any element of $\ell^1(\Omega)$ can be written as $\alpha p - \beta q$, where $\alpha, \beta \in \mathbf{R}_+$, $p, q \in \Sigma(\Omega)$ even if $|\Omega| = \infty$.

**Exercise 2.49.** A *norm* on a real or complex vector space $\mathcal{X}$ is a non-negative function $\mathcal{X} \to \mathbf{R}$, such that

(1) $\|X + Y\| \le \|X\| + \|Y\|$
(2) $\|\lambda X\| = |\lambda| \|X\|$
(3) $\|X\| > 0$ if $X \ne 0$.

We say that a normed space is *complete in the norm* if every Cauchy sequence converges: this means that whenever a sequence $\{X_n\}$ is such that $\|X_m - X_n\| \to 0$ as $m$ and $n$ go to infinity, then there exists $X \in \mathcal{X}$ such that $X_n \to X$ as $n \to \infty$.

Show that $\| \bullet \|_1$ is a norm, and that $\ell^1(\Omega)$ is complete in this norm.

**Exercise 2.50.** Let $\Omega = \Omega_1 \times \Omega_2$, and let $\Sigma(\Omega_1)$ and $\Sigma(\Omega)$ be the sets of states on $\Omega_1$ and $\Omega$ respectively; show that the marginal map $\mathcal{M}_1 : \Sigma(\Omega) \to \Sigma(\Omega_1)$, given in (14.9), is continuous in the norm $\| \bullet \|_1$.

**Exercise 2.51.** Show that the mean and variance of a random variable $F$ on a sample space $\Omega$ furnished with a probability $p$ are given by

$$\langle F \rangle = \sum_f f p_F(f)$$

$$\mathcal{V}F = \sum_f (f - \langle F \rangle)^2 p_F(f).$$

**Exercise 2.52.** Show that the joint probability distribution $p_{F,G}$ must satisfy the inequality

$$\left| \sum_{f,g} f g \, p_{F,G}(f,g) \right|^2 \le \mathcal{V}F.\mathcal{V}G.$$

**Exercise 2.53.** Show that the characteristic function of a random variable $F$ is the Fourier transform of $p_F(x)$. Show that the characteristic function of $F_1, \dots, F_n$ is the Fourier transform of the joint probability distribution.

**Exercise 2.54.** Show that any random variable with zero variance is sure.

**Exercise 2.55.** Show that the variance is given by

$$\mathbf{E}[F^2] - (\mathbf{E}[F])^2 = \frac{\partial^2}{\partial s^2} \log \mathcal{G}(s); \tag{2.56}$$

show that in two variables, the covariance is given by

$$\mathbf{E}[FG] - \mathbf{E}[F]\mathbf{E}[G] = \frac{\partial^2}{\partial s \partial t} \log \mathcal{G}(s,t). \tag{2.57}$$

**Exercise 2.58.** Show that $\|F\|_\infty$ is a norm obeying $\| |F|_\infty^2 \| = \|F\|_\infty^2$.

Exercise 2.59. Show that $\mathcal{C}(\Omega)$ is the topological dual of $\ell^1(\Omega)$. If $|\Omega| < \infty$, show that $\mathcal{C}(\Omega)$ and $\ell^1(\Omega)$ are mutual duals, and obtain the equations

$$\|p\|_1 = \sup_F \{|p \cdot F| : F \in \mathcal{A}, \|F\|_\infty = 1\} \text{ for any } p \in \ell^1$$
$$\|F\|_\infty = \sup_p \{|p \cdot F| : p \in \ell^1, \|p\|_1 = 1\} \text{ for any } F \in \mathcal{A}.$$

Exercise 2.60. Show that $\Sigma(\Omega)$ and the cone of positive bounded random variables on $\Omega$ are dual cones.

Exercise 2.61. Show that the function 1 on $\Omega$ cannot be a uniform limit of functions of finite support if $|\Omega| = \infty$.

Exercise 2.62. Show that two functionally dependent random variables can be statistically independent only if they are both sure.

Exercise 2.63. Let $\Omega = \prod_{x=1}^n \Omega_x$. Show that a function $F(\omega_1, \ldots, \omega_n)$ depends only on $\omega_1, \ldots, \omega_j$ if and only if $F$ is measurable relative to the $\sigma$-ring generated by the cylinder sets based on all gates defined by sets in $\Omega_1, \ldots, \Omega_j$.

Exercise 2.64. Show that any conditional probability is a probability.

Exercise 2.65. Show that the map $G \mapsto \mathbf{E}[G|F]$ is the orthogonal projection of $G \in \ell^2(\Omega, p)$ onto the subspace $\ell^2(\Omega, p, \mathcal{B}(F))$.

Exercise 2.66. Show that $S_\alpha(p|q) \geq \frac{1}{2}\|p - q\|_2^2$.

Exercise 2.67. Show that for all positive $p$ and $q$,

$$(2p/3 + 4q/3)\,(p\log(p/q) + q - p) \geq (p - q)^2. \tag{2.68}$$

Show that for probabilities $\{p_i\}$, $\{q_i\}$, we have

$$\sum_i p_i \log(p_i/q_i) \geq \frac{1}{2}\|p - q\|_1^2.$$

Hint: take the square root of (2.68) and use Cauchy-Schwarz; or look at [Gzyl (1995)].

Exercise 2.69. Let $(\Omega, p)$ be a countable probability space. Define the Shannon entropy of a random variable $F$ to be

$$S(F) := -\sum_x p_F(x) \log p_F(x),$$

where $p_F(x) = p\{\omega : F(\omega) = x\}$. Show that $S(F) \leq S(p)$, with equality if and only if for each $x$, the inverse image $F^{-1}\{x\}$ has only one element almost surely.

Hint: let $D_x = \{\omega : F(\omega) = x\}$; show that if $p(\omega) \neq 0$ for all $p$, then

$$S(F) - S(p) = -\sum_x \sum_{\omega \in D_x} p(\omega) \log\left[1 + \frac{\sum_{\omega' \in D_x : \omega' \neq \omega} p(\omega')}{p(\omega)}\right]. \qquad (2.70)$$

Exercise 2.71. Shannon interpreted $S(F)$ as the amount of information (about $\omega$) that can be obtained by measuring $F$, given that we have a noisy source $(\Omega, p)$. Show that $S(F) = S(p)$ if and only if the $C^*$-algebra generated by $F$ is $L^\infty(\Omega, p)$.

Exercise 2.72. Let $\{F_1, \ldots, F_n\} = \mathcal{F}$ be bounded random variables on $(\Omega, p)$ as in (2.69), and define the channel entropy to be

$$S_{\mathcal{F}} := -\sum_{x_1, \ldots, x_n} p_{F_1, \ldots, F_n}(x_1, \ldots, x_n) \log p_{F_1, \ldots, F_n}(x_1, \ldots, x_n).$$

Show that $S_{\mathcal{F}} \leq S(p)$, with equality if and only if the $C^*$-algebra generated by $\mathcal{F}$ is $L^\infty(\Omega, p)$. (If so, we say that $\mathcal{F}$ is a *sufficient statistic*.)

# Chapter 3

# Linear Dynamics

## 3.1 Reversible Dynamics

By dynamics in classical mechanics, we mean the motion of a point in phase space as time goes by; this is determined by Newton's laws. In a field theory, the field equations of motion are expressed as partial differential equations. In our much simplified models with a countable configuration space we cannot formulate differential equations. We choose the time $t$ to be discrete, labelled by the group $\mathbf{Z} = \{\ldots, -1, 0, 1, 2, \ldots\}$. Then one time step in the dynamics is specified by giving a bijection of $\Omega$, that is, a permutation, denoted by $\tau$. The set of permutations of a set $\Omega$ is a group, denoted $\Gamma = AUT\,\Omega$. A permutation is invertible, and $\tau$ is thus non-dissipative. We shall write the action as a right multiplication, thus: $\omega \mapsto \omega\tau$. The effect of this is that the product $\tau_1\tau_2$ means the permutation in which $\tau_1$ is done first, followed by $\tau_2$. The invariance of the system under translations in time is expressed by choosing $\tau$ to be independent of time. Then the time-evolution through $n$ steps is given by the composition of $n$ single time steps, given by the permutation

$$\tau(n) = \underbrace{\tau \circ \tau \circ \ldots \circ \tau}_{n}.$$

We express this by saying that the dynamics is given by a group homomorphism $\tau : \mathbf{Z} \to AUT\,\Omega$. This means that $\tau(m) \circ \tau(n) = \tau(m+n)$ and $\tau(-n)$ is the inverse permutation $(\tau(n))^{-1}$. These relations are obvious from the definition of $\tau(n)$.

Given a dynamics $\tau$, and a point $\omega_0$, the set of points got by moving $\omega_0$, backwards as well as forwards in time, is called the *orbit* of $\omega_0$. It is denoted by $\omega_0\mathbf{Z}$, that is, we have

$$\omega_0\mathbf{Z} = \{\omega \in \Omega : \omega_0\tau(n) = \omega \text{ for some } n \in \mathbf{Z}\}.$$

43

Being on the same orbit is an equivalence relation, (3.12) and so $\Omega$ is divided up as the union of disjoint orbits. We may also move subsets around by time evolution: if $\Omega_0$ is an event, then the orbit of $\Omega_0$ is the set

$$\Omega_0 \mathbf{Z} = \{\omega \in \Omega : \omega = \omega_o \tau(n) \text{ for some } n \in \mathbf{Z} \text{ and some } \omega_o \in \Omega_0\}.$$

We have chosen the measurable structure so that every set is measurable, but we might be interested in a Boolean subring $\mathcal{B}_1$. We say that a map $T : \Omega \to \Omega$, whether invertible or not, is $\mathcal{B}_1$-measurable if the inverse image $\Omega_1 T^{-1}$ of every element $\Omega_1 \in \mathcal{B}_1$ is also in $\mathcal{B}_1$. We say that a subset $\Omega_0$ is invariant under a time evolution $\tau$ if $\Omega_0 \mathbf{Z} = \Omega_0$. Given a state $p \in \Sigma(\Omega)$, some people say that a dynamics $\{\tau(n)\}_{n=\ldots,-1,0,1\ldots}$ is ergodic if there are no invariant subsets except those of measure 0 or 1. Alternatively, if $\tau$ is given, and the same property holds, we say that the state $p$ is ergodic. The word is constructed out of the erg, the unit of energy, and should really be related to an energy function in the theory. Thus let $\mathcal{E}$ be a random variable, chosen to represent the energy of the system. In order to get a theory obeying the first law of thermodynamics (the conservation of energy), we must choose the dynamics $\tau$ so that it leaves all the level sets of $\mathcal{E}$ invariant. This still leaves many choices of $\tau$. This contrasts with Hamiltonian mechanics, where the dynamics is uniquely determined by the Hamiltonian.

In order for the canonical state to exist, each energy-shell must be finite (Exercise 2.39). To allow for energy-conserving dynamics, we do not require the above definition of ergodicity, but limit the notion to energy-conserving maps, and use the following

**Definition 3.1 (Ergodicity).** *Let $\mathcal{E}$ be a random variable on a countable set $\Omega$, giving rise to energy-shells $\Omega_E$. Let $\tau$ be one time-step of an energy-conserving dynamics. We say that $\tau$ is ergodic if for each $E$, there is no subset of $\Omega_E$ that is invariant except the empty set and $\Omega_E$ itself.*

This definition does not mention a measure; we have implicitly used the counting measure, which assigns the measure 1 to each point. For any finite set (with $n$ elements) it is easy to construct an ergodic permutation: we just choose any $n$-cycle such as

$$\begin{pmatrix} 1\,2\,\ldots\,n-1\,n \\ 2\,3\,\ldots\quad n\quad 1 \end{pmatrix}$$

It follows that for any $\mathcal{E}$ with a finite partition function we can find many ergodic time-evolutions. Historically the idea of ergodicity on energy-shells was introduced in classical Hamiltonian mechanics in an attempt to explain irreversibility; unfortunately, Hamiltonian systems are generally not ergodic. This is the upshot of the deep work of Kolmogorov, Arnol'd and Moser, known as the *KAM* theorem. A few cases of ergodic Hamiltonian motion have been established, but not anything interesting. In response to this, in the attempt to explain statistical mechanics, the ergodic *theorem* has given way to the ergodic *hypothesis*, which says that whatever the unknown Hamiltonian is, it leads to an ergodic action on phase space. This enables one to show that the infinite time average of an observable is the same as the average over phase space using the microcanonical state. One can then argue that, since we take a large amount of time to measure an observable (large compared to atomic times), we really measure the time average, which is the same as the mean in the microcanonical state. This idea is not helpful to us, as it is entirely geared to explaining equilibrium statistical mechanics, and there it still leaves open the question of what happens in a particular model. The fact is that the ergodic programme for explaining irreversibility in classical Hamiltonian systems is moribund and has ground to a halt. For our countable space $\Omega$, split into finite energy-shells, on the contrary, it is easy to obtain examples for which the ergodicity can be proved.

In many models of interest, such as those that occur in the theory of chemical reactions, the dynamics is not ergodic in exactly the sense given here, for the simple reason that there is one or more further conserved quantities, typically charge and atomic number. If we denote these random variables by $\mathcal{N}_i$, then the sample space is split into the disjoint union of energy-number-shells $\Omega_{E, N_i}$. The time-evolution must conserve all these, and will be said to be ergodic (in the general sense) if no non-trivial subset of any energy-number shell is mapped to itself.

For any set $\Omega$, the permutation group $\Gamma(\Omega)$, has been taken to act on the right. We can obtain from this an associative left multiplication, also called $\tau$, by the definition

$$\tau\omega := \omega\tau^{-1}.$$

This gives us an anti-isomorphism from $\Gamma$ into itself, namely the map to the *opposite* group, denoted $\overset{o}{\Gamma}$, which is the same set $\Gamma$ with the product written in the opposite order. By associativity, we mean that $\sigma(\tau\omega) = (\sigma\tau)\omega$; see (3.16). In terms of transformations, this means that $\tau$ is done first, followed by $\sigma$; this is the opposite order from the group law in $\Gamma$, which is defined by right multiplication. The point $\omega \in \Omega$ can be identified with the random variable which is 1 at $\omega$ and zero elsewhere. We can therefore extend this left action to the whole of $\mathcal{C}(\Omega)$ by linearity. Now, in the natural basis of Span $\Omega$, a permutation is represented by an orthogonal matrix, whose inverse is its transpose. The left action defined by this extension therefore agrees with that defined in (2.31), in which we write $F(\omega)$ as $\omega \cdot F$. This defines an automorphism $F \mapsto \tau F$ of the unital Banach algebra $\mathcal{C}$; see Example 3.13. The properties of an automorphism are these; let $\sigma : \mathcal{C}(\Omega) \to \mathcal{C}(\Omega)$. Then we say that $\sigma$ is an automorphism if $\sigma$ is:

(1) linear: $\sigma(\lambda_1 F_1 + \lambda_2 F_2) = \lambda_1 \sigma F_1 + \lambda_2 \sigma F_2$.
(2) unital: $\sigma 1 = 1$.
(3) multiplicative: $\sigma(F_1 F_2) = (\sigma F_1)(\sigma F_2)$.
(4) bijective: $\sigma$ maps onto and is invertible.

The set of automorphisms of a $C^*$-algebra $\mathcal{A}$ is a group under composition; it is denoted $AUT\,\mathcal{A}$. An automorphism automatically preserves the norm: $\|\sigma F\|_\infty = \|F\|_\infty$; it also obviously preserves the norms $\| \ \|_1$ and $\| \ \|_2$. It follows from (3) that any automorphism is a positive map, since it takes any element of the positive cone into the positive cone: $\sigma(F^2) = (\sigma F)^2 \geq 0$. We shall see that in more general dynamics, including dissipation, we replace (3) by positivity, and drop (4). There is a converse: if $|\Omega| < \infty$, any automorphism of $\mathcal{C} = \mathcal{C}(\Omega)$ is given by some permutation of $\Omega$; (3.15). Thus, $\overset{\circ}{\Gamma}(\Omega)$ and $AUTC(\Omega)$ are isomorphic groups.

If we specify an energy $\mathcal{E}$, and a dynamics generated by $\tau$ is energy-preserving, then $\tau\mathcal{E} = \mathcal{E}$, at least if $\mathcal{E}$ is bounded. For, we can write

$$\omega \cdot \mathcal{E} = \sum_E E\omega \cdot \chi_E$$

where $\chi_E$ is the indicator function for $\Omega_E$, the energy-shell of energy $E$, and all these sets are invariant under $\tau$. If $\mathcal{E}$ is not bounded, then all finite partial sums in this expression are invariant. Similarly, any function of $\mathcal{E}$ is invariant under $\tau$. We shall need a partial converse to this: if the energy-shells are finite and in addition $\tau$ is ergodic, then any invariant element $F$ of $\mathcal{C}$ is a function of $\mathcal{E}$. For, if it were not, it would not be measurable with respect to the Boolean ring $\mathcal{B}_1$ generated by $\mathcal{E}$, and so its level sets would not all lie in $\mathcal{B}_1$. A non-empty level set of $F$ is invariant, and its intersection with the level sets of $\mathcal{E}$ give us a part of an energy-shell that is invariant, contrary to the ergodicity.

The time-evolution of the observables generated by the left multiplication by $\tau$ is the analogue of the Heisenberg picture in quantum mechanics.

The analogue of the Schrödinger picture is the dual of this. Thus given $p \in \Sigma$, define the right action by $p \mapsto p\tau$ defined by $p\tau \cdot \omega := p \cdot \tau\omega$. To see that this right action takes states to states, note that $\tau F$ is positive if and only if $F$ is positive. Then $p\tau \cdot F = p \cdot \tau F$ is positive if and only if $F$ is positive, which ensures that $p\tau$ lies in the positive cone. The normalisation of $p$ is the statement that $p \cdot 1 = 1$. This gives $p\tau \cdot 1 = p \cdot \tau 1 = p \cdot 1$, which is 1 if $p$ is normalised. Being a right action, $p \mapsto p\tau$ gives us an isomorphism of $\Gamma$ onto the group of reversible dynamical laws on the states. The assumption that $\tau$ preserves the energy-shells, which are finite, ensures that the dual action maps $\ell^1$ to $\ell^1$, and does not take us to non-normal states. Indeed the algebra of functions on an energy-shell $\Omega_E$ is the dual of the linear span of the set of probabilities on the same energy-shell; the question reduces to the finite-dimensional case. Then the dynamics conserves the mean value of the energy (the first law of thermodynamics).

$$p\tau \cdot \mathcal{E} = p \cdot \mathcal{E}. \tag{3.1}$$

For, $p\tau \cdot \mathcal{E} = p \cdot \tau\mathcal{E}$ and this is equal to $p \cdot \mathcal{E}$ as $\mathcal{E}$ is invariant under $\tau$. If $\tau$ is ergodic, then the only invariant states are constants on the energy shells. Thus they are mixtures of microcanonical states of various energies; see Exercise 3.17.

The transformation law of states is compatible with the writing of $p$ as a vector in the span of $\Omega$: in a formal way, $p = \sum_\omega p(\omega)\omega$, and the permutation $\tau$ acts on the $\omega$, but not on the $p(\omega)$, which are just numerical coefficients. Thus the action of $\tau$ is

$$p = \sum_\omega p(\omega)\omega \mapsto \sum_\omega p(\omega)\omega\tau$$
$$= \sum_{\omega'} p(\omega'\tau^{-1})\omega'$$

from which we read off the result (because the elements of $\Omega$ are linearly independent) that $p\tau(\omega) = p(\omega\tau^{-1})$. This convention, with the argument of $F$ transforming with $\tau$, and $p$ with $\tau^{-1}$, is the correct way round. When we come to consider dissipative dynamics, we shall allow endomorphisms, $T$, obeying (1), (2) and (3) of the axioms of automorphisms, but not necessarily being invertible. Then we could not define $F(\omega T^{-1})$ but we can give a meaning to $p(\omega T^{-1})$ as the probability of the inverse image of $\omega$: $pT(\omega) = p(\omega T^{-1}) = \mathrm{Prob}\{\omega_1 : \omega_1 T = \omega\}$.

The dynamics $\tau(n) = \tau \circ \ldots \circ \tau$ traces out an orbit in $\Sigma(\Omega)$, which is thus split into the union of disjoint orbits. All the points on the same orbit have the same entropy. For, $S$ is a sum over all points of $\Omega$, and

$\tau$ just rearranges the order. Thus, reversible dynamics obeys the second law of thermodynamics, but in the trivial sense that while entropy does not decrease, it does not increase either. The partial order given by the concept of 'more chaotic' [Alberti and Uhlmann (1981)] is also preserved along the orbit.

We now turn to some examples of dissipative dynamics.

## 3.2   Random Dynamics

Let $\Omega$ be a countable sample space, with state space $\Sigma$, and suppose that the dynamics depends on some parameters that are unknown. We can treat this using probability. Let us assign probabilities $\lambda_1, \lambda_2, \ldots, \lambda_n$ to the dynamics $\tau_1, \tau_2, \ldots, \tau_n$ respectively. We then compute a quantity of interest using each of the possible choices of dynamics, and then compute the mean over the $\lambda$'s. This works well; its success is either due to the fact that a small system cannot be shielded from random influences, or to the chaotic nature of the actual dynamics over the time-interval represented by one time-step. The mean dynamics can be defined as

$$\tau = \sum_{i=1}^{n} \lambda_i \tau_i, \tag{3.2}$$

where $\tau$ acts on the algebra on the left and on the states on the right. If $|\Omega| = n < \infty$, in the natural basis in $\mathrm{Span}\,\Omega$, we can represent $\tau$ by a *stochastic* matrix, $T$. This is one whose rows add up to 1. One can see that the transpose matrix, $T^d$, is also stochastic. Such a matrix is called *bistochastic* or doubly stochastic. We can also deal with the case where $n = \infty$ and $\{\lambda_i > 0\}$ is a sequence whose sum is 1. It is clear that $\tau$ maps $\mathcal{C}$ into itself, and also that $\tau$ maps $\Sigma$ to $\Sigma$, since each $\tau_i$ does this, and $\Sigma$ is convex.

**Example 3.2.** Take $\Lambda = \{1, 2, \ldots, n\}$, representing the points on a segment of a line, and take $\Omega_x = \{0, 1\}$ for all $x$, with 0 representing an empty site and 1 an occupied site. As always, we take $\Omega = \prod_x \Omega_x$. Let $\tau_x$ denote the flip operator, which exchanges the situation at the sites $x$ and $x + 1$.

Thus

$$(\omega_1, \ldots, \omega_x, \omega_{x+1}, \ldots, \omega_n)\tau_x = (\omega_1, \ldots, \omega_{x+1}, \omega_x, \ldots, \omega_n).$$

The part of the space involving $x$ and $x + 1$ has four points,

$$(0,0), \ (0,1), \ (1,0), \ (1,1),$$

and in this basis $\tau_x$ is given by the matrix

$$\begin{pmatrix} 1 & 0 & 0 & 0 \\ 0 & 0 & 1 & 0 \\ 0 & 1 & 0 & 0 \\ 0 & 0 & 0 & 1 \end{pmatrix}.$$

The last row expresses that the two particles at the two sites are indistinguishable; their exchange does not alter the sample point. The operator on the whole algebra $\mathcal{C}(\Omega)$ is the tensor product of this matrix with the unit operator on $\mathcal{C}(\prod' \Omega_x)$ where the $\prod'$ omits $x$ and $x + 1$. The map $\tau_x$ conserves the random variable representing the number of particles, $\mathcal{N}(\omega) = \sum_x \mathcal{N}(x)$, where $\mathcal{N}(x)$ is the random variable $\mathcal{N}(x)(\omega) = \omega_x$. Its mean per site is the particle density. It is clear that the uniform distribution on a number-shell (the microcanonical state) is a fixed point of each $\tau_x$, and therefore of the convex sum

$$\tau = (n-1)^{-1} \sum_{x=1}^{n-1} \tau_x.$$

This gives rise to a symmetric matrix in the natural basis. The action on the random field $\mathcal{N}(x)$ follows at once: only $\tau_x$ and $\tau_{x-1}$ affect $\mathcal{N}(x)$, and these shift it to the right and left respectively. So we get, if $x$ is not an end-point:

$$\tau(\mathcal{N}(x)) = (n-1)^{-1} \left( \mathcal{N}(x+1) + \mathcal{N}(x-1) + (n-3)\mathcal{N}(x) \right)$$
$$= \mathcal{N}(x) + (n-1)^{-1} \left( \mathcal{N}(x+1) - 2\mathcal{N}(x) + \mathcal{N}(x-1) \right). \tag{3.3}$$

We recognise the increment in the density in one time-step to be proportional to a finite difference approximation to minus the second derivative. When the elements $x$ of $\Lambda$ are taken as a basis, we can express $\mathcal{N}$ as a column matrix $\mathcal{N} := \left( \mathcal{N}_1, \ldots, \mathcal{N}_{|\Lambda|} \right)^d$, and put the change in the mean values of $\mathcal{N}(x)$ during one time-step (3.3) in matrix form: $N(t+1) = TN(t)$, where $T$ is the matrix

$$T = \begin{pmatrix} 1-\lambda & \lambda & 0 & \ldots & 0 \\ \lambda & 1-2\lambda & \lambda & \ldots & 0 \\ 0 & \ldots & \ldots & \ldots & 0 \\ 0 & \ldots & \ldots & 1-2\lambda & \lambda \\ 0 & \ldots & \ldots & \lambda & 1-\lambda \end{pmatrix}. \tag{3.4}$$

Here, $\lambda = (n-1)^{-1}$.

In (3.18) you will show that at the end points the operator obeys Neumann boundary conditions, corresponding to no flow into or out of the region. One shows that if the initial state is a point measure with a certain value of total number, then the system converges to the corresponding microcanonical state. In fact, the dynamics is not only ergodic but has a spectral gap, a concept that will be met in the next section. This explains why it converges.

We get a variant of this model by considering the $n$ points of $\Lambda$ to lie on a ring. That is, we take $\Lambda = \mathbf{Z}_n$ the cyclic group on $n$ elements. Then 1 and $n$ become neighbours and if we add the permutation exchanging these, making $n$ altogether, and add them with equal weights we get, instead of (3.4) the following transition matrix, in which $\lambda = 1/n$:

$$
T = \begin{pmatrix}
1 - 2\lambda & \lambda & 0 & \ldots & \lambda \\
\lambda & 1 - 2\lambda & \lambda & \ldots & 0 \\
\ldots & \ldots & \ldots & \ldots & \ldots \\
0 & 0 & \ldots 1 - 2\lambda & \lambda \\
\lambda & 0 & \ldots & \lambda & 1 - 2\lambda
\end{pmatrix}
\tag{3.5}
$$

which is the finite-difference approximation to the second derivative with periodic boundary conditions.

The two matrices, (3.4) and (3.5), both have the property that the sum of the elements in each column is equal to 1. This reflects that the transformation $p \mapsto pT$ takes a probability to a probability. In these examples the transition matrices are symmetric; this reflects the fact that each of the permutations $\tau_i$ making up the dynamics is equal to its own inverse. More generally, we would get a symmetric matrix if, whenever the sum $\tau$ contained a permutation $\tau_i$ it also contained its time-inverted permutation, and with the same probability $\lambda_i$. This property is justified in classical mechanics by invoking invariance under time-reversal: to each possible dynamics, its time-reversed dynamics is possible. Let us add the probabilistic assumption that the reversed process is equally likely. This time-reversal symmetry leads, for observables that are themselves time-reversal invariant, to a symmetric matrix $T$ in the dynamics. However, not all observables are invariant under time-reversal. In fluid dynamics, the velocity field is an important exception. This is the subject of Chapter 7. In biological systems such as neural nets a more general form of the transition matrix is sometimes needed; for example, when the transition matrix element $T_{jk}$ represents the probability that the firing of neuron $j$ causes the firing of neuron $k$, there is no reason why $T$ should be symmetric. To allow for this, we prefer not to require that the symmetry of $T$ be a fundamental part of the theory. The following simple result is quite useful. Let us say that a sample space $\Omega$ has *time-reversal symmetry* if there exists a permutation $\gamma_0$ of $\Omega$ with the interpretation of time reversal. Let us say that a map $T$ is invariant under $\gamma_0$ if

$$T_{\omega,\omega'} = T_{\omega'\gamma_0,\omega\gamma_0} \qquad \text{for all } \omega, \omega' \in \Omega.$$

A matrix $T : \mathbf{M}_n \to \mathbf{M}_n$ of non-negative numbers is said to be *stochastic* if the sum of elements of each column is 1. We say $T$ is *bistochastic* if in addition the sum of each row is also 1. Then we have:

**Lemma 3.3.** *If a stochastic map $T$ is invariant under a time-reversal, then it is bistochastic.*

For,

$$\sum_i T_{\omega_i,\omega_j} = \sum_i T_{\omega_j\gamma_0,\omega_i\gamma_0} = 1$$

since as $\omega_i$ runs over $\Omega$ so does $\omega_i\gamma_0$.

The next example has two conserved quantities and is ergodic in the generalised sense.

**Example 3.4.** The space $\Lambda = \{1, 2, \ldots, n\}$ is the same as in example 3.2, but now $\Omega_x$ has three points, called 0, $A$ and $B$. We will interpret 0 as saying that $x$ is an empty site, and the others as saying that it is occupied, by a molecule of type $A$ or $B$. Then $\Omega$ is the space of 'words' of length $n$, and alphabet $\{0, A, B\}$. It is split into the number-shells $\Omega_{N_1,N_2}$, where $N_1$ is the number of $A$'s in the word $\omega$ and $N_2$ is the number of $B$'s. There are three types of exchange permutations of $\Omega$ that conserve the number of $A$-molecules and the number of $B$-molecules. We can exchange $A$ at $x$ with an empty site at $y$: call this $\tau(A, x, y)$; we can exchange $B$ at $x$ with an empty site at $y$: call this $\tau(B, x, y)$; or we can exchange an $A$ at $x$ with a $B$ at $y$: call this $\tau(A, x; B, y)$. Any mixture of these permutations, as we vary $x$ and $y$, gives rise to a left map $T$ on $\mathcal{C}(\Omega)$ that leaves the indicator functions of the shells $\Omega_{N_1,N_2}$ fixed. It is easy to ensure that the action is ergodic in the generalised sense, if we mix all these permutations with non-zero coefficients. If we only use exchanges of neighbours, where $x = y \pm 1$, this model will possess only short-ranged forces. We could express that $B$ is more mobile than $A$ by choosing the weight of the permutation $\tau(B, x, y)$ to be larger than the weight of the permutation $\tau(A, x, y)$. The same effect could be achieved if we included some permutations $\tau(B, x, y)$ with a large distance between $x$ and $y$. This model differs from two copies of the (3.2) since $A$ and $B$ interfere with each other; thus, if we did not use $\tau(A, x; B, y)$, and merely exchanged a particle with a neighbouring hole, then the action would not be ergodic, even in the generalised sense using the two conserved numbers; the initial order of the letters $A$ and $B$ would not changed by the dynamics. This order would give us more conserved quantities, labelled by a partition of $N_1 + N_2$. That the particles $A$ and $B$ cannot both sit at the same site effectively gives them an infinite repulsion, called a *hard core*.

Our third model allows linear chemical reactions

$$A \rightleftharpoons B \tag{3.6}$$

as well as diffusion of each species.

**Example 3.5.** We take $\Lambda$ to have $n$ points as in the previous examples, and consider all permutations $\tau(A, x, y)$, $\tau(B, x, y)$ and $\tau(A, x; B, y)$, which cause diffusion, and include the permutation causing a transition from $A$ to $B$ and *vice versa*. It is said to be a local transition if it occurs at a site $x$. Thus define $\tau(A, B, x)$ to be the permutation

$$(\omega_1, \ldots, \omega_x = A, \omega_{x+1}, \ldots, \omega_n)\tau(A, B, x)$$
$$= (\omega_1, \ldots, \omega_x = B, \omega_{x+1}, \ldots, \omega_n)$$

and the inverse, expressed as $\tau(A, B, x) = \tau(B, A, x) = \tau(A, B, x)^{-1}$. Let us mix the permutations giving diffusion with one weight, and those giving reactions with another. We get a linear reaction-diffusion system, in which $\mathcal{N}_1 + \mathcal{N}_2$ is conserved, but not each separately.

This example is generalised to $m$ species of chemical in Section 4.3.

Let us now return to the general theory. We take $\Omega$ to be countable, and consider a mixture $T$ as in (3.2) of permutations $\tau_i$ each of which is a finite product of finite cycles. Since $T$ maps $\Sigma$ to $\Sigma$, it follows that the entropy of $pT$ is well defined (with infinity as a possible value if $\Omega$ is infinite). In fact, entropy can increase under the action of $T$ if it is a proper mixture. Thus if two different permutations $\tau_1$ and $\tau_2$, are involved, then there exists a state $p$ and $\delta > 0$ such that $\|p\tau_1 - p\tau_2\|_2 = \delta$. Then

$$S(pT) - S(p) = S(\lambda p\tau_1 + (1 - \lambda)p\tau_2) - S(p)$$
$$\geq \lambda S(p\tau_1) + (1 - \lambda)S(p\tau_2)$$
$$+ \frac{\lambda(1 - \lambda)}{2}\|p\tau_1 - p\tau_2\|_2^2 - S(p)$$
$$\text{by theorem (2.9)}$$
$$= \lambda(1 - \lambda)\delta^2/2$$

since the permutations do not alter the entropy. We conclude that irreversibility generally holds if we postulate random dynamics. You might wonder how our not knowing the true dynamics can possibly lead to a physical law such as the increase of entropy with time. The answer is that we have not computed the entropy gain (zero) for each of the possible $\tau_i$, and then found the mean of these (zero), but found the entropy gain of the mean dynamics. This is a clever model of the observed fact, the second law. You could ask, why do it this way? Well, I truly believe in the usefulness of probabilistic methods in science, and doing it this way gives the answer we need. Koseki [Koseki (1993)] has remarked that probability is successful in thermodynamics just because the properties of the theory, such as convexity, are so similar to those of thermodynamics.

In order to get a model satisfying the first law as well, we recall the examples (3.2, 3.4, 3.5) and generalise: choose an energy $\mathcal{E}$, with level sets $\Omega_E$, and some maps $\tau_i$ mapping each $\Omega_E$ to itself. We consider the case where $|\Omega_E| < \infty$ for each $E$. Then the mean energy is conserved by the mean dynamics, since it is conserved by each $\tau_i$ (3.1). Moreover, any microcanonical state is a fixed point of $T$. To see this, note that $\tau^{-1}$ also leaves each $\Omega_E$ invariant. See (and do) (3.20). It follows that $T$ leaves the state $p_E := \chi_E/|\Omega_E|$ invariant. This is true for all $E$. We thus have in place many elements of a theory of non-equilibrium statistical mechanics: the first law, the second law and the stationarity of the microcanonical states. But even if $\tau$ is ergodic, the system need not converge to equilibrium. It could move a point measure $\omega$ around and around. We need a stronger concept, that of a spectral gap. In the traditional approach to statistical mechanics via the ergodic theorem, it is argued that an actual observable is not a random variable at time $t$, since a measurement takes some time, indeed a very, very long time on the time-scale of the atomic variables. So we measure not $F$ but

$$\overline{F} := (1 + \tau + \tau^2 + \ldots, \tau^n)F/(n + 1)$$

for some $n$; this is a special case of the mixing of permutations studied just now; a typical ergodic theorem (3.21) says that if $\tau$ is ergodic, then this converges, to the *ergodic mean* of $F$, as $n \to \infty$. It is this limit that is supposed to describe the macroscopic observable. We do not follow this traditional approach, in which $n \to \infty$, as it is only useful for the equilibrium theory. If we keep $n$ finite, then the sum represents the average measurement over the time-interval $n$, which can still be small relative to macroscopic time. If $F$ is slow to change, then $\overline{F}$ is close to $F$; the dual action on the states hardly changes the entropy if $p$ is nearly invariant under $\tau$. But if $p\tau$ differs from $p$ by a large amount, then the entropy increases by a large amount. Thus, the dissipation depends on the dynamics, and is large for fast variables, without the need to discuss the occurrence of chaos in the technical sense. This approach, then, is close to the ergodic programme, but is adapted to the non-equilibrium case. Our random dynamics, (3.2) given by any mixture of permutations, is more general than the equal mixtures of powers of a single permutation suggested by the ergodic programme. In fact, it gives us any bistochastic map; this is the Birkhoff-von Neumann theorem. We shall prove this later, when $|\Omega| < \infty$ (3.11). This has recently been extended to some infinite $\Omega$ [Kendal (1960); Safarov (2005)].

More holds if the energy-shells are mapped to themselves: if for all $E$, a bistochastic map leave $\chi_E \in \text{Span}\,\Omega$ invariant under left and right multiplication, then it is a mixture of energy-conserving permutations. Similarly, if there are one or more conserved atomic numbers as well as energy, then any map increasing entropy is a mixture of permutations each of which conserves the same numbers. Renyi proved that the set of bistochastic maps consists of those stochastic maps not decreasing the entropy of any state [Renyi (1970)]. Thus all linear dynamics obeying the first and second laws in fact have the form of random dynamics, (3.2). To explain these results, we develop the general theory a bit more, to cover also the case of infinite $\Omega$.

Although the map $T$, a mixture of permutations, might not be invertible, it is linear and positive, and maps 1 to 1. This is an easy exercise (3.22). These are all the properties of a stochastic map:

**Definition 3.6 (Stochastic Maps).** *Let $\mathcal{C}(\Omega)$ be the usual algebra of random variables on the countable space $\Omega$. A map $T : \mathcal{C}(\Omega) \to \mathcal{C}(\Omega)$, $F \mapsto TF$, is said to be* stochastic *if*

- *$T$ is linear;*
- *$T$ is positive; that is, $T$ maps the positive cone to itself;*
- *$T1 = 1$.*

It is usual to add the condition of $W^*$-continuity to the definition, in order that its dual map leave $\ell^1$ invariant. We shall soon impose a much stronger condition which will ensure this. We get an important class of stochastic maps as follows. Let $\tau : \omega \mapsto \omega\tau$ be a mapping of $\Omega$ into itself, not necessarily bijective. For $p \in \Sigma$, let $(p\tau)(\omega) = p(\omega\tau^{-1})$, where $\omega\tau^{-1}$ denotes the inverse image of $\omega$. This means that the measure of any set $S \subset \Omega$ in the state $p\tau$ is $p(S\tau^{-1})$. More, the map on $\mathcal{C}(\Omega)$ given by $(\tau F)(\omega) = F(\omega\tau)$ is an *endomorphism* of $\mathcal{C}$. This is explained and proved in Example 3.23.

If $|\Omega| < \infty$, a stochastic map $T$ induces by duality a right multiplication $T$ on $\ell^1$: to a given $p \in \ell^1$, $pT$ is the functional which gives the value $p \cdot TF$ for the random variable $F$. The dual map has the following properties:

(1) $p \mapsto pT$ is linear;
(2) $T$ is positive, that is, takes non-negative functionals to non-negative functionals;
(3) $\|pT\|_1 = 1$ for all $p \in \Sigma$.

Thus, $T$ takes $\Sigma$ to itself. The first one is obvious. To prove property (2), suppose $p$ is in the positive cone. If $F$ is in the positive cone of $\mathcal{A}$, then so is $TF$, and so $p \cdot TF \geq 0$. Since this is true for all $F$ in the positive cone, we have that $pT$ is in the positive cone, since the cones are dual. Property (3) is also easy: since $pT$ is in the positive cone,

$$\|pT\|_1 = pT \cdot 1 = p \cdot T1 = p \cdot 1 = \|p\|_1.$$

The same result follows immediately if $|\Omega| = \infty$, but is divided up as the union of finite sets $\Omega_i$ (say, the energy-shells) and the indicator function of each $\Omega_i$ is invariant under the action of $T$.

For any $\Omega$, it follows from positivity that a stochastic map is bounded, in the sense of an operator on the Banach space $\mathcal{C}$. For, $F$ can be written in terms of its positive and negative parts: $F = F_+ - F_-$; where $F_\pm \geq 0$. Then $TF_\pm \geq 0$, and

$$|(TF)(\omega)| = |(TF_+)(\omega) - (TF_-)(\omega)|$$
$$\leq (TF_+)(\omega) + (TF_-)(\omega)$$
$$= (T|F|)(\omega).$$

This is true for all $\omega$, so we get the same result for the sup-norm: $\|TF\|_\infty \leq \|T|F|\|_\infty$. Now, for each $\omega$, $|F(\omega)| \leq \|F\|_\infty$, so $\|F\|_\infty 1 - |F|$ is in the positive cone. Hence $T(\|F\|_\infty 1 - |F|)$ is a non-negative function, so

$$(T|F|)(\omega) \leq (T\|F\|_\infty 1)(\omega) = \|F\|_\infty,$$

since $T1 = 1$. Take the sup over $\omega$, to get $\|T|F|\|_\infty \leq \|F\|_\infty$, which combined with the result above, $\|TF\|_\infty \leq \|T|F|\|_\infty$, gives the following important

**Theorem 3.7.** *A stochastic map $T$ on $\mathcal{C}$ is a contraction in the norm, with norm* 1.

The norm of $T$ is not smaller than 1, since $T1 = 1$. Also, when acting on the positive cone, a stochastic map does not increase the maximum of the function, nor decrease the minimum. If $T$ maps a subset $\Omega_0$ of $\Omega$ to itself, then likewise this result is localised to $\Omega_0$.

The set of stochastic maps is a convex set (3.24). This raises the question as to what the extreme points are; the answer is nice: if $\Omega$ is countable, a stochastic map is extremal if and only if it is induced by an endomorphism (3.25). This result is implicit in [Alberti and Uhlmann (1981)] p. 13. There is a general result (the Krein-Milman theorem) giving conditions under which a general point in a convex set is a mixture of the extreme points of the set. Usually it involves the integral over the extreme points, rather than just a sum. Since the number of endomorphisms of $\Omega$ is uncountable, even when $\Omega$ is countable, this is outside the scope of this book. In (3.25) you are asked to prove these results when $|\Omega| < \infty$. Simon [Simon (2005)] gives an accessible proof of the Krein-Milman theorem. Now suppose that $T$ maps $\ell^1(\Omega)$ to itself. That $T$ is bounded follows from the inequality

$$|pT \cdot F| = |p \cdot TF| \leq \|p\|_1 \|TF\|_\infty \leq \|p\|_1 \|F\|_\infty,$$

so $pT$ is a continuous functional of norm $\leq \|p\|_1$. Thus right multiplication by $T$ is a contraction (in the $\ell^1$-norm).

On iterating a contraction map we generate a dissipative dynamics which quite often converges to a fixed point, which could be called equilibrium. For example, if $T$ is a proper contraction, with norm less than 1, then $T^n p$ converges to zero as $n$ becomes large. Such maps, however, are excluded from our theory by the condition $T1 = 1$, which leads to the conservation of probability. Many models of the reaction-diffusion equation have been assiduously studied which violate this condition, in that the dynamics does not preserve the $\ell^1$-norm of the density. Such equations could not be realistic models of dissipative systems.

The right action of a general stochastic map $T$ does not necessarily increase the entropy of the state; this has puzzled some writers, because stochastic maps are widely used by physicists and chemists to represent one time-step. This has been elevated to an axiom by Penrose [Penrose (1970)], a book that, if not the father of the present one, is at least an uncle. The answer to the puzzle is that the dynamics generated by $T$ (with a strictly positive fixed point $p_0$) is not the dynamics of an isolated system, but that of a system in contact with a heat bath. Entropy flows in and out with the heat-flows, and it is another thermodynamic function, sometimes called $\Psi$, or Massieu's function [Bailyn (1995)], that increases. Penrose shows that the relative entropy $S(p|p_0)$, given in (2.10), decreases along the orbit. It is noted in [Thirring (1980)] p. 68 that this is related to the free energy. We shall obtain a sharp version of this result on the decrease in free energy in (5.30). In thermodynamics a system must be isolated (also called adiabatic) before we can be certain that the entropy increases, and so by Renyi's theorem, (3.29), the dynamics must be bistochastic, at least, if $|\Omega| < \infty$.

**Definition 3.8 (Bistochastic Map).** *Let $\Omega$ be a countable sample space, and $T$ a stochastic map on $\mathcal{C}(\Omega)$, acting on the left. Let $T$ map the subspace $\ell^1$ of $\mathcal{C}$ to itself, and denote by $T_1$ the restriction of $T$ to $\ell^1$. Then we say $T$ is bistochastic if and only if $T_1$ is trace-preserving.*

The set of bistochastic maps themselves form a convex set, (3.24). It is obvious that if $T$ is a bistochastic matrix, then any matrix obtained from $T$ by permuting the rows and columns is also bistochastic. Since a bistochastic map is *a fortiori* a stochastic map, it can be written as a convex sum of stochastic endomorphisms. The celebrated theorem of Birkhoff and von Neumann [Birkhoff (1946); Safarov (2005)] states more: the set of extreme points of the set of bistochastic maps is exactly the set of permutations. The hard part of the proof, as explained by Alberti and Uhlmann [Alberti and Uhlmann (1981)], p. 15, is not that permutations are extreme points (3.26), but that there are no extreme points of the set of bistochastic maps other than these. We now prove the Birkhoff-von-Neumann theorem, following [Bapat and Raghavan (1997)]; see also Ando's paper [Birkhoff (1946)].

Given an $n \times n$ matrix $A = [a_{ij}]_{1 \leq i, j \leq n}$ and a permutation $\pi \in S_n$, the *diagonal associated to* $\pi$ is the set $\{a_{1,\pi(1)}, a_{2,\pi(2)}, \ldots, a_{n,\pi(n)}\}$. A diagonal is said to be *positive* if each entry is positive. The product $\prod_i^n a_{i,\pi(i)}$ is called the *diagonal product* of $A$ associated with $\pi$. The sum of all the diagonal products of a matrix is called its *permanent*:

$$\operatorname{per} A := \sum_{\pi \in S_n} \prod_{i=1}^n a_{i,\pi(i)}. \tag{3.7}$$

Two matrices which are related by permuting the rows and columns, have the same permanent. Permanents have the same (Laplace) expansions, with change of sign, as hold for determinants. It is easy to see that if $A$ has block form,

$$A = \begin{pmatrix} B & 0 \\ C & D \end{pmatrix},$$

then per $A = $ per $B$ per $D$. The next theorem is known as the Frobenius-König lemma.

**Theorem 3.9.** *Let $A$ be a matrix with non-negative entries. Then per $A = 0$ if and only if some permutation of the rows and columns of $A$ gives a matrix having an $r \times s$ zero submatrix with $r + s = n + 1$.*

**Proof.**    The theorem is obviously true if $n = 1$, and if $A = 0$.
(1) Suppose per $A = 0$ and as inductive hypothesis, that the theorem holds for square matrices of size $\leq n - 1$. Suppose $A \neq 0$, and that $a_{ij} > 0$. Let $A(i, j)$ denote the matrix $A$ with the $i$-th row and the $j$-th column deleted. As per $A = 0$, every diagonal product is zero. Hence per $A(i, j) = 0$. By the inductive hypothesis, $A(i, j)$ and hence $A$ has a zero $u \times v$ submatrix such that $u + v = n$. So there exist permutation matrices $\Pi_1, \Pi_2$ such that

$$\Pi_1 A \Pi_2 = \begin{pmatrix} B & 0 \\ C & D \end{pmatrix},$$

where $B$ and $D$ are square matrices of size $u$ and $n - u$, respectively. Since per $A = 0$, either per $B = 0$ or per $D = 0$. Suppose that per $B = 0$. By induction, a permutation of the rows and columns of $B$ gives a matrix having a $p \times q$ zero submatrix with $p + q = u + 1$. Without loss in generality, we can assume it to be the matrix formed by the first $p$ rows and the last $q$ columns of $B$. Thus $A$ has a $p \times (q + n - u)$ zero submatrix, and $p + (q + n - u) = u + 1 + n - u = n + 1$ as required. A similar argument holds when per $D = 0$. So the induction hypothesis holds for matrices of size $\leq n$.

(2) For the converse, assume that the matrix of the first $r$ rows and the first $s$ columns is zero, where $r + s = n + 1$. Let $\pi \in S_n$. We must show that $a_{i, \pi(i)} = 0$ for some $i$. Suppose not. Then we must have

$$\{1, 2, \ldots, r\} \cap \{\pi(1), \pi(2), \ldots, \pi(s)\} = \emptyset,$$

which contradicts the fact that $r + s = n + 1$.                                   □

**Lemma 3.10.** *If $T \in \mathbf{M}_n$ is bistochastic, then it has a positive diagonal.*

**Proof.**    It is sufficient to show that per $T > 0$. If per $T = 0$, then by (3.9), $T$ has an $r \times s$ zero submatrix with $r + s = n + 1$. After permuting the rows and columns, we may assume that

$$T = \begin{pmatrix} B & 0 \\ C & D \end{pmatrix},$$

where the zero matrix is $r \times s$ with $r + s = n + 1$. Thus $B$ is $r \times (n - s)$ and $D$ is $(n - r) \times s$. Since the rows of $T$ add up to 1, the rows of $B$ add up to 1, and the entries of $B$ add up to $r$. The columns of $T$ add up to 1, so the entries of $B$ and $C$ together add up to $n - s$. Therefore $r \leq n - s$, which contradicts $r + s = n + 1$. Thus per $T > 0$ is proved. $\square$

We now give the famous Birkhoff-von-Neumann theorem.

**Theorem 3.11.** *Any bistochastic matrix $T$ lies in the convex hull of the permutation matrices.*

**Proof.** This is obvious if $T$ is a permutation matrix. So suppose not; by (3.10), $T$ has a positive diagonal, say $\{a_{1,r(1)}, a_{2,r_2}, \ldots, a_{n,r(n)}\}$. Let $\Pi_1$ be the permutation matrix corresponding to this, thus:

$$\{\Pi_1\}_{ij} = \begin{cases} 1 \text{ if } (i,j) \in \{(1, r(1)), \ldots, (n, r(n))\} \\ 0 \qquad \text{otherwise.} \end{cases}$$

Let

$$\lambda_1 = a_{t,r(1)} := \min\{a_{1,r(1)}, a_{2,r(2)}, \ldots a_{n,r(n)}\}.$$

Since $T$ is not a permutation, not all elements are 1, so $\lambda_1 < 1$. Then we have

$$T = \lambda_1 \Pi_1 + (1 - \lambda_1)W,$$

where

$$W := \frac{1}{1 - \lambda_1}(T - \lambda_1 \Pi_1)$$

is obviously bistochastic. If $W$ is a permutation, we are home; if not we note that $W$ has one more zero than $T$, and so proceeding as above, we can write it as $W = \lambda_2 \Pi_2 + (1 - \lambda_2 X)$, where $X$ is bistochastic with two more zeros than $T$. Proceeding, we write $T$ as the sum of at most $n^2 - n + 1$ permutations, since no bistochastic matrices exist with more than $n^2 - n$ zeros. It follows that any extreme point of the set of bistochastic matrices is a permutation: the extreme point, when written as the mixture of permutations, can only involve one term, and so is itself a permutation. $\square$

Obviously a stochastic map that is symmetric (in the natural basis, $\Omega$), is bistochastic. They generate the special class of *symmetric* Markov chains, in which entropy increases in time. The generalisation to bistochastic maps is both natural and necessary for applications; we have already seen (3.3) that any stochastic map that transforms reasonably under time-reversal must be bistochastic.

In Exercise 3.27 you will show that if a bistochastic map conserves energy, then the permutations making it up must also conserve energy. Together with (3.29), this gives the main result of this section: any linear map obeying both the first and second laws of thermodynamics must be a random dynamics. The next bit of theory shows that under a condition on the spectrum of the matrix $TT^d$, such dynamics leads to the convergence of the system to a mixture of microcanonical states (but not to the canonical state) for any initial state.

## 3.3   Convergence to Equilibrium

In this section we find a criterion, called the spectral gap, which ensures that a random conservative dynamics converges to equilibrium. We shall show that there are large classes of tractable examples. We shall also find that the limit state, at large times, is a microcanonical state, which explains why these occur. The occurrence of canonical and grand canonical states will be explained in the next Chapter.

We begin with a sharp estimate for the entropy gain under a bistochastic map.

**Lemma 3.12.** *Let $w_k \geq 0$, $\sum w_k = 1$ with $\{w_k\}$ a sequence in $\ell^1$ and let $x = \sum w_k x_k$ with $x_k$ a sequence with $0 \leq x_k \leq 1$. Let $s(y) = -y \log y$. Then*

$$\sum_k w_k s(x_k) \leq s(x) - \frac{1}{2} \sum_k w_k (x - x_k)^2. \tag{3.8}$$

**Proof.**   By Taylor's theorem,

$$s(x_k) = s(x) + (x_k - x)s'(x) - (x - x_k)^2/(2\xi_k)$$
$$\leq s(x) + (x_k - x)s'(x) - (x - x_k)^2/2.$$

Multiply by $w_k$ and sum; the middle term drops out and we get the result. $\square$

In the following, the measure on $\Omega$ is taken to be counting measure.

**Theorem 3.13 (Contraction property of bistochastic maps).**
*Suppose that $\|\Omega\| < \infty$, and $T$ be a bistochastic map on $\ell^\infty(\Omega)$. Then $T$ is a contraction in $\ell^\infty$, $\ell^1$ and $\ell^2$.*

**Proof.**   By the Birkhoff-von-Neumann theorem, $T$ is a mixture of permutations, $\tau_i$, say. A permutation preserves each of these norms, denoted $\| \ \|$.

Then

$$\|pT\| = \| \sum_i \lambda_i p\tau_i \| \leq \sum_i \lambda_i \|p\tau_i\| = \sum_i \lambda_i \|p\| = \|p\|.$$

$\square$

The theorem is proved for $\ell^2$ without the assumption of finiteness, and without using the Birkhoff-von-Neumann theorem, in the course of the next theorem.

We now come to the main estimate. In this theorem, a *semiprobability* is an element of $\ell^1$ with components in $[0, 1]$.

**Theorem 3.14.** *Let $T$ be a bistochastic map on $\mathcal{C}(\Omega)$, where $\Omega$ is countable. Let $p$ be a semi-probability. Then*

$$S(pT) - S(p) \geq \frac{1}{2} \left( \|p\|_2^2 - \|pT\|_2^2 \right). \tag{3.9}$$

**Proof.** Choose the natural basis, the points in $\Omega$, and write $T$ as an infinite matrix of non-negative numbers with rows and columns in $\ell^1$. The condition that $T$ is stochastic says that the columns of $T_{jk}$ sum to 1, and the condition that it is the dual of a stochastic map says that the rows of $T_{jk}$ sum to 1. The action of $T$ on the states is given by $p \mapsto q$, where $q_j = \sum_k p_k T_{kj}$ is a convergent sum. Fix $j$ and put $x_k = p_k$ and $w_k = T_{kj}$ for all $k$ in the lemma. Then $x = \sum_k T_{kj} p_k = q_j$. By Eq. (3.8) we get for each $j$:

$$-\sum_k T_{kj} p_k \log p_k \leq -q_j \log q_j - \sum_k T_{kj}(q_j - p_k)^2/2.$$

Sum over $j$ and use $\sum_j T_{kj} = 1$ and $\sum_k T_{kj} = 1$:

$$-\sum_j q_j \log q_j + \sum_k p_k \log p_k \geq \frac{1}{2} \sum_{k,j} T_{k,j}(q_j - p_k)^2$$

$$= \sum_{k,j} T_{kj}(q_j^2 - 2q_j p_k + p_k^2)/2$$

$$= \left( \sum_k p_k^2 - \sum_j q_j^2 \right) /2$$

$$= \left( \|p\|_2^2 - \|pT\|_2^2 \right)/2$$

since $\sum_k p_k T_{kj} = q_j$.

$\square$

The right-hand side is non-negative, since it is equal to $\frac{1}{2} \sum T_{jk}(q_j - p_k)^2$. Incidently, this shows that a bistochastic map is a contraction in $\ell^2$. This estimate gives us our first result for the long-time behaviour of random dynamics. Let us take the case of interest, when the bistochastic map $T$ maps the energy-shells into themselves. Let $E$ be chosen. Since $|\Omega_E| = n$ say, the restriction $T_E$ of $T$ to $\Omega_E$ is an $n \times n$ matrix. Then $TT^d$ is a positive semi-definite matrix, with 1 as an eigenvalue and with the microcanonical state, denoted $e$, as eigenvector. Let us say that $T$ has a *spectral gap* if 1 is a simple eigenvalue of $T_E T_E^d$. The space $\ell^2(\Omega_E)$ is a real $n$-dimensional Hilbert space with scalar product $p \cdot q = pq^d$, (recall that $q^d$ is a column matrix). For any $n \times n$ matrix $T$, we have $pT \cdot q = p \cdot qT^d$. We note that if $p$ is any probability, then its part on the $\Omega_E$ is a semiprobability with $n$ components. We can therefore apply Eq. (3.9) for an $n \times n$ matrix with spectral gap, for which we have the following improvement:

**Theorem 3.15.** *Let $|\Omega| = n < \infty$ and let $T$ be a bistochastic map with a spectral gap, $\gamma > 0$, so that $TT^d$ has no eigenvalues between $1 - \gamma$ and 1. Let $e$ denote the uniform probability on $\Omega$. Then for any semi-probability $p$ we have*

$$S(pT) - S(p) \geq \frac{1}{2}\gamma \left\| p - \|p\|_1 e \right\|_2^2. \tag{3.10}$$

*Proof.* We note that

$$\|p\|_2^2 - \|pT\|_2^2 = p(1 - TT^d) \cdot p = (p - \|p\|_1 e)(1 - TT^d) \cdot (p - \|p\|_1 e)$$

since $(1 - TT^d)$ annihilates $e$. Also, $p - \|p\|_1 e$ is orthogonal to $e$ since

$$(p - \|p\|_1)e \cdot e = p \cdot e - \|p\|_1 e \cdot e = \frac{1}{n}\|p\|_1 - \|p\|_1 \frac{1}{n} = 0.$$

Now $1 - TT^d$ is bounded by $\gamma$ on the space orthogonal to $e$, so

$$(p - \|p\|_1 e)(1 - TT^d) \cdot (p - \|p\|_1 e) \geq \gamma\|p - \|p\|_1 e\|_2^2.$$

Combining this with (3.9) gives the theorem.                        □

This shows that unless we are at the fixed point of the map the entropy is increased by a non-zero amount when we apply $T$. This is a very useful property, in non-linear as well as linear dynamics: it allows us to prove a fundamental law of chemistry called (by Tolman [Tolman (1938)]) 'the law of detailed balance'. It is the following. Suppose that dynamical law $\tau$ is a mixture of two maps, $\tau_1$ and $\tau_2$, and that $S$ is increased by $\tau_1$ unless we are at a fixed point of $\tau_1$, and by $\tau_2$ unless we are at a fixed point of $\tau_2$. Then a fixed point of $\tau$ must be a simultaneous fixed point of both $\tau_1$ and $\tau_2$. The proof is easy, using the concavity of the entropy: if $p$ is a fixed point of $\tau$, then

$$S(p) = S(p\tau) \geq \lambda S(p\tau_1) + (1 - \lambda)S(p\tau_2)$$
$$> \lambda S(p) + (1 - \lambda)S(p)$$
$$= S(p),$$

a contradiction unless $p$ is a simultaneous fixed point. Tolman interprets this as saying that in a complex chemical reaction (where several processes can occur together), chemical equilibrium is not reached until all the constituent processes are separately in balance, meaning at equilibrium. There is no perpetual motion, where $A \to B \to C \to A$. Our proof of this, above, needed the inequality rather than equality in the estimate (3.10), and depends on $\gamma > 0$.

Our inequality also allows us to apply Lyapunov's 'direct method' to show that the system converges at large times. We now give a simple version of this theory.

**Definition 3.16 (Lyapunov function).** *Let $\tau$ be a continuous map of a separable compact space $\Sigma$ to itself. Let $S$ be a continuous function on $\Sigma$ such that $S(p\tau) > S(p)$ unless $p$ is a fixed point. Then we say that $S$ is a Lyapunov function for the dynamics given by $\tau$.*

We now give the main theorem, which applies to both linear and non-linear dynamics.

**Theorem 3.17 (Lyapunov).** *Let $\Sigma$ be a compact separable space and let $\tau$ be a continuous map from $\Sigma$ to $\Sigma$. Suppose that there is a unique fixed point $p_0$ under $\tau$. Let $S$ be a Lyapunov function for $\tau$. Then for any initial point $p$, we have*

$$p\tau^n \to p_0 \text{ as } n \to \infty.$$

***Proof.*** The sequence $p\tau^n$ has a convergent subsequence $\{p\tau^{n_i}\}_{i \in \mathbf{N}}$, by the Bolzano-Weierstrass theorem. Let $p_\infty$ be its limit. Since $S$ is continuous, $S(p\tau^{n_i}) \to S(p_\infty)$ as $i \to \infty$. Also, as $\tau$ is continuous,

$$(p(\tau^{n_i})\tau \to p_\infty \tau. \tag{3.11}$$

Note that $\tau^{n_i}\tau = \tau^{(n_i+1)}$ might not be equal to $\tau^{n_{i+1}}$ so we cannot immediately say that $p\tau^{n_i}\tau \to p_\infty$. We must use the existence of the Lyapunov function, $S$, which is bounded as it is a continuous function on a compact set. Then $S(p\tau^n)$ is bounded and non-decreasing, and so converges. This implies that each of its subsequences $S(p\tau^{n_i})$ and $S(p\tau^{(n_i+1)})$ converges to the same limit. The first converges to $S(p_\infty)$ as we have seen; so the second must too. But it is $S(p\tau^{n_i}\tau)$ which converges to $S(p_\infty\tau)$ by (3.11) since $S$ is continuous. Hence $S(p_\infty) = S(p_\infty\tau)$. Since the entropy is strictly increasing, except at a fixed point, we have that $p_\infty$ is a fixed point. Since there is only one fixed point, $p_0$ and the above argument applies to any convergent subsequence of the orbit, we have proved that *every* convergent subsequence of the orbit converges to $p_0$. This means that the orbit itself converges to the fixed point, by the following curious theorem, (3.28): if $\{x_n\}$ is a sequence in a compact space such that every convergent subsequence has the same limit, then $\{x_n\}$ itself is convergent.                  □

It is necessary for the function $S$ to strictly increase along an orbit, to get the result. For example, if $\Omega = \{0, 1\}$ and $\Sigma$ is the set of states on $\Omega$, then the permutation $\tau$ which exchanges the two elements of $\Omega$ is ergodic, since there is no proper non-trivial invariant subset. The right action on $\Sigma$, is the permutation matrix

$$\tau = \begin{pmatrix} 0 & 1 \\ 1 & 0 \end{pmatrix}.$$

This conserves entropy but does not increase it. The sequence $p\tau^n$ has two subsequences, those for odd and even values of $n$, which are constant and so convergent, but not to the same limit unless $p$ is a fixed point. In this case, $\tau\tau^d$ is the unit matrix, of which 1 is a double eigenvalue. This means that the spectral gap $\gamma$ of (3.10) is zero. The proof breaks down since 1 is a repeated eigenvalue and $\tau\tau^d$ has a fixed point orthogonal to $e$.

In our application of this theorem in this Chapter, the entropy is a Lyapunov function if $\gamma$ is positive; in finite dimensions, this is the same as saying that 1 is a simple eigenvalue of $TT^d$. To ensure this it is enough for $T$ to be a matrix with positive entries. This is part of the Perron-Frobenius theorem. Let us say that a vector is *positive* if it is non-zero and its components are non-negative, and *strictly positive* if its components are positive. Also a matrix $T^d$ is *positively improving* if $pT$ is strictly positive whenever $p$ is positive. Clearly, a matrix with positive entries is positively improving.

**Theorem 3.18 (Uniqueness part of Perron-Frobenius).**     *Let $T$ be a positively improving $n \times n$ matrix, and having 1 as an eigenvalue with a positive eigenfunction $p$. Then 1 is simple.*

***Proof.*** Suppose if possible that there are two linearly independent eigenvectors $p_1$ and $p_2$ where $p_1$ is strictly positive. If $p_2$ is complex, then the real and imaginary parts are real eigenfunctions, since $T$ is real. So we may assume that $p_2$ is real with some negative entries (if it is positive, then take $-p_2$ instead, in the following argument). For real $\lambda$, consider the eigenvector $p(\lambda) = p_1 + \lambda p_2$, which has eigenvalue 1. For $\lambda$ sufficiently close to zero, this is strictly positive, but as we increase $\lambda$ a value is reached when $p(\lambda)$ has some zero entries, and the rest positive. Then $p(\lambda) = p(\lambda)T$ must be strictly positive, contradicting the presence of zero components. $\qquad\square$

The theorem has a useful corollary; if the matrix $T$ obeys the conditions of the theorem, except that only some power $T^M$ has positive entries, then 1 is simple.

A non-negative matrix $T$ is *irreducible* if there exists $M$ such that, given any two points $\omega_0$, $\omega$, there is a path of transitions $\{\omega_0, \omega_1, \ldots, \omega_M = \omega\}$ such that $T_{\omega_{i-1},\omega_i} > 0$, $i = 1, \ldots, M$. This has the property that its $M^{\text{th}}$ power is strictly positive. Such a matrix, then, has 1 as a simple eigenvalue, and if $\Omega$ is finite, the spectral gap $\gamma$ is positive. A useful special case arises when the diagonal entries of $T$ are positive; the irreducibility of the matrix follows if for each pair $(\omega_0, \omega)$ there is a chain connecting them of length $\leq M$; we then get a chain of length $M$ by inserting enough trivial hops $\omega_i \to \omega_i$. For example, the $j$-th power of the diffusion matrix with Neumann boundary conditions (3.4), with $\lambda < 1$ has non-zero elements making a band of width $2j + 1$ down the diagonal, which causes spatial diffusion of a particle at the rate of $\lambda$ units per unit time. Clearly, some power of it is strictly positive on each number-shell. Another easy case is the bistochastic $n \times n$ matrix obtained by mixing the identity matrix with a cycle of length $n$.

We are now in a position to prove the main theorem for random conservative dynamics:

**Theorem 3.19.** *Let $\Omega$ be a countable space and let $\mathcal{E}$ be a random variable (the energy) such that each energy-shell is finite. Let $T$ be a bistochastic map on $\mathcal{C}(\Omega)$, made up of permutations, each leaving every energy-shell invariant. Suppose that $T$ restricted to each energy-shell $\Omega_E$ is irreducible. Then for any initial state $p$, we have $pT^n \to p_\infty$ as $n \to \infty$. The convergence is in $\ell^1$, and the limit state is a mixture of microcanonical states*

$$p_\infty = \sum_E |\Omega_E|^{-1} p(E) \chi_E, \text{ where } p(E) = \sum_{\omega \in \Omega_E} p(\omega).$$

*The mean energy of the state is constant in time, and entropy properly*
*increases unless the initial state is a mixture of microcanonical states.*

**Proof.**    Restrict the map $T$ to an energy-shell $\Omega_E$.    It is a finite-
dimensional problem and the bistochastic matrix representing the map $T_E$
is irreducible.  Hence there is a non-zero spectral gap $\gamma_E$, and by (3.10),
entropy is a strict Lyapunov function.  Also, as the eigenvalue 1 is simple
(for the restricted map), we can apply Lyapunov's theorem, which says that
iterates of the map $T_E$ applied to $p$ converge, preserving the $\ell^1$-norm.  Since
the fixed point is the uniform distribution, we get the result.          $\square$

In particular, if all the probability is initially in one energy-shell, then
the system converges to the microcanonical state with that energy. This
tendency to move to the uniform distribution is called the *equipartition of
energy*.

    This is a partially satisfactory result. We have not had to use the ergodic
*hypothesis*, since we have a good criterion allowing us to prove the stronger
property of having a spectral gap. For example, the diffusion model satisfies
this, and so converges to the uniform distribution (3.19). But we are not
able to set up an energy-conserving model which redistributes probability
between energy-shells, and so are still unable to understand why systems
move to the canonical state. This is achieved in the next Chapter.

## 3.4   Markov Chains

Let $\Omega_0 = \{\omega_1, \ldots, \omega_N\}$ and $\mathcal{A} = \mathcal{C}(\Omega_0)$. Let $T$ be a stochastic map on
$\mathcal{A}$, acting on Span $\Omega$ on the left and right, obeying (2.31). Thus the sure
random variable 1 is represented by $\sum_j \omega_j$. Then we define the *stochastic
matrix* $T_{ij}$ of $T$ to be

$$T_{ij} = \omega_i \cdot T\omega_j.$$

It has the property that $\sum_j T_{ij} = 1$ for all $j$. For

$$\sum_{j=1}^{N} T_{ij} = \omega_i \cdot T \sum_j \omega_j$$

$$= \omega_i \cdot \sum_j \omega_j = 1.$$

We can interpret $T_{ij}$ as the probability that the system is in state $\omega_j$ at
time $t + 1$ given that it is in state $\omega_i$ at time $t$. It is then a conditional

probability. Suppose that the initial probability is $\overset{\circ}{p} \in \Sigma$. We can use the definition (2.5) of conditional probability to define the probability that at times $t = 0$ and $t = 1$ the sample points are $(\omega(0), \omega(1))$:

$$\text{Prob}\{\omega(0), \omega(1)\} = \overset{\circ}{p}(\omega(0))P(\omega(1)|\omega(0)) = \overset{\circ}{p}(\omega(0))T_{01}.$$

Proceeding at successive times, we can define the probability that the sample point $\omega(t)$ trace out a given walk $(\omega(0), \omega(1), \ldots, \omega(n))$ in $n$ time-steps to be

$$\text{Prob}\{\omega(0), \omega(1), \ldots, \omega(n)\} = \overset{\circ}{p}(\omega(0))T_{01}T_{12} \ldots T_{n-1,n}.$$

If we are not interested in the intermediate states, but just want the probability that at time $t = n$ the state is $\omega(n)$, we sum over the intermediate positions at all previous times, to get

$$\text{Prob}\{\omega(n)\} = \sum_{\omega(n-1),\ldots,\omega(0)} \overset{\circ}{p}(\omega(0))T_{0,1} \ldots T_{n-1,n} = \overset{\circ}{p} T^n \cdot \omega(n)$$

since this is just matrix multiplication; so our rule for calculating probabilities agrees with the given dynamics on $\Sigma$ when limited to the arrival point.

In the theory of the random walk we are interested in the path by which $\omega(n)$ was arrived at, not only in $\omega(n)$ itself. This extra detail can be incorporated at the cost of a much larger sample space. Suppose for simplicity that we know that the sequence started at $\omega(0)$. Then for a sequence of indefinite length we need the space

$$\Omega = \{\omega(0)\} \times \Omega_0 \times \Omega_0 \ldots$$

(one copy of $\Omega_0$ for each time $n \geq 1$). If the sequence is infinite, care is needed in the definition of measurable sets, measures and so on. We cannot assign a non-zero probability to each sample point, a sequence of points in $\Omega_0$, since such points are uncountable in number. Instead, this is how we proceed; let $\mathcal{B}_n$ be the Boolean ring generated by all cylinder sets having as base any subset of $A \subseteq \Omega_0$ located at time $n$; that is, any set of the form

$$\Omega_0 \times \Omega_0 \times \ldots \times A \times \Omega_0 \ldots \qquad \text{with } A \subseteq \Omega_0 \text{ in the } n\text{-th place.}$$

Various choices of $A$ represent events that can be measured at time $n$. Physically, we can measure the present, and might have a record of past measurements, but we cannot know the future. It makes sense to define a Boolean ring $\mathcal{B}_{\leq n}$ as that generated by $\{\mathcal{B}_j : j \leq n\}$. For example, a measurable set (relative to $\mathcal{B}_{\leq n}$) could be a cylinder set defined by gates

$A_1, A_2, \ldots, A_n$, in which event $\Omega_j$ happened at time $j$, with $1 \leq j \leq n$. Clearly, $\mathcal{B}_{\leq n+1} \supseteq \mathcal{B}_{\leq n}$. The increasing family of Boolean rings $\{\mathcal{B}_{\leq n}\}$ is called a *filtration*. We can define the probability of any cylinder set, as the sum of the probabilities of all random walks passing through the (finite number of) gates. This is called a cylinder measure. We shall denote it by $\mu$. Books on probability theory [Neveu (1965)] prove the Kolmogorov extension theorem, which says that this measure on the cylinder sets can be uniquely extended to a countably additive measure on the $\sigma$-ring $\mathcal{B}$ generated by the cylinder sets. We call the extension $\mu$ as well; it is the Kolmogorov measure of the chain. The semi-group of time-evolution acts naturally on the sample-space, by left-shift: $(\omega\tau)(t) = \omega(t+1)$, where the sample point $\omega$ is the sequence $\{\omega(t) \in \Omega_0\}_{t=0,1,\ldots}$. This is $\mathcal{B}$-measurable. The process we have obtained has the Markov property: the future, conditioned by the present, is independent of the past. This can be expressed in every-day terms by saying that the random walk starts afresh at each time, from where it's at. In terms of conditional expectations, the Markov property can be expressed neatly:

**Theorem 3.20.** *Let $F : \Omega \to \mathbf{R}$ be $\mathcal{B}$-measurable. Then*

$$\mathbf{E}[F|\mathcal{B}_{\leq n}] = \mathbf{E}[F|\mathcal{B}_n] \qquad \text{for each } n.$$

We shall not prove this. For further details on Markov chains, see [Revuz (1975)].

One of the problems of the non-linear dynamics, which we meet in the next Chapter, is that, unlike in a Markov chain, the time-evolution of $p$ in $n$ steps is not obtained by summing the probabilities of all walks leading to the final point; so the traditional theory of stochastic processes needs modifying. If the dynamics $p \mapsto pT$ is linear, as in a Markov chain, we can extend it to act on all functions in $\ell^1(\Omega)$. We can equally well consider $t$ steps in the dynamics to act on the algebra of random variables, $F \mapsto T^t F$. Let $t_1 < t_2 \ldots < t_n$ be $n$ integers denoting times, and let the initial state be $p$. The many-time correlation functions of random variables $F_1, \ldots, F_n$ are defined to be

$$p \cdot T^{t_1} F_1 \ldots T^{t_n} F_n.$$

This can be expressed in terms of the Kolmogorov measure $\mu$ on the Markov chain:

$$\int d\mu(\omega) F_1(\omega(t_1)) F_2(\omega(t_2)) \ldots F_n(\omega(t_n)).$$

In the next chapter we shall consider dynamics where the map on the states is non-linear, and cannot easily be used to define a 'Heisenberg' picture $F \mapsto TF$. In this case the definition of the many-time correlation functions is not obvious.

## 3.5   Exercises

Exercise 3.12. Show that the property of being on the same orbit under a group is an equivalence relation.

Exercise 3.13. Let $\tau$ be a permutation of $\Omega$. Show that the map $F \mapsto \tau F$ is an automorphism of $\mathcal{C}(\Omega)$.

Exercise 3.14. Show that a group and its opposite are isomorphic. Hint: two groups $G_1$ and $G_2$ are said to be isomorphic if there exists a bijective map $\phi : G_1 \to G_2$ such that $\phi(g \circ_1 h) = \phi(g) \circ_2 \phi(h)$, where $\circ_i$ is the multiplication law in the group $G_i$, $i = 1, 2$.

Exercise 3.15. Suppose that $|\Omega| < \infty$. Show that any automorphism of $\mathcal{C}(\Omega)$ is of the form $F \mapsto \tau F$, where $\tau$ is a permutation of $\Omega$. Show that the group of automorphisms of $\mathcal{C}(\Omega)$ is isomorphic to the permutation group of $\Omega$.

Exercise 3.16. Show that the left action by a stochastic map $T$, $F \mapsto TF$, is associative, thus $(T_1 T_2)F = T_1(T_2 F)$ for all random variables $F$ and all stochastic maps $T_1, T_2$.

Exercise 3.17. Let $\tau$ be a permutation of $\Omega$ leaving invariant the energy-shells defined by a random variable $\mathcal{E}$. Suppose that $\tau$ is ergodic relative to energy. Prove that the only states invariant under $\tau$ are mixtures of microcanonical states.

Exercise 3.18. Show that the finite difference equation for the density $N(x)$, coming from the matrix in (3.4) corresponds to the Neumann boundary condition $\partial \mathcal{N}/\partial x = 0$. Write the marginal distribution of $p \in \Sigma(\Omega)$ onto the factor $\Omega_x$ as $(q_x, p_x)$. Show that the time evolution of $p_x$ along the orbit of $\tau$ obeys the same equation. This is the Fokker-Planck equation for this dynamics.

Exercise 3.19. Show that the dynamics given by the matrix in (3.4) takes any initial state to the uniform state at large times.

Exercise 3.20. Let $\Omega$ be a countable set, and $\tau$ a permutation of $\Omega$. Let $\Omega$ be the disjoint union of $\Omega_1$ and $\Omega_2$, and suppose that $\Omega_1$ and $\Omega_2$ are invariant under $\tau$. Show that each is invariant under $\tau^{-1}$.

Exercise 3.21. A sample space $\Omega$ is the union of energy-shells $\Omega_E$, with $|\Omega_E| < \infty$, and $\tau$ is an ergodic permutation of $\Omega$. Let $p$ be a point measure with support in $\Omega_E$. Show that

$$p(1 + \tau + \tau^2 + \ldots + \tau^n)p/(n+1) \to |\Omega_E|^{-1}\chi_E \text{ as } n \to \infty$$

in $\ell^1$-norm.

Exercise 3.22. Show that the mapping $\tau = \sum \lambda_i \tau_i$, where $\tau_i$ are permutations of $\Omega$ and the $\lambda_i$ are probabilities on $\{1, 2, \ldots, n\}$, is a bistochastic map.

Exercise 3.23. Let $\tau : \Omega \to \Omega$ be any map, and define $\tau F$ by $\omega \cdot \tau F := \omega\tau \cdot F$. Prove that $\tau$ is a stochastic endomorphism, that is, it is a stochastic map, and obeys the two properties of an endomorphism

(1) For $F_i \in \mathcal{C}$ and real $\lambda_i$, we have linearity:

$$\tau(\lambda_1 F_1 + \lambda_2 F_2) = \lambda_1 \tau F_1 + \lambda_2 \tau F_2.$$

(2) $\tau(F_1 F_2) = \tau F_1 \tau F_2$.

Also prove the converse, that any stochastic endomorphism of $\mathcal{C}(\Omega)$ is induced from a unique map $\tau$. Hint: construct $\tau$ from the action of the endomorphism on functions with support at a single point.

Exercise 3.24. Show that the set of stochastic maps on $\mathcal{C}(\Omega)$ is convex. Show that the subset of bistochastic maps is also convex.

Exercise 3.25. Suppose that $|\Omega| < \infty$; show that a stochastic map $T$ is an extreme point of the convex set of stochastic maps if and only if it is induced by an endomorphism. Show that any stochastic map is a convex mixture of endomorphisms. [Hint: use the Krein-Milman theorem]

Exercise 3.26. Show that a permutation is an extreme point of the set of bistochastic maps.

Exercise 3.27. Let $T$ be a bistochastic map on $\mathcal{C}(\Omega)$ that conserves energy: $T\chi_E = \chi_E$ and $\chi_E T = \chi_E$ for all $E$, where $\Omega = \bigcup \Omega_E$ is the division of $\Omega$ into energy-shells. If each $\Omega_E$ is finite, show that $T$ is a mixture of energy-conserving permutations.

Exercise 3.28. Let $\{x_n\}$ be a sequence of points in a compact space such that every convergent subsequence has the same limit. Show that $\{x_n\}$ is convergent.

Exercise 3.29. Show that a stochastic matrix which increases the entropy of every state must be bistochastic.

Exercise 3.30. Let $\Omega_x = \{0, 1\}$ and $\Lambda = \{1, 2, \ldots, n\}$, $\Omega = \prod_x \Omega_x$, and $\mathcal{E}(\omega) = \sum_x \omega_x$. Let $\tau_x$ denote the permutation which exchanges site $x$ if occupied, with the site $x + 1$, if unoccupied, and makes no other changes to $\omega$; here $\tau_n$ exchanges 1 and $n$. Let $T$ be any mixture of all these $\tau_x$, for $x = 1$ to $x = n$, with non-zero weights. Show that the system converges to a distribution with uniform energy-density in space from any initial state.

Exercise 3.31. Show that if $\| \ \|$ is a seminorm on Span $\Omega$ that is contracted by any permutation, then it is contracted by any bistochastic map. Hence show that all norms

$$\|p\|_\alpha := \left( \sum_j |p_j|^\alpha \right)^{1/\alpha}$$

on the sequence space are contracted by any bistochastic map.

# Chapter 4

# Isolated Dynamics

## 4.1 The Boltzmann Map

We have seen that the random mixture of reversible maps can lead to thermodynamics, including the convergence to equilibrium; however in this way we get only the microcanonical states and not the canonical states. Within the linear theory there is no mechanism for mixing up the various energy-shells, or the number-energy-shells if there is more than one conserved quantity, if we wish to retain the conservation laws. The extra randomness needed to mix the shells can be achieved by a construction hidden in Boltzmann's equation, and called the Stosszahlansatz by him. This is called the hypothesis of molecular chaos in early English text-books [Fowler (1936); Tolman (1938)]. We shall express this in a general but very simple form. Suppose that $\Omega = \Omega_1 \times \Omega_2$, and also that the energy of the whole system is the sum of the energies of the two subsystems, thus:

$$\mathcal{E}(\omega) = \mathcal{E}_1(\omega_1) + \mathcal{E}_2(\omega_2) \text{ where } \omega = (\omega_1, \omega_2).$$

This expresses that there is no interaction between the two systems. Then for a general state $p \in \Sigma(\Omega)$, the mean energy is

$$p \cdot \mathcal{E} = \sum_{\omega_1, \omega_2} p(\omega_1, \omega_2) \left( \mathcal{E}_1(\omega_1) + \mathcal{E}_2(\omega_2) \right)$$

$$= \sum_{\omega_1} \mathcal{E}_1(\omega_1) \sum_{\omega_2} p(\omega_1, \omega_2) + \sum_{\omega_2} \mathcal{E}_2(\omega_2) \sum_{\omega_1} p(\omega_1, \omega_2)$$

$$= p_1 \cdot \mathcal{E}_1 + p_2 \cdot \mathcal{E}_2.$$

$p_1$ and $p_2$ are the marginal probabilities. From this it follows that if we replace $p$ by $p_1 \otimes p_2$, which is the stoss map, then the mean energy is unchanged. In (2.14) we saw that the entropy of $p_1 \otimes p_2$ is not less than that of $p$. We can then define the dynamics as follows; let $T$ be an energy-conserving bistochastic map on $\Sigma(\Omega)$, and define one time-step by

$$p \mapsto pT \mapsto pT\mathcal{M}_1 \otimes pT\mathcal{M}_2 := pB_T. \tag{4.1}$$

We call this the Boltzmann map defined by $T$. Since each step conserves the average energy and does not decrease the entropy, our map $B_T$, which is non-linear, obeys both the first and the second laws of thermodynamics. The stoss map introduces a bit of mixing between energy-shells, while preserving the mean energy, and quite often is enough to drive the system to the canonical or grand canonical state. The discrete Boltzmann equation ([Monaco and Preziosi (1991)]) is actually of this form: after a scattering, the marginal distributions of the two scattered particles are calculated, and the particles then re-enter the population as independent particles. This was the 'hypothesis'. Since in general $pT$ is actually not equal to $pB_T$, it is fruitless to try to prove the hypothesis. We shall use it as a part of the model; it expresses random effects such as exterior noise that is not included in the mapping $T$. This noise does not inject any energy, but destroys the correlations set up between the two systems by the scattering $T$. The system for which the Boltzmann map is ideal is a pair of 'well stirred' chemical reactors, joined by a semi-permeable membrane. Chemicals in pots (i) and (ii) diffuse into the membrane, where they react or diffuse to one side or the other. After each time step, each reactor is stirred by a paddle, which does not change the amount of each chemical, or add a significant quantity of heat. All it does is to destroy the correlation between the two reactors. The Boltzmann map then should not be regarded as a possibly bad approximation to the true Hamiltonian dynamics, but as theory in its own right. Thus, the next example is a model of two well-stirred reactors containing three chemicals, and both pots are densely occupied, as in a liquid rather than a gas.

**Example 4.1.** Let $\Omega_1 = \{A, B, C\} = \Omega_2$, and $\Omega = \Omega_1 \times \Omega_2$. This falls into the field picture if $\Lambda$ has two points and $\Omega_x = \Omega_1 = \Omega_2$.

We wish to describe the chemical process

$$A + C \rightleftharpoons 2B.$$

For this to obey the first law, our energy function must satisfy $\mathcal{E}(A) + \mathcal{E}(C) = 2\mathcal{E}(B)$. Assume this, and that there are no accidental relations, such as all being zero. Then $\Omega$, which has nine points, splits into the energy-shells

$$\{AA\}, \quad \{AB, BA\}, \quad \{AC, BB, CA\}, \quad \{BC, CB\}, \quad \{CC\}.$$

An energy-conserving $T$ must be the identity on the one-point shells $\{AA\}$ and $\{CC\}$. Let us define $T$ on the sets $\{AB, BA\}$ and $\{BC, CB\}$ to be the symmetric matrix

$$T = \begin{pmatrix} 1 - \kappa & \kappa \\ \kappa & 1 - \kappa \end{pmatrix}. \tag{4.2}$$

The number $\kappa$ is the probability that the particles $A$ and $B$ exchange places in one time-step, and so $0 < \kappa < 1$. We have chosen the same rate for the exchange of $B$ and $C$. The action of $T$ on the remaining energy-shell is chosen to be the $3 \times 3$ matrix

$$\begin{array}{c} \\ AC \\ BB \\ CA \end{array} \begin{array}{ccc} AC & BB & CA \\ \begin{pmatrix} 1 - \lambda - \kappa & \lambda & \kappa \\ \lambda & 1 - 2\lambda & \lambda \\ \kappa & \lambda & 1 - \lambda - \kappa \end{pmatrix}. \end{array} \tag{4.3}$$

So $\kappa$ is the rate that $A$ and $C$ exchange places, and $\lambda$ is the rate of reaction. We get a strictly positive bistochastic matrix on the energy-shell if $\kappa > 0$, $\frac{1}{2} > \lambda > 0$ and $\kappa + \lambda < 1$. The action of $T$ on $\Omega$ is the product of the actions on the various energy-shells; these commute, so the order of the product makes no difference. The restriction of $T$ to each energy-shell is positively improving. Thus the map $T$ increases the entropy unless $p$ is constant on each of these energy-shells. The stoss map increases entropy unless the probability $p$ is a product, so any fixed point has the form $p_1 \otimes p_2$, and whether at equilibrium or not, this is the form of the state just after the stoss map. One step in the dynamics starts with $p = p_1 \otimes p_2$, then transforms this to $pT = p'$ which then leads to new marginals $p_1'$ and $p_2'$. For our example, the first step is

$$p'(AA) = p(AA)$$
$$p'(AB) = (1 - \kappa)p(AB) + \kappa p(BA)$$
$$p'(BA) = (1 - \kappa)p(BA) + \kappa p(AB)$$
$$p'(CC) = p(CC)$$
$$p'(BC) = (1 - \kappa)p(BC) + \kappa p(CB)$$
$$p'(CB) = (1 - \kappa)p(CB) + \kappa p(BC)$$
$$p'(AC) = (1 - \lambda - \kappa)p(AC) + \lambda p(BB) + \kappa p(CA)$$
$$p'(BB) = (1 - 2\lambda)p(BB) + \lambda p(AC) + \lambda p(CA)$$
$$p'(CA) = (1 - \lambda - \kappa)p(CA) + \lambda p(BB) + \kappa p(AC).$$

From this the marginals $p'_1$ and $p'_2$ can be computed. Thus

$$
\begin{aligned}
p'_1(A) &= p'(AA) + p'(AB) + p'(AC) \\
&= p(AA) + (1 - \kappa)p(AB) + \kappa p(BA) \\
&\quad + (1 - \lambda - \kappa)p(AC) + \lambda p(BB) + \kappa p(CA) \\
&= p_1(A)p_2(A) + (1 - \kappa)p_1(A)p_2(B) + \kappa p_1(B)p_2(A) \\
&\quad + (1 - \lambda - \kappa)p_1(A)p_2(C) + \lambda p_1(B)p_2(B) + \kappa p_1(C)p_2(A) \\
&= p_1(A) + \kappa\,(p_2(A) - p_1(A)) - \lambda\,(p_1(A)p_2(C) - p_1(B)p_2(B))
\end{aligned}
$$

where we have simplified using $p_i(A)+p_i(B)+p_i(C) = 1$, $i = 1, 2$. Similarly the other five marginals $p'_1(B) \ldots p'_2(C)$ can be computed; see (4.122).

This example shows that the resulting dynamics is non-linear; the state after one time step is a quadratic function of the components of the state before the step. Because of this, the time evolution of a mixed state $p$ is not the mean of the evolution of the pure states into which $p$ decomposes. We have replaced the random path through the sample space with a non-linear, non-random path through the state space $\Sigma(\Omega)$. There is a kind of coherence to the mixture $p$, which we contrast with the incoherence of the every-day mixture; in the every-day mixture the states being mixed represent systems in different boxes.

The example describes a model with a diffusion term, $\kappa(p_1 - p_2)$, giving a flow proportional to the gradient of the density. If we extend the model to the case where $\Lambda = \{1, 2, \ldots, n\}$ then the gradient term builds up to the second derivative, and we get a discrete approximation to the heat equation. The non-linear term $\lambda\,(p_1(A)p_2(C) - p_1(B)p_2(B))$ causes the reaction. This is of the form used by chemists under the name 'the law of mass action'. The limit to continuous time is called the reaction-diffusion equation. If $\kappa$ is much bigger than $\lambda$ then $p_1$ and $p_2$ rapidly become equal, and the system behaves as two particles in a single container.

From the equations of motion, one can see that the following are conserved quantities:

$$
\begin{aligned}
p_1(A) + p_1(B) + p_1(C) &= 1 \\
p_2(A) + p_2(B) + p_2(C) &= 1 \\
2\,(p_1(A) + p_2(A)) + p_1(B) + p_2(B) &= 2N_A + N_B \\
2\,(p_1(C) + p_2(C)) + p_1(B) + p_2(B) &= 2N_C + N_B \ .
\end{aligned}
$$

The first two say that there is always exactly one particle of some sort at each point of $\Lambda$. The third says that the removal of one $A$ from the total number of $A$ is accompanied by the increase by two in the total number of $B$. The fourth is not independent of the others, since $N_A + N_B + N_C = 2$. The value of $2N_A + N_B$ on the orbit is fixed by the initial state. Since each of the restrictions of $T$ to the energy-shells is positively improving, the law of detailed balance proves that the only fixed points are fixed points of all the maps simultaneously. The diffusion terms require $p_1 = p_2$ at equilibrium, and, putting this in the reaction term, we get $p(B)^2 = p(A)p(C)$ at equilibrium. If both $2N_A + N_B$ and $2N_C + N_B$ are initially non-zero, then either some $B$ is present, or if there is no $B$, then there is a non-zero amount of both $A$ and $C$. The chemical interpretation of this situation is that the reaction can proceed. A quite different situation arises if either, $2N_A + N_B$ or $2N_C + N_B$ is zero; then there is no $B$ present, and one of $A$ and $C$ is absent. That is, the configuration is either $(A, A)$ or $(C, C)$. Then the reaction cannot proceed in either the forward or backward direction. This configuration is on the boundary of the space $\Sigma(\Omega)$ of states, and is an unstable fixed point, in that a small addition of the missing ingredient (the 'catalyst') leads to the first case, where both $2N_A + N_B$ and $2N_C + N_B$ are non-zero, and the reaction proceeds. It follows by Lyapunov's theorem that the system converges to equilibrium for large times. The fixed point can be interpreted as the canonical state for some beta. Let $\mathcal{N}_A(\omega)$ denote the number of $A$-particles in the configuration $\omega$; it is the random variable taking values 0, 1, or 2, whose mean is $N_A$. Similarly $\mathcal{N}_B$ and $\mathcal{N}_C$ are defined. Then, the state $p_\beta \in \Sigma(\Omega)$ given by

$$p(\omega) = Z^{-1} \exp -\beta(E_A \mathcal{N}_A + E_B \mathcal{N}_B + E_C \mathcal{N}_C)$$

obeys all the equilibrium conditions, where $\beta$ is determined by the initial conditions, and $E_A + E_C = 2E_B$. The conservation law of $2\mathcal{N}_A + \mathcal{N}_B$ is equivalent to the conservation of energy $\mathcal{E} = E_A \mathcal{N}_A + E_B \mathcal{N}_B + E_C \mathcal{N}_C$, provided that $E_A + E_C = 2E_B$. Thus the stoss map adds enough entropy to take the system to thermal equilibrium, provided that the initial point is not one of the unstable equilibrium points on the boundary of $\Sigma$, such as $(A, A)$ or $(C, C)$.

**Example 4.2.** This models cotransport through a membrane. Suppose that we have two cells containing chemicals $\{A, B, C, D\}$ which can undergo the reaction

$$A + B \rightleftharpoons C + D. \tag{4.4}$$

Suppose that the cells are separated by a membrane, which is permeable to $C$ and $D$ but not to $A$ and $B$. What are the possible equilibria?

We model this by choosing $\Lambda$ to have four points, two lying in one cell and two in the other. Each point holds exactly one of the four types of molecule. The sample space is thus

$$\Omega = \Omega_1 \times \Omega_2, \text{ where } \Omega_1 = \Omega_2 = \{A, B, C, D\} \times \{A', B', C', D'\}.$$

We place the second factor of $\Omega_1$ and the first factor of $\Omega_2$ next to the membrane. Introduce a bistochastic map on $\Omega_1$ which allows the reaction (4.4) to occur in one cell with rate $\lambda$, in which the configuration $(C, D')$ produces with rate $\lambda/2$ each of the configurations $(A, B')$ and $(B, A')$. Similarly, we allow (4.4) to occur in the other cell with rate $\lambda$. We introduce diffusion terms allowing all the molecules to move inside each cell, but include diffusion between the cells for $C$ and $D$ only, since the membrane is not permeable to $A$ and $B$. Let $p_1(A)$, $p_1(A')$, $p_2(A)$, $p_2(A')$ denote the marginal probabilities that the four sites in $\Lambda$ are occupied at time $t$ by an $A$, with $p_1'(A)$ and so on denoting these probabilities at time $t + 1$. We use a similar notation for the others. This leads by a calculation similar to that in (4.1) to the following equations for the marginal probabilities:

$$p_1'(A) = p_1(A) + \kappa \left(p_1(A') - p_1(A)\right)$$
$$+ \lambda/2 \left(p_1(C)p_1(D') + p_1(D)p_1(C') - 2p_1(A)p_1(B')\right)$$
$$p_1'(C') = p_1(C') + \kappa \left(p_1(C) - 2p_1(C') + p_2(C)\right)$$
$$+ \lambda/2 \left(p_1(A)p_1(B') + p_1(B)p_1(A') - 2p_1(D)p_1(C')\right)$$

with similar equations for the other variables. The fixed points of this map must be fixed points of each of the reactions taking part, by the theorem on detailed balance. The existence of diffusion means that at equilibrium, $A$ and $B$ must be of uniform density in each cell, and $C$ and $D$ must be uniform throughout:

$$p_1(A) = p_1(A'); \ p_2(A) = p_2(A'); \ p_1(B) = p_1(B'); \ p_2(B) = p_2(B').$$

$$p_1(C) = p_1(C') = p_2(C) = p_2(C');$$

$$p_1(D) = p_1(D') = p_2(D) = p_2(D').$$

The chemical balance of the reaction then means that the law of mass action

$$p_1(A)p_1(B) = p_1(C)p_1(D) = p_2(C)p_2(D) = p_2(A)p_2(B)$$

holds at equilibrium. Eliminating $C$ and $D$ gives us the relation at equilibrium

$$\frac{p_1(A)}{p_2(A)} = \frac{p_2(B)}{p_1(B)}.$$

This ratio is called the 'static head'. The imbalance between the concentrations of $A$ on the two sides of the membrane, $p_1(A) \neq p_2(A)$ is balanced at equilibrium by the opposite ratio for $B$. It is thought that living bodies make use of this to efficiently transport desired chemicals, such as sugar, through a membrane, such as the wall of the gut. To push the chemical $B$ say from cell 1 to cell 2, we add chemical $A$, say sodium, to cell 1 in large quantities. In this way all but a few percent of the chemical $B$ can be taken out of cell 1, a much better system than relying on diffusion, which would lead at equilibrium to equal concentrations on the two sides. It should be said that equilibrium may not be reached, as the products in cell 2 could be swept away by flowing blood. Nevertheless, the mechanism of cotransport is a useful addition to simple diffusion. More elaborate models are described in [Streater (1993d)].

These two examples illustrate a quite reasonable theory of chemical kinetics, which can be extended to include any number of points in $\Lambda$ and any number of chemicals. However, it has some limitations; only balanced reactions, where the number of reactants is the same on each side, are included; there is no concept of heat, in that the energy is a linear sum of particle-numbers, and all the energy is 'chemical', measured by the chemical potentials that enter the equilibrium state; it is not possible to discuss the heat produced in an exothermic reaction. Finally, the rate of the forward reaction is the same as the reverse rate, contrary to experiment. These criticisms led to the following developments of the theory.

To allow unbalanced reactions we borrow the idea of Fock space from quantum mechanics. Instead of allowing only one particle at a site $x \in \Lambda$, the sample space $\Omega_x$ is

$$\mathcal{F} := \{0, 1, 2, \ldots\}, \tag{4.5}$$

for each type of particle. $\mathcal{F}$ is called 'classical Fock space'. Molecules take up some space, and we cannot have more than one at each position in space; thus if we choose Fock space for $\Omega_x$, the site $x$ should represent a macroscopic volume, which can contain a large number of particles. In this case we are omitting all microscopic information, and will get only phenomenological laws of dynamics. Thus the entropy will not include that of the omitted microscopic degrees of freedom. When we want a more microscopic description, that takes account of the impossibility of more than one particle within a certain distance of another, we choose instead of classical Fock space the hard-core space $\Omega_x = \{\emptyset, A, B, \ldots, C\}$, which says that we can have either no particles at $x$, or one of the named particles.

The hard-core is more severe than using Fermi Fock space, (11.3), which would allow two different particles to sit at the site. If we use the hard-core space it is natural to allow diffusion only from an occupied site to a nearby unoccupied site. The exchange of particles, as in the example, might also be allowed, but might be given a much smaller rate than the rate of moving to an empty site. In either case we can set up dynamical laws for reactions that are not balanced. In the case of the hard-core space, it just amounts to calling one of the particles in a balanced reaction by the symbol $\emptyset$, and giving it no energy or atomic number. In (4.123) you will consider the reaction

$$A + C \rightleftharpoons B$$

using the hard-core model $\Omega = \Omega_1 \times \Omega_2$, where $\Omega_{1,2} = \{\emptyset, A, B, C\}$. The reaction goes forward only when one site is occupied by an $A$ and the other is occupied by a $C$, in which case a $B$ is produced with equal probability, $\lambda$, at each site. The back reaction goes, with rate $\lambda$, when either site is occupied by $B$, and the other is empty; it produces the state $(A, C)$ and $(C, A)$ with equal probability. Omitting the possibility of exchange and diffusion, you will get the rate equation

$$p_1'(A) = p_1(A) + \lambda \left[ p_1(\emptyset)p_2(B) \right.$$
$$\left. + \ p_1(B)p_2(\emptyset) - p_1(A)p_2(C) - p_1(C)p_2(A) \right]. \tag{4.6}$$

This differs from the equations often seen in chemistry books, and called the law of mass action; our theory leads to the inclusion of the extra factors $p_i(\emptyset)$. These represent an inhibition of the reaction caused by the hard-core: unless one site is empty, we cannot produce two particles, $A$ and $C$, from a $B$ at the other site. This leads to the slowing down of the rate per molecule of reactions producing an increase in the number of molecules, as the density increases. This is because the pressure increases.

The second criticism, summarised in the question 'where is the heat?', has been answered in studies of the classical Boltzmann equation; the sample space (of a single particle) is taken to be $\mathbf{R}^6$, phase space, a point of which gives us the position and velocity of the test particle. The distribution function $f(\boldsymbol{x}, \boldsymbol{v})$ allows us to define the density of kinetic energy per particle as

$$T(\boldsymbol{x}) = \int f(\boldsymbol{x}, \boldsymbol{v}) \boldsymbol{v}.\boldsymbol{v}/2 \, d^3 v. \tag{4.7}$$

If we postulate that the kinetic energy thermalises almost instantly, because the time-scale of particle exchange is very fast compared with most chemical rates, then this energy is identified with heat, and the velocity distribution is taken to be of the Maxwell-Boltzmann form

$$f(\boldsymbol{x}, \boldsymbol{v}) = N \exp\{-\beta m \boldsymbol{v}.\boldsymbol{v}/2\}$$

where $\beta$ and $N$ are functions of $\boldsymbol{x}$. In this way we can track the production and flow of heat [Monaco and Preziosi (1991)]. So far our model is not able to do this. As a start, we can enlarge the hard-core sample space to include a label telling us the kinetic energy of the particle as well as its type; this is less than telling us its velocity, but is enough to keep track of the heat. It is possible to keep track of momentum in a scattering process, as in the Boltzmann equation. To remain within our theory, which has a countable sample space, we need discrete velocities. This is also needed for numerical calculation. An early model is that of Broadwell [Chentsov (1982)]; a modern account can be found in [Boon and Rivet (2001)]. In Chapter 7 we shall introduce a model which does keep track of momentum conservation; it provides us with a new theory of gases, not found in traditional books such as [Chapman and Cowling (1970)].

Even with the simplification brought about by replacing the velocities by the kinetic energy, the dynamics $p \mapsto pT$ carries at any time the detailed distribution of the energy among the energy-levels, and this is far more information than is usual in non-equilibrium theories in physics and chemistry. In fact, we might guess that very little difference will be made to any macroscopic property by a coarser picture. It has been remarked by Fowler [Fowler (1936)] that the transfer of heat within a local region $x \in \Lambda$ happens $10^8$ times faster than most chemical reactions. This is also true of the exchange of molecules between neighbouring small regions. So we might expect that a state at $x$, the marginal state, is almost instantly thermalised. This is known as the *hypothesis of local thermodynamic equilibrium*, or *LTE*. This can be implemented mathematically as follows. Given the state $p$ at time $t$, and the random dynamics $T$, we first find the marginal map $pT\mathcal{M}_x$, and then find the grand canonical state with the same mean energy and atomic numbers as this. This state will be specified by the beta and one chemical potential for each additional atomic number. We call this the local thermalised state. Then we form the product of the local thermalised states over the space $\Lambda$. This is taken to be the map for one time-step. Since each stage in the map can only increase the entropy, or leave it the same, the map obeys the second law of thermodynamics. By construction, it will conserve the energy and just those atomic numbers that are conserved by $T$. Whether or not it gives nearly the same dynamics we would get from the same $T$ without $LTE$ does not really matter; we are not trying to prove the hypothesis, but to construct a model in which it is true. There are many situations, argues Fowler, when $LTE$ replaces unspecified fast variables, and gives a better fit to experiment than the theory without $LTE$ and without the fast variables.

The following model [Streater (1994b)] shows these features.

**Example 4.3.** This models a free particle in gravity.
Let us take $\Lambda = \{0, 1, \ldots, N\}$, which we interpret as points in a vertical direction. We introduce one type of molecule with a hard core, whose kinetic energy can be any non-negative multiple of a unit.

Thus, the sample space at a site $x$ is

$$\Omega_x = \{\emptyset, 0, 1, 2, \ldots\} \tag{4.8}$$

where $\omega_x = \emptyset$ means that no particle is present at $x$, and $\omega_x = j$ means that there is one particle present at $x$ and its kinetic energy is $j$. We only allow one state with each kinetic energy; this is for simplicity. For a more realistic model, see (4.124). In taking this choice of sample space, we also assume in (4.8) that the molecule has only one spin state. Let us write $p_x(\emptyset) = q_x$, and $p_x(j)$ for the marginal probabilities of the state $p$. The number of molecules at $x$ is the random variable $\mathcal{N}_x(\omega) = 0$ if $\omega_x = \emptyset$, and $= 1$ otherwise. The energy of a molecule at $x$ is the sum of its kinetic energy and its potential energy. Thus $\mathcal{E}(\omega) = \sum_x \mathcal{E}_x(\omega_x)$ where

$$\mathcal{E}_x(\omega_x) = \begin{cases} 0 & \text{if } \omega_x = \emptyset \\ 2\pi\hbar\nu j + mgdx & \text{otherwise} \end{cases} \tag{4.9}$$

where $\nu$ is the frequency of a photon with the same kinetic energy, $g$ is the acceleration due to gravity, $m$ is the mass of the molecule and $d$ is the distance between the sites. The factor $2\pi\hbar\nu$ is just to give the energy the right dimensions. We assume that $g$ is independent of $x$ for simplicity. The grand canonical state at $x$ with beta $\beta_x$ and chemical potential $\mu_x$ is by definition

$$q_{\beta_x, \mu_x} = Z^{-1}(\beta_x, \mu_x) \tag{4.10}$$

$$p_{\beta_x, \mu_x}(j) = Z^{-1} \exp\{-\beta_x(2\pi\hbar\nu j + mgdx - \mu_x)\} \tag{4.11}$$

where $Z$ is the grand partition function for the system at $x$:

$$Z(\beta, \mu) = 1 + e^{-\beta(mgdx-\mu)} \sum_j e^{-\beta 2\pi\hbar\nu j} = 1 + \frac{e^{-\beta(mgdx-\mu)}}{1 - e^{-2\pi\beta\hbar\nu}}. \tag{4.12}$$

Every state has density bounded by $1/d$, since there can be at most one particle at each point. There are, however, some states for which the mean kinetic energy at a point $x$, $p \cdot T_x = \sum_{j=0}^{\infty} j p_x(j) 2\pi \nu \hbar$, is infinite. Let $\mathcal{D}$ be the affine subspace of $\Sigma(\Omega)$ consisting of states having finite mean kinetic energy at each $x \in \Lambda$. Let us denote the grand canonical state at $x$, the pair $(q_{\beta_x, \mu_x}, p_{\beta_x, \mu_x}(j))$ for all $j$, by $p_{\beta_x, \mu_x}$. We define the $LTE$ map $Q : \mathcal{D} \to \mathcal{D}$ by

$$pQ = \otimes_x p_{\beta_x, \mu_x} \qquad (4.13)$$

where $\mu_x$ and $\beta_x$ are uniquely determined by the requirement that $(pQ)_x$ have the same mean density and kinetic energy as $p_x$. We know that $(pQ)_x$ is the state of greatest entropy among all states with this property, and that the state $pQ$, which is independent over $\Lambda$, is the state of greatest entropy among all states with the given marginals. It follows that $Q$ is entropy-nondecreasing. Note that the entropy is finite for any state $p$ in $\mathcal{D}$, since it is not greater than

$$S(pQ) = \sum_x S(p_{\beta_x, \mu_x}).$$

The linear part of the dynamics is given by a convex sum of local terms, denoted $T_x(j)$. This acts on $\Sigma(\Omega_x \times \Omega_{x+1})$, and transfers a particle at $x$ with energy $2\pi \nu \hbar j$ into a hole at $x + 1$, using up kinetic energy $mgd$ to overcome gravity. It also transfers in the reverse direction, and with the same probability. In order to conserve energy, the ratio of the parameters $2\pi \nu \hbar$ and $mgd$ must be an integer. We shall assume they are equal, for simplicity. Then in going up the ladder from $x$ to $x + 1$ a particle loses one unit of kinetic energy, and in going down the same step, it gains one unit. So $T_x(j)$ is the symmetric stochastic matrix

$$T = \begin{array}{c} \\ \omega_1 \\ \omega_2 \end{array} \begin{array}{cc} \omega_1 & \omega_2 \\ \begin{pmatrix} 1-\lambda & \lambda \\ \lambda & 1-\lambda \end{pmatrix} \end{array}. \qquad (4.14)$$

In this matrix, the first row and column are labelled by a point $\omega_1 \in \Omega$ such that $\omega_1(x) = j + 1$, $\omega_1(x + 1) = \emptyset$ and the second row and column are labelled by a point $\omega_2 \in \Omega$ such that $\omega_2(x) = \emptyset$, $\omega_2(x + 1) = j$. For simplicity we have chosen $\lambda$ to be independent of $j$, although we might expect that the rate of hopping to increase for larger $j$, and this is easy to arrange. The sample point $\omega_1$ is connected to only finitely many points by such matrices. So by forming a convex sum of all the matrices involving $\omega_1$ we get a stochastic matrix taking care of all its possible transitions.

Similarly for any other element of $\Omega$. Thus we get an infinite stochastic matrix, any of whose rows or columns has only finitely-many non-zero elements. It therefore acts on Span $\Omega$ (the algebraic sum) in a well-defined way, and preserves the mean total energy and total particle number. This follows from the fact that $T$ connects only states with the same number of particles and the same energy. We shall see it explicitly in a while. It follows that $T$ maps $\mathcal{D}$ into $\mathcal{D}$. Since the hopping particle carries energy with it, the model describes the *convection* of heat. To find the fixed points of $\tau := QT$ note that entropy is a strict Lyapunov function for each $T_x(j)$, and is strictly concave, so a fixed point of a convex sum of these stochastic maps is a fixed point of each of them (Tolman's principle of detailed balance is true here). So a fixed point of $\tau$ is a grand canonical state that is left unchanged by all the $T_x(j)$.

Take $p \in \Sigma(\Omega)$ to be independent over $\Lambda$, and consider the action of $T_x(j)$ on the relevant part of $p$, namely $p_x \otimes p_{x+1} \in \Sigma(\Omega_x \times \Omega_{x+1})$. As before, (see 4.8 *et seq*) put

$$p_x = (q_x, p_x(j)) \qquad p_{x+1} = (q_{x+1}, p_{x+1}(j)) \ .$$

Let $p' \in \Sigma(\Omega_x \times \Omega_{x+1})$ be the relevant part of $pT_x(j)$. Then $p'$ differs from $p$ at only two points, as we see from (4.14):

$$p'(j+1, \emptyset) = (1 - \lambda)p_x(j+1)q_{x+1} + \lambda q_x p_{x+1}(j)$$
$$p'(\emptyset, j) = \lambda p_x(j+1)q_{x+1} + (1 - \lambda)q_x p_{x+1}(j) \ .$$

Now

$$p'_x(j+1) = p'(j+1, \emptyset) + \sum_{l=0}^{\infty} p'(j+1, l)$$

$$= p'(j+1, \emptyset) + \sum_{l=0}^{\infty} p_x(j+1)p_{x+1}(l) \qquad (4.15)$$

since $T_x(j)$ does not alter $p(j+1, l) = p_x(j+1)p_{x+1}(l)$. Thus

$$p'_x(j+1) = p'(j+1, \emptyset) + p_x(j+1) \sum_{l=0}^{\infty} p_{x+1}(l)$$

$$= (1 - \lambda)p_x(j+1)q_{x+1} + \lambda q_x p_{x+1}(j)$$
$$+ p_x(j+1)(1 - q_{x+1})$$
$$= p_x(j+1) - \lambda\left(p_x(j+1)q_{x+1} - p_{x+1}(j)q_x\right). \qquad (4.16)$$

This could have been guessed, since the change in $p_x(j+1)$ can come about from the absence of a particle at $x$ (probability: $q_x$) and the presence of one at $x+1$ of energy $j2\pi\hbar\nu$ (probability: $p_{x+1}(j)$), or from a hole at $x+1$ and a particle of energy $(j+1)2\pi\hbar\nu$ at $x$ (probability: $p_x(j+1)q_{x+1}$). Our calculation shows that this intuitively natural dynamical law comes from applying $T$ followed by the stoss map, and is therefore entropy-increasing. Similarly we get

$$p'_{x+1}(j) = p_{x+1}(j) + \lambda\left(p_x(j+1)q_{x+1} - p_{x+1}(j)q_x\right). \tag{4.17}$$

We remark that $p_0(0)$, the ground-state occupation density, does not appear in either equation of motion, since $j \geq 0$ and $x \geq 0$, and so is a constant of the motion under the Markov chain generated by $T$. It is $Q$ that causes the time-dependence of this variable, and leads us to equilibrium. We can now see explicitly that $Q_T$ conserves the mean total energy; it is enough to show this for $T_x$ acting on the two sites $x$ and $x+1$. For, the energy is equal to

$$((j+1)2\pi\hbar\nu + dmgx)p_x(j+1) + (j2\pi\hbar\nu + dmg(x+1))p_{x+1}(j)$$
$$= (j+x+1)2\pi\hbar\nu(p_x(j+1) + p_{x+1}(j))$$

and since

$$p_x(j+1) + p_{x+1}(j) = p'_x(j+1) + p'_{x+1}(j),$$

this is conserved; this also shows that the mean total number of particles is conserved.

We conclude that $\{\tau^n p\}_{n=1,2,\dots}$ is an orbit in $\Sigma(\Omega)$, such that the mean total energy and particle number is fixed, moving through local grand canonical states for which entropy is a Lyapunov function. By Lyapunov's theorem the proof of convergence to a fixed point will be complete when we show that there is a unique fixed point. This we now do. To be stationary, a state must be a fixed point of each $T_x(j)$, and so must satisfy

$$p_x(j+1)q_{x+1} = q_x p_{x+1}(j) \quad \text{for all } x, j \geq 0 \tag{4.18}$$

as well as the $LTE$ conditions, (4.11):

$$p_x(j+1)/q_x = \exp\left(-\beta_x\left((j+1)2\pi\hbar\nu + mdgx - \mu_x\right)\right) \tag{4.19}$$

$$p_{x+1}(j)/q_{x+1} = \exp\left(-\beta_{x+1}(j2\pi\hbar\nu + mgd(x+1) - \mu_{x+1})\right). \tag{4.20}$$

We note that these conditions do not depend on $\lambda$, or on whether $\lambda$ depends on $x$ or $j$, as long as $\lambda > 0$. This is usual in statistical dynamics, in that the equilibrium state does not depend on $T$. Combine (4.18) with (4.19) and (4.20) for all $j$ to give $\beta_x = \beta_{x+1}$ and $\mu_x = \mu_{x+1}$, since $mgd = 2\pi\hbar\nu$. Thus our model of convection is fully mixing, in that it ensures a uniform beta and chemical potential at equilibrium. The value of $\mu$ is fixed by the total mean number of particles, provided that it is not greater than $N + 1$. As usual, beta is fixed by the mean energy, which must be non-negative. We conclude that there is only one fixed point in the set of local grand canonical states with a given mean energy and particle number. This is enough, by Lyapunov's direct method, to establish that the system converges at large time, since the set of local grand canonical states with given total mean energy and particle number is compact with the usual topology on the parameters $\mu_x, \beta_x$. At equilibrium the density $\rho_x = p_x/d$ varies with height $x$; we see that

$$p_x = \sum_j p_x(j) = Z^{-1} \exp{-\beta\{mdgx - \mu\}} \left(1 - \exp\{-2\pi\beta\hbar\nu\}\right)^{-1}.$$

Putting in the grand partition function (4.12) and rearranging, gives (4.125)

$$p_x = \frac{\exp{-\beta\{mdgx - \mu\}}}{1 + \exp{-\beta\{mdgx - \mu\}} - \exp\{-\beta mdg\}} . \qquad (4.21)$$

Putting $x = 0$ gives us a relation between the chemical potential $\mu$ and the density at $x = 0$, which can be solved to give

$$\exp\{\beta\mu\} = p_0 \left(1 - \exp\{-\beta mgd\}\right) \left(1 - p_0\right)^{-1}.$$

Substituting, we get (4.125)

$$p_x = \frac{p_0 \exp\{-\beta mdgx\}}{1 - p_0 \left(1 - \exp\{-\beta mdgx\}\right)} . \qquad (4.22)$$

For small values of $p_0$ and large values of $x$ this shows an exponential thinning of the density with increasing $x$; this is Laplace's barometric formula. If $p_0$ is not small there is a correction due to the hard core.

This example allows us to understand how to compute the pressure in statistical dynamics. Often, in equilibrium statistical mechanics, the pressure is defined in terms of the grand partition function. But here the pressure is not an intensive variable, the same throughout the system, but must depend on $x$. What is holding up the molecules at height $x$, against the gravitational field? Let $z = xd$ be the height in dimensional units. The difference in pressure at $z$ and $z + \delta z$ provides the force to hold up the weight $mg\rho\delta z$. So we get the equation

$$-\frac{\partial P}{\partial z}\delta z = mg\rho\delta z \qquad (4.23)$$

where we write $\rho$ as a function of $z$. In addition, the pressure must vanish at $z = \infty$. You might not like that remark if $\Lambda$ is finite; in that case, we note that the pressure must vanish when the density is zero. Either condition fixes the constant of integration. To solve the differential equation, note that

$$Z(z) = 1 + \frac{\exp\{\beta\mu\}}{1 - \exp\{-\beta mdg\}} \exp\{-\beta mgz\} \qquad (4.24)$$

from which we can compute

$$\frac{\partial \log Z}{\partial z} = \frac{\beta mg \exp \beta\{\mu - mgz\}}{\exp \beta\{\mu - mgz\} + 1 - \exp\{-\beta mdg\}} = -\beta mgd\rho . \qquad (4.25)$$

Since $\log Z$ satisfies the boundary conditions, we conclude that

$$P(z) = (\beta d)^{-1} \log Z(z). \qquad (4.26)$$

This relation (with $d=1$) is usually derived for homogeneous systems within the equilibrium theory.

The third criticism of this theory of chemistry, that the forward and backward rates of reaction are equal, contrary to experiment, will be addressed in the next section.

## 4.2 The Heat-Particle

The sample space chosen in the model of the gas in a gravitational field (4.3) attributes the kinetic energy to the particular particle at $x$; however, the interaction between close particles is very strong, and it is reasonable, as well as simpler, to share the kinetic energy between neighbours. There is another thing to bear in mind, that in thermal systems the electromagnetic field is also thermalised, and some energy of motion of the charged particles should be attributed to the photon field. We shall attribute all the kinetic energy of the particles to some extra degrees of freedom, the quanta of which are called heat-particles. Like the electromagnetic field, these will be located on the chemical bonds between molecules, and can be regarded as carrying the energy of bond vibration. This should not be counted twice, once as the kinetic energy of the particles, and a second time as a field vibration. The introduction of the heat-particle instead of the space of kinetic energy of specified particles expresses the idea that heat does not belong to any particular particle. In the early days, the idea that the kinetic energy is quickly thermalised was incorporated in the Smoluchowski approximation. We express the same physical idea in a different way, through the presence of the heat-particle. When we view a body at positive temperature, actual photons are observed as black body radiation, so they are really there.

We now add to the general structure of the space $\Lambda$ a set $\mathcal{B}$ of pairs $\{x, y\}$ specifying the bonds, so that $\{\Lambda, \mathcal{B}\}$ is a finite graph. The vertices of the graph are the points of $\Lambda$, and the edges of the graph are the bonds. Denote a typical bond by $b \in \mathcal{B}$, so that $b = (x, y)$. We choose a sample space $\Omega_{cx}$ at each site, which can be classical Fock space, (4.5), of one or more chemical types, or the hard core space of several chemicals, or a mixed product of these. The suffix $c$ indicates that this is the space of chemicals. To each bond $b$ we associate a sample space $\Omega_{\gamma b}$, which is a product of classical Fock spaces, one for each heat-particle; the suffix $\gamma$ indicates that the space concerns the photon field. For simplicity we usually only introduce heat-particles with energy equal to that of one of the energy transfers in the process, just as the kinetic energy $2\pi\hbar\nu$ in (4.3) was chosen equal to $mgd$ to allow transitions to conserve energy. The total sample space thus has the form

$$\Omega = \prod_x \Omega_{cx} \times \prod_b \Omega_{\gamma b}. \tag{4.27}$$

For example, if there are two chemicals $A$ and $B$ and they have hard cores, and can reach every site, then each $\Omega_{cx}$ could be chosen to be $\{\emptyset, A, B\}$ and if there is one type of heat-particle we choose $\Omega_{\gamma b} = \mathcal{F} = \{0, 1, 2, \ldots\}$; the integer here specifies the number of quanta present. Then a sample point is the pair $\omega = (\omega_c, \omega_\gamma)$; here, $\omega_c$ is an ordered set of symbols $\emptyset, A$ or $B$, one for each $x$, and $\omega_\gamma$ is an ordered set of non-negative integers, one for each bond.

The energy of the system is a random variable which is a sum of a potential energy and a kinetic energy. The potential energy $\mathcal{E}_c$ is a function only of the variables $\omega_c$, and represents the chemical energy bound up in the molecule, as well as the physical potential energy such as that of gravity. Electrostatic energy is also an important example in chemistry and biology. The kinetic or heat energy, $\mathcal{E}_\gamma$, is a function only of $\omega_\gamma$; thus

$$\mathcal{E}(\omega_c, \omega_\gamma) = \mathcal{E}_c(\omega_c) + \mathcal{E}_\gamma(\omega_\gamma). \tag{4.28}$$

We do not postulate any contribution to the energy that depends on both $\omega_c$ and $\omega_\gamma$; this is reasonable if we interpret the two parts as potential and kinetic energy (and we omit magnetic forces) but with the new idea that $\mathcal{E}_\gamma$ is the energy of the electromagnetic field, it might seem that molecules should exhibit some interaction. Nevertheless, we stick by (4.28), and interpret it as the energy of free, asymptotic molecules and free photons, after they have come apart from their scattering. The scattering is achieved by the stochastic map $T$, which is bistochastic and so is a random mixture of reversible maps, by Theorem 3.11. We interpret these as scattering by a random short-ranged potential. It is known that the scattering operator conserves the total energy of the asymptotic states. (Of course, it does mix up the various contributions.) This is why we still get reactions occurring even though there is no 'interaction' term in (4.28). Thus the dynamics is determined by a stochastic map $T$ mapping the energy-shells to themselves; mixing between energy-shells is achieved by the stoss map, which replaces $pT$ by $pT\mathcal{M}_c \otimes pT\mathcal{M}_\gamma$. Here, $\mathcal{M}_c$ and $\mathcal{M}_\gamma$ are the marginal maps onto the chemical space and the heat space respectively. The stoss map is quite within the spirit of statistical methods; more is true here; since heat is, by definition, the thermalised part of the energy, any disturbance of its thermal state rapidly reverts back to a new thermal state. So it is reasonable to follow the stoss map by the map that replaces $pT\mathcal{M}_\gamma$ by the canonical state with the same mean energy. This is the $LTE$-map for heat, say $Q_\gamma$. Thus we are coming round to defining the *isolated dynamics with heat-particles* as the map

$$p \mapsto pT \mapsto pT\mathcal{M}_c \otimes pT\mathcal{M}_\gamma Q_\gamma .$$

This map conserves mean energy and obeys the second law. Then the aim of the theory is to show that, for a suitable choice of $T$, the system converges to equilibrium for large time; the equilibrium state will be a canonical state for the heat, and a grand canonical state for the chemicals. The beta and chemical potentials are determined by the initial mean values of the energy and the atomic numbers, not by an external heat-bath. We express this as saying that an isolated system finds its own equilibrium.

It is often hard to handle a model of the above sort, because the state at time $t$ continues to carry details of the distribution of all the chemicals taken together; that is, it carries the information about the correlations between different points of $\Lambda$. Great simplifications take place if we add more randomness, namely, we replace the state $pT\mathcal{M}_c$ by the grand canonical state with the same energy and atomic numbers. Let us call this map $Q_c$. One time-step is then

$$p\tau = pT\mathcal{M}_c Q_c \otimes pT\mathcal{M}_\gamma Q_\gamma . \tag{4.29}$$

This expresses that a great stirring has rapidly occurred within the time-step, within the region of space represented by $x$. The resulting map $\tau$ is called 'isolated dynamics with $LTE$'. The map $\tau$ is highly non-linear, since

$p_c Q_c$ is a product of degree $|\Lambda|$ in the grand canonical states at each $x$. We shall see that the resulting equations are among those postulated by chemists on the basis of the law of mass action.

**Example 4.4.** Let us reconsider the problem of a particle in a gravitational field, but let us assign the kinetic energy to a heat-particle rather than to the chemical. Thus $\Lambda = \{0, 1, \ldots, N\}$, and

$$\mathcal{B} = \{(0,1), (1,2), \ldots, (N-1, N)\}$$

is the set of bonds. Also, we can have at most one particle with a hard core at each $x \in \Lambda$, and any number of heat-particles on each bond, so we choose $\Omega_{cx} = \{\emptyset, 1\}$ and $\Omega_{\gamma b} = \{0, 1, \ldots\}$. We interpret $x \in \Lambda$ as the height above ground. The sample space for the system is

$$\Omega = \prod_{x \in \Lambda} \Omega_{cx} \times \prod_{b \in \mathcal{B}} \Omega_{\gamma b}, \qquad\qquad \omega = (\omega_c, \omega_\gamma). \qquad (4.30)$$

The energy $\mathcal{E}$ differs from that in (4.3) in that a site can possess energy even if it contains no particles; this is just what we want: it is the radiation in the cavity. Thus

$$\mathcal{E}(\omega) = \mathcal{E}_c(\omega_c) + \mathcal{E}_\gamma(\omega_\gamma) = \sum_{x=0}^{N} mgxd\mathcal{N}(\omega_{cx}) + \sum_{b \in \mathcal{B}} 2\pi\hbar\nu\omega_{\gamma b} \, .$$

Here, as defined above, $\mathcal{N}(\omega_{cx})$ is either 0 or 1, and $\omega_{\gamma b}$ is one of $0, 1, 2, \ldots$. As before, we need compatibility between the kinetic energy and the energy jumps in the potential; the simplest choice is the one we made before: $mgd = 2\pi\hbar\nu$. Then convection can occur if a particle at height $x$ absorbs $j$ quanta from the bond $b = (x-1, x)$, jumps to the empty site at height $x+1$, and emits $j-1$ quanta to the bond $(x, x+1)$. One quantum is used up to overcome the gravitational potential. It could equally well absorb $j$ quanta from the bond $(x, x+1)$, jump from $x$ to $x+1$ and release $j-1$ quanta to the bond $(x+1, x+2)$. To get a symmetric process, and hence a bistochastic map, we choose the reverse of these processes to be equally likely. The simplest case is $j = 1$, and we shall give the details for this.

Choose $x \in \Lambda$ and concentrate on the factors $\Omega_{cx} \times \Omega_{c,x+1} \times \Omega_{\gamma b}$, where $b = (x, x+1)$. Introduce the bistochastic map $T$ which takes the pure state $\{\omega_{cx} = 1, \omega_{c,x+1} = \emptyset, \omega_{\gamma b} = j\}$ to the state $\{\omega_{cx} = \emptyset, \omega_{c,x+1} = 1, \omega_{\gamma b} = j-1\}$, with probability $\lambda$; the inverse is taken to have the same probability. All other components of $\omega$ are unchanged. Following $T$, we apply the stoss map with *LTE*. Now, any state $(q_x, p_x)$ on a two-point set is already a grand canonical state with some local value of $\mu$, so we can omit $Q_c$, and the *LTE* map is

$$p \mapsto \bigotimes_x pT\mathcal{M}_x \otimes \bigotimes_b pT\mathcal{M}_\gamma Q_{\gamma b} .$$

We stay with the convention introduced in [Streater (1993d)] to write $s$ for the state of the heat particle, and $s_\beta$ for the canonical state with beta $\beta$, while retaining $p_x$ for the marginal state of $p$ on the set $\Omega_{cx}$, and $q_x = 1 - p_x$. Then $p_x$ is the probability of occupation of the site $x$, so it is the density. Since $T$ acts only on $x$, $x + 1$ and $b$, we can ignore the other factors and take $p \in \Sigma(\Omega_{cx} \times \Omega_{c,x+1} \times \Omega_{\gamma b})$, of the form

$$p = p_x \otimes p_{x+1} \otimes s.$$

Put $p' = pT$. Then one time-step gives

$$p'(0,0,j) = p(0,0,j) \text{ for all } j \geq 0$$
$$p'(0,1,j) = (1 - \lambda)p(0,1,j) + \lambda p(1,0,j+1) \text{ for all } j \geq 0$$
$$p'(1,0,j) = (1 - \lambda)p(1,0,j) + \lambda p(0,1,j-1) \text{ for all } j \geq 1$$
$$p'(1,0,0) = p(1,0,0)$$
$$p'(1,1,0) = p(1,1,0) . \tag{4.31}$$

In addition, it is reasonable to add a diffusion of the heat, with a probability $\kappa$ that an occupied bond sends a heat-particle to a neighbouring bond. The full dynamics is a convex mixture of all these. Let $p'_x$ be the density at $x$ after the time-step. Then

$$p'_x = \sum_{j \geq 0} p'(1,0,j) + \sum_{j \geq 0} p'(1,1,j)$$
$$= p(1,0,0) + p_x p_{x+1} \sum_j s(j)$$
$$+ \sum_{j \geq 1} \{(1 - \lambda)p(1,0,j) + \lambda p(0,1,j-1)\}$$
$$= p_x q_{x+1} s(0) + (1 - \lambda)p_x q_{x+1}(1 - s(0)) + \lambda q_x p_{x+1} + p_x p_{x+1}$$
$$= p_x - \lambda p_x q_{x+1}(1 - s(0)) + \lambda q_x p_{x+1} . \tag{4.32}$$

This is no surprise: the transition probability from $x + 1$ to $x$ is $\lambda$ times $p_{x+1} q_x$, which is the probability that there is a particle at $x + 1$ and a hole at $x$. The inverse probability is similar, multiplied by the factor $(1 - s(0))$. Now, we have applied the *LTE* map $Q_\gamma$ to the state, so $s$ is a canonical state, with beta $\beta$ say, the instantaneous local beta of the bond $(x, x+1)$. Then $1-$

$s(0) = 1 - Z_\beta^{-1} = e^{-2\pi\beta\hbar\nu}$. Thus the forward and backward rates are related by the factor $e^{-\beta E}$, where $E$ is the energy difference between the states. This is known as the Boltzmann factor to physicists, and less accurately as the principle of detailed balance; it follows from a simple form of microscopic reversibility, which in our case is that $T$ is symmetric. It will be shown to arise in all reactions that absorb or emit heat. In chemistry it follows by applying the Arrhenius law to the forward and backward processes; however, it is very much more accurate than the Ahrrenius law. It gives the temperature-dependence of the ratio of the forward and backward rates, and is well confirmed by measurements.

There is a unique fixed point for any initial values of the total energy and total number of molecules. For, the heat-particles cannot be at equilibrium, because of the direct diffusion, unless the temperature is uniform. Then the condition for equilibrium of the molecules

$$p_{x+1}/q_{x+1} = e^{-\beta E} p_x/q_x \qquad (4.33)$$

determines the density in sequence, starting at $x = 0$. It can be summarised as saying that the chemical potential is independent of $x$ at equilibrium. Thus we get the same variation of density with height as in (4.3). Now try (4.128).

In this example suppose that we omit the *LTE* map on the space $\Omega_c$, while keeping the map $Q_\gamma$ which thermalises the heat-particles. By the principle of detailed balance, any fixed point must be a fixed point of the diffusion of the heat particles as well as the convection of the molecules. So there must be a uniform temperature, and the state $s_\beta$ of the heat-particle must be unchanged by the dynamics. The factors $s$ in the dynamics of the chemicals just appear as constants. In this case the dynamics of the chemicals reduce to a linear stochastic matrix, which is irreducible on each number shell, since each pair of states is connected by a finite chain of non-zero transition probabilities. The fixed point must therefore be the uniform distribution on each number-shell, by the uniqueness part of the Perron-Frobenius theorem. By (3.17), the system converges to a mixture of uniform distributions on each number-shell, the fraction of probability on each shell being the same as in the initial state; the actual number of molecules is conserved, and not only the mean number. Thus the system fails to converge to the grand canonical state; there is not enough mixing for that. We conclude that the chemical *LTE*-map is needed to get the grand canonical state.

The chemical *LTE* map produces from a state with a definite number of molecules a state which is a mixture of states with various numbers, but with the same mean number. This means that if the initial state has a definite number, then after one time-step, there is a non-zero probability of finding a different number. This seems to say that there is a violation of the law of conservation of atomic number. Some people are more worried by this than the same problem that arises in the thermalisation of energy. I was for a time, and I asked 't Hooft about it. His reply was worth keeping: 'The theorems and results in mathematical physics are not theorems and results about Nature, but about our description of it.'

In other words, a model (in this case the *LTE*-map) is only a partial description, and the interpretation of the model might not agree with experiment in all detail. We must treat the number of particles statistically; only the mean values, variance and other higher moments, are predicted by the model. The exact number of atoms is not one of the predictions of the model. If the volume is large, the local states in equilibrium are almost the same, whichever ensemble is arrived at. This is the result of the important theorem on the equivalence of ensembles. The text-books just say ([Fowler (1936)]) we use the grand canonical state for computational convenience, and rely on the fact that the limit as $|\Lambda| \to \infty$ is the same for all ensembles. We state a version of this in (5.34).

Our theory aims to be more ambitious, and to describe dynamics, not only the equilibrium state. We also want to apply it to *mesoscopic systems*, in which $\Lambda$ is not infinite. In this case the dynamics with or without chemical *LTE* is different. Undoubtedly, the dynamics with $Q_c$ is simpler, and converges to a different equilibrium. To avoid the problem of non-conservation of particles, we could just keep the Boltzmann map and leave out *LTE*; then we get convergence to the canonical state, and do not mix the number-shells. But then we are left with the same philosophical problem with the mixing of energy-shells: a state with a sharp energy can move to a state with mixed energy. I offer the following idea: to say that we have an isolated system is an idealisation, as no system is completely isolated. The walls of the jar are not part of the system, and are designed to conserve the mean number of molecules; we cannot prevent the rapid exchange of adhering molecules with those of the system; this does not alter the mean number, but any given measurement will reveal that the number has changed from that in the initial state (but with small chance of being very different). We choose to model this by the *LTE* map, on the assumption that the exchange of molecules is one of the fast variables, and so all correlations are destroyed in one time-step.

In the next section we show how to model any chemical reaction, and how the heat-particle represents the energy of reaction.

## 4.3   The Hard-Core Model of Chemical Kinetics

*There are no photons in chemistry*

From the referee's report to [Streater (1993d)]

Physical chemistry started with the law of mass action, formulated by Guldberg and Waage in 1867 (Guldberg and Waage, 1864). This stated that the rate of a reaction

$$A + B + C + \ldots \rightarrow \text{products} \tag{4.34}$$

at a given temperature is proportional to the product of the concentrations of $A$, $B$, $C$, …. The same was claimed for the back reactions, with a different rate constant. Let us denote these concentrations by $N_A, N_B$, and so on. Then the reaction, say

$$A + B \rightleftharpoons C \tag{4.35}$$

should have forward rate $K_+ N_A N_B$ and backward rate $K_- N_C$, with dynamics given by

$$\frac{dN_A}{dt} = K_- N_C - K_+ N_A N_B \tag{4.36}$$

$$= \frac{dN_B}{dt} = -\frac{dN_C}{dt}.$$

Equilibrium holds when $\frac{dN_A}{dt} = 0$, giving

$$\frac{N_A N_B}{N_C} = \frac{K_-}{K_+}, \tag{4.37}$$

a formula taught to generations of students of physical chemistry. For many dilute solutions, (4.36) is found to be approximately true; it does not fit all experiments very well, especially at high concentrations. Moreover, there are theoretical reasons why (4.37) cannot be exact: the grand canonical state does not satisfy this condition. This suggests that the dynamics (4.36) violates the second law of thermodynamics.

Many reactions go faster at high temperatures than at low temperatures. This suggests that some energy must be supplied before a reacting molecule can penetrate an initial repulsive barrier. We saw that Arrhenius suggested that at a temperature $\Theta$ the rate of a reaction should be proportional to

$e^{-\beta E}$, where $\beta = (k_B \Theta)^{-1}$ and $E$ is an energy characteristic of the reaction (now called the activation energy by chemists, and the energy-threshold by physicists). Arrhenius's 'law' is true for some reactions but not so good for others. This idea does predict another law, which is accurate over many orders of magnitude in the temperature. Namely, let $E_+$ and $E_-$ be the activation energies of the forward reaction $A + B + \ldots \to C + D + \ldots$ and the backward reaction $C + D + \ldots \to A + B + \ldots$ respectively. Then it is observed that for most processes

$$\frac{K_+}{K_-} = e^{-\beta(E_+ - E_-)} \tag{4.38}$$

holds. This is what we would get if the Arrhenius law were true, and $K_+ = Ke^{-\beta E_+}$, $K_- = Ke^{-\beta E_-}$. We therefore seek a theory of chemistry which shows these features:

(1) The law of mass action is nearly true at low concentrations.
(2) Arrhenius's law is nearly true, sometimes.
(3) (4.38) is exactly true for all temperatures.

Let us start with the simplest system, located at a single point in space, which can be occupied by one but not more than one of the chemicals $A_1, A_2, \ldots, A_n$; these transform into each other at various rates. This process can be described by the sample space $\Omega = \{A_1, A_2, \ldots, A_n\}$ and the probability $p = (p_1, p_2, \ldots, p_n)$. The dynamics is given by a bistochastic map $p \mapsto pT$, leading to a Markov chain in $n$ variables. This type of model includes the extreme case when $T$ is a permutation, and the reaction rotates around the space, as well as the ergodic cases, when the probability converges to the uniform distribution on $\Omega$. We saw that this occurs when the diagonal of $T$ has no zeros, and every state has a chain of non-zero probabilities leading to any other.

The next simplest case is in isomer formation.

### 4.3.1 *Isomers and Diffusion in a Force-Field*

Suppose that two isomers $A_1, A_2$ of a molecule $A$ differ in energy by $E$. Let $\gamma_E$ be a heat particle with energy $E$. Then the reaction

$$A_1 + \gamma_E \rightleftharpoons A_2 \tag{4.39}$$

conserves energy. We can expect this reaction to occur if $A_1$ needs energy $E$ to activate it, while $A_2$ can decay into $A_1$ with no stimulation. The energy of activation is drawn from the ambient heat, modelled by a heat particle.

The sample space for this is classical Fock space, 4.5, in which each photon has energy $E$. The sample space for an isomer process is therefore taken to be

$$\Omega = \{A_1, A_2\} \times \mathcal{F} = \Omega_c \times \mathcal{F}. \tag{4.40}$$

In a reaction with an energy threshold, the temperature of the system changes with time. The temperature is defined by the beta in the canonical state $s$, a measure on $\mathcal{F}$ given by $s_\beta$. The state of the chemicals is described by $p_c = (p_1, p_2)$, with $p_1 + p_2 = 1$. Thus normalised, the conservation law $\mathcal{N}_1 + \mathcal{N}_2 = $ constant becomes just the normalisation of probability on $\Omega_c$, which will be conserved if one step of the dynamics is given by a suitable bistochastic map. To get a plausible choice for this map, we must decide what happens when the state is any point in $\Omega$. If the initial state has $n$ heat particles, it is natural to assume that any one of them can be absorbed to change $A_1$ into $A_2$, with equal probability. Einstein argued [Einstein (1917)] that the rate of the process

$$A_1 + n\gamma \rightleftharpoons A_2 + (n-1)\gamma, \qquad\qquad n \geq 1 \tag{4.41}$$

should be proportional to $n$. We seek a symmetric stochastic matrix (as the simplest example of a bistochastic matrix) on $\Sigma(\Omega)$ with this property. Let us rewrite (4.41) slightly, to the reaction

$$A_1 + (n+1)\gamma \rightleftharpoons A_2 + n\gamma, \qquad\qquad n \geq 0, \tag{4.42}$$

in which the forward rate is now proportional to $n+1$. With extraordinary prescience, Einstein saw that the decay of $A_2 + n\gamma$ to $A_1 + (n+1)\gamma$ can be explained by two processes: the 'spontaneous' decay of $A_2$ to $A_1 + \gamma$, and a further decay 'stimulated' by the presence of the $n$ photons, proportional to $n$. This suggests that the transitions between the states $(A_1, n+1)$ and $(A_2, n)$ be given by the bistochastic matrix

$$T = \begin{matrix} \\ A_1, n+1 \\ A_2, n \end{matrix} \begin{matrix} A_1, n+1 & A_2, n \\ \begin{pmatrix} 1 - (n+1)\lambda & (n+1)\lambda \\ (n+1)\lambda & 1 - (n+1)\lambda \end{pmatrix} \end{matrix}. \tag{4.43}$$

Here the stimulated and spontaneous transitions come together, and the need for a symmetric transition matrix forces their rates, $\lambda > 0$ per time-step to be the same. This was taken for granted by Einstein. To make sense, $T$ must not contain any negative elements, so we require $(n+1)\lambda \leq 1$. We temporarily limit the number of photons present by a cut-off: $T$ is the identity map if $n \geq N$; we take $\lambda$ to be proportional to the time-step. Then if the time-step is small enough, no negative matrix elements arise.

In the state at time $t$, the isomers are present in the ratio given by $p_c = (p_1, p_2)$. There is some ambient heat, say given by the mean energy in the canonical temperature state $s_\beta$. We always take the heat particle before the reaction to be independent of the chemicals, so the initial state is $p = p_c \otimes s_\beta$. After one time-step, the state is $p' := Tp$, from which we can calculate the amount of $A_1$ and $A_2$ by taking the chemical marginal map, $\mathcal{M}_c$:

$$p'_c := (p'_1, p'_2) = \mathcal{M}_c p' = \sum_n p'(\omega_c, n) = (Tp)(\omega_c, n). \qquad (4.44)$$

What is the probability that the state at time $t+1$ is $(A_1, n)$ with $0 \leq n < N$ photons? The state $(A_1, 0)$ can occur only when this is the state at time $t$; this has probability $p_1 s_\beta(0)$. The case $n \geq 1$ can arise in two ways: (a) the state at time $t$ is the same, $(A_1, n)$ and nothing happens; or (b), the state at time $t$ is $(A_2, n - 1)$, and a transition occurs. The probability of (a) is

$$p_1 s_\beta(n)(1 - n\lambda)$$

and the probability of (b) is

$$p_2 s_\beta(n - 1)(n\lambda).$$

The marginal is therefore

$$p'_1 = \sum_{n \geq 0} p'(A_1, n)$$

$$= p_1 s_\beta(0) + \sum_{n \geq 1} [p_1 s_\beta(n)(1 - n\lambda) + p_2 n s_\beta(n - 1)\lambda]$$

$$= p_1 \left[ s_\beta(0) + \sum_{n \geq 1} s_\beta(n) \right] + \sum_n \lambda(n)(n s_\beta(n))(e^{\beta E} p_2 - p_1)$$

since $s_\beta(n - 1) = e^{\beta E} s_\beta(n)$; we have put $\lambda(n) = \lambda$ up to $n = N$ and zero after. Now, $s$ is a normalised distribution, so the first bracket [...], the coefficient of $p_1$, sums to 1. More, $\lambda$ is proportional to the time-step, say $\lambda(n) = \kappa(n)dt$; we can thus rearrange the equation to get

$$\frac{p'_1 - p_1}{dt} = \sum_{n \geq 0} \kappa(n)(n s_\beta(n)) \left[ e^{\beta E} p_2 - p_1 \right]. \qquad (4.45)$$

As $dt \to 0$, the left side converges to $dp_1/dt$, after which we can let $N$ go to infinity. The right side converges, using

$$\sum_{n \geq 0} n s_\beta(n) = \sum_{n \geq 0} n(1 - e^{-\beta E}) e^{-n\beta E} = e^{-\beta E}(1 - e^{-\beta E})^{-1}$$

to get

$$\frac{dp_1}{dt} = \frac{\kappa}{1 - e^{-\beta E}}[p_2 - e^{-\beta E}p_1] = -\frac{dp_2}{dt}. \tag{4.46}$$

The second equation can be derived in the same way, or from the remark that $A_1$ is destroyed exactly when $A_2$ is produced, and *vice versa*.

We see that the rate at which $A_1$ is converted into $A_2$ is proportional to the mean occupation number of the oscillator, and hence to Planck's mean energy (2.42). This was Einstein's starting point, based perhaps on commonly known experiments at the time (Einstein does not give any references to such results). We also see that the law of Arrhenius is approximately true, if $\beta E \gg 1$: the loss rate for $A_1$, with activation energy $E$, is proportional to $e^{-\beta E}$, and the loss rate for $A_2$, with zero activation energy, it is independent of the temperature. More, the ratio of the forward to backward rates is exactly $e^{-\beta E}$.

The dynamics just derived assumes that the heat particle is always in its thermal state. After the application of the bistochastic $T$, the heat bath is knocked out of equilibrium. We postulate that it rethermalises, as part of the assumption that photon processes are much quicker than the chemical processes of interest. Thus to the state $p'$ we apply not only the marginal map onto $\mathcal{F}$, but we follow this by replacing the state of the photon (whatever it is) by the equilibrium state with the same mean energy; we can then repeat the map $T$, and argue as above for the second step of the dynamics, and so on. Since the number of heat particles + the number of $A_2$ is conserved, we must have for the mean number of photons

$$\frac{d\bar{n}}{dt} = \frac{dp_1}{dt} = \frac{\kappa}{1 - e^{-\beta E}}[p_2 - e^{-\beta E}p_1].$$

From this, you can find the equation of motion of the temperature (4.120).

More generally, both isomers might have a non-zero activation energy; in that case there are two active modes of the ambient heat source, and the reaction would be

$$A_1 + \gamma(1) \rightleftharpoons A_2 + \gamma(2) \tag{4.47}$$

where the energies of the heat particles obey $E(1) - E(2) = E$. The ambient heat is then described by the classical Fock space over two states, $\mathcal{F}_1 \times \mathcal{F}_2$, at a common beta; after each step of the reaction, the new measure on this space thermalises to a new common beta, without loss of mean energy. We leave the analysis of this model to an exercise, (4.121).

The five inert gases have molecules with van der Waals radius differing at most by a factor of 1.5, the smallest being neon. The compound molecule

$C_6H_6$ is less than twice as large as a neon atom: [Atkins and de Paula (2006)], Table 1.1, p. 41. The simple theory given here does not cover large molecules such as *DNA*. We adopt the rough approximation, in which all the ions and molecules are of the same size and can be represented as a hard sphere of diameter $a$. A point in space will be represented by $x \in \Lambda \subseteq (\mathbf{Z}a)^3$. Since two molecules cannot occupy the same point, we take the sample space to be

$$\Omega := \prod_{x \in \Lambda} \Omega_x, \tag{4.48}$$

$$\text{where} \quad \Omega_x := \{\emptyset, 1\}. \tag{4.49}$$

A sample point $\omega$ is the collection of *configurations* $\{\omega_x \in \Omega_x\}_{x \in \Lambda}$. When $\omega_x = \emptyset$ we say that the site $x$ is empty (or, has a hole) in the configuration $\omega$. When $\omega_x = 1$ we say that the site $x$ is occupied. A *state* is, as usual, a probability $p$ on $\Omega$. For chemistry, the most important special case is when $p$ in 'independent over $\Lambda$', that is when

$$p(\omega) = \prod_{x \in \Lambda} p_x(\omega_x), \tag{4.50}$$

$$\text{where} \quad p_x(\omega_x) = \sum_{\omega' : \omega'_x = \omega_x} p(\omega') \tag{4.51}$$

is the marginal of $p$ at $x$. There is no correlation between different point in a product state. In Exercise 4.119, you are asked to prove that such a $p$ is the state of maximum entropy, among all states with the given marginals, $\{p_x\}_{x \in \Lambda}$. Now, each $p_x$ is determined by its probability of occupation, $p_x\{\omega_x : \omega_x = 1\}$, since $\Omega_x$ has only two points, $\emptyset$ and 1. Let us write $p(x)$ for this occupation probability. The probability of a hole at $x$ is obviously $1 - p(x)$. If $0 < p(x) < 1$, we define the partition function by

$$Z(x) := (1 - p(x))^{-1} \tag{4.52}$$

and the *canonical field* $\xi(x)$ by

$$\xi(x) := -\log Z(x) - \log p(x). \tag{4.53}$$

Then $p(x)$ takes the form of a Gibbs state

$$p_x = Z^{-1} \exp\{-\xi(x)\mathcal{N}_x\} \tag{4.54}$$

where $\mathcal{N}_x$ is the random variable giving the number of particles at $x$:

$$\mathcal{N}(\emptyset) = 0$$
$$\mathcal{N}(1) = 1. \tag{4.55}$$

In thermodynamics it is usual to write $\xi(x) = -\beta(x)\mu(x)$, in terms of the inverse temperature $\beta$ and the chemical potential $\mu$. In this simple model there is no energy, and so no temperature. So we may interpret the canonical field as the negative chemical potential, up to a constant factor, and the marginal states $p(x)$ as being in *local thermodynamic equilibrium, LTE*. Since $p(x)$ is exactly the mean field $\mathbf{E}_p[\mathcal{N}_x]$, we can say that the mean field determines the state, if it is in *LTE*. The set of states in *LTE* makes up the *information manifold*, $\mathcal{M}$.

### 4.3.2  Markov Dynamics

We now introduce a class of Markovian dynamics for this model; we then get the dynamics in terms of the mean fields by projecting the state, after each time-step, onto the information manifold, $\mathcal{M}$. The resulting (discrete time) dynamics will obey both the first and second laws of thermodynamics. The first law here pertains to the conservation of number of particles: the only transitions are between states having the same number of particles. The simplest such is a hop between neighbouring sites, and will be termed *diffusion*; say $0 \in \Lambda$ and $e \in \Lambda$, where in three dimensions $e$ is one of the 6 unit vectors of the lattice. The sample space for the two points is $\Omega_{0,e} = \omega_0 \times \Omega_e$. This has four points,

$$\{(\emptyset, \emptyset), (\emptyset, 1), (1, \emptyset), (1, 1)\}. \tag{4.56}$$

A hop can occur only when one site is empty and the other is occupied, and all the remaining components of $\omega$ and $\omega'$ are equal. It is natural to ask that the forward and the backward hops occur with the same probabilty, $\kappa$ say. This ensures that the resulting transition matrix is bistochastic:

$$T_{0,e} := \begin{pmatrix} 1 & 0 & 0 & 0 \\ 0 & 1-\kappa & \kappa & 0 \\ 0 & \kappa & 1-\kappa & 0 \\ 0 & 0 & 0 & 1 \end{pmatrix}. \tag{4.57}$$

This is written in the basis given by (4.56). This map cannot decrease the entropy, as it is bistochastic. Now let $T_{x,e}$ be the similar operator, with $x$ replacing 0. If $x$ is one of the points in the even lattice, $(2\mathbf{Z})^3$, then all the $T_x$ commute. Let $T_+$ be the product of all operators $T_{x,e}$ with $x$ on the even lattice. This acts simultaneously on all even points $x \in \Lambda$, with the same value of $\kappa > 0$. It is a symmetric stochastic map: for suppose first that $\omega_x, \omega'_{x+e}$ is either of the pairs $(\emptyset, 1)$ or $(1, \emptyset)$. Then

$$\sum_{\omega'} T(\omega, \omega') = \sum_{\omega'} T_{x,e}\left((\omega, (...\omega'_{x-2e}, \omega'_{x-e}, \omega'_x, \omega_{x+e}, ...)))\right)$$

and the entries are $\delta_{\omega_y,\omega'_y}$ unless $y = x$ or $x+e$. The sum over all these is 1. So the test reduces to checking that the second and third rows of $T(x,e)$ sum to one (which they do). In the remaining two cases, $\omega_x, \omega'_{x+e}$ is either $\emptyset, \emptyset$ or $1, 1$. Then the matrix element of $T(x,e)$ is 1, and the role of $x$ is played by another point, not a neighbour of $x$, with the same result. Clearly, $T_+$ is a symmetric operator, being the product of commuting symmetric operators. It follows that the map increases the entropy of any state (or leaves it the same).

Let us find the dynamics given by $T_+$, starting with a probability $p \in \mathcal{M}$. The dynamics is taken to be $p \mapsto pT_+ \mapsto pT_+Q = p'$, where $Q$ is the projection of a state (here $pT_+$) onto $\mathcal{M}$. By definition, $Q$ does not change the mean field at any $x$. The dynamics is thus a path in $\mathcal{M}$, and can be defined as a path in the coordinates describing points of $\mathcal{M}$. Let us take the mean field, $N(x)$ as the coordinate of the current state. In one time step, say $\Delta t$, in which $p$ changes to $p'$, the field $N$ changes to $N'$, where

$$N'(x) := p'.\mathcal{N}_x = pT_+Q.\mathcal{N}_x = pT_+.\mathcal{N}_x. \tag{4.58}$$

So it is sufficient to find the change in $N$ due to the linear part of the dynamics. We may express this as saying that the difference between $pTQ$ and $pT$ cannot be observed by measuring the mean field $N_x$. Since we start on the manifold, $p = \otimes_y p_y$ is a product state of the marginal probabilities; the only two that enter the calculation are

$$p_x(\emptyset) = 1 - N_x; \qquad p_x(1) = N_x;$$
$$p_{x+e}(\emptyset) = 1 - N_{x+e}; \qquad p_{x+e}(1) = N_{x+e}.$$

Then the hop $T_{x,e}$ leads to

$$pT_{x,e}(1) = N_x N_{x+e} + (1 - \kappa)N_x (1 - N_{x+e}) + \kappa (1 - N_x) N_{x+e}$$
$$= N_x + \kappa (N_{x+e} - N_x). \tag{4.59}$$

The quadratic terms have cancelled. We consider the terms $\pm e$ together, so as to include the odd points, by forming the convex mixture (which is also bistochastic)

$$T_e := \frac{1}{2} (T_{x,e} + T_{x-e,e}),$$

which gives the dynamics

$$N(x, t + \Delta t) = N(x, t) + \frac{\kappa}{2} (N_{x+e} - 2N_x + N_{x-e}),$$

which is the finite-difference approximation to

$$N(x, t + \Delta t) - N(x, t) = \frac{\kappa |e|^2}{2} \frac{\partial^2 N}{\partial x^2}.$$

Notice that the quadratic term has cancelled. In three dimensions, we have three directions to account for; so we add two similar terms, dividing by 3 to retain the bistochasticity of $T := 1/3\,(T_{e_1} + T_{e_2} + T_{e_3})$. Now divide both sides by $\Delta t \to 0$; the dynamics is then given by the heat equation with parameter $\lambda$:

$$\frac{\partial N}{\partial t} = \lim_{\Delta t \to 0} \frac{\kappa |e|^2}{6\Delta t} \nabla^2 N = \lambda \nabla^2 N. \tag{4.60}$$

We can turn this round; the diffusion equation can be approximated by a Markov chain with time-step $\Delta t$ and space grid of size $|e|$, provided that $6\Delta t \lambda / |e|^2$ is not so big that $\kappa$ lies outside the range $0 < \kappa < 1$. As $\kappa$ approaches 1, the finite difference approximation becomes worse, and for $\kappa > 1$ we get quasi-periodic motion with increasing amplitude, which approaches chaotic dynamics as $\kappa$ increases further [Rondoni (1991)]. Note that whereas the diffusion equation for $N(x, t)$ is the same as that of a random walk on a lattice of a single particle, the state space and thus the entropy, is not the same.

### 4.3.3  *Entropy Production*

Since our $T$ is bistochastic, the discrete linear model increases entropy at each step; since $Q$ is also entropy-increasing, the dynamics increases entropy along the orbit on $\mathcal{M}$. This property persists in the continuum limit.

The entropy of the product state in $\mathcal{M}$ is the sum of the entropy at each site. Thus

$$S(p) = -\sum_x p_x \log p_x = -\sum_x N_x \log N_x.$$

If the lattice size is $a$, then for systems of finite density, with $\rho = N/(a^3)$, we may replace $\sum_x$ by $a^{-3} \int ...dx$. Then the entropy of the continuum

$$S(\rho) = \lim_{a \to 0} a^{-3} \sum_x N_x \log N_x \to \int (\rho(x) \log \rho(x) - 3 \log a) dx$$

diverges. The information contained in a continuum field is infinite. However, since the space integral of the density is conserved, the divergent term is time-independent. The rate of entropy production is finite, and is positive:

$$\begin{aligned}
\frac{dS}{dt} &= -\lambda \int d^3x \frac{\partial \rho}{\partial t} \log \rho - \int d^3x \frac{\rho}{\rho} \frac{\partial \rho}{\partial t} \\
&= -\lambda \int d^3x \log \rho \, \partial_i \partial_i \rho \\
&= \lambda \int d^3x \, \partial_i \rho \frac{1}{\rho} \partial_i \rho \geq 0. \tag{4.61}
\end{aligned}$$

This rate happens to coincide with the Fisher information (15.19) in $p$, with the factor $\lambda$.

### 4.3.4 *Osmosis*

Suppose that $\Lambda$, a cubic lattice in dimension 2 or 3 with base-vector $e$, is divided by a semi-permeable membrane, which allows the solvent $W$ to pass through a hole to the other side, but does not allow any solute $A$ to pass. So, if $x$ is on one side not containing any $A$ and $x + e$ is on the other, in a region with some $A$, $W$ will pass through until the chemical potential of $W$ is the same on both sides of the membrane. The only dynamics through the membrane is the $(W, \emptyset)$ exchange diffusion, obeying the equation

$$N'_W(x) - N_W(x) = Z_x^{-1} Z_{x+e}^{-1} \left(a_W(x + e) - a_W(x)\right). \qquad (4.62)$$

Suppose that at time $t = 0$ the pressure, and thus $Z$, is the same on both sides. So initially,

$$1 + a_W(x) = 1 + a_W(x + e) + a_A(x + e)$$

from which we conclude that

$$a_W(x) > a_W(x + e).$$

Therefore, $W$ will flow from the side without $A$ to the side with some $A$, and the pressure will rise on that side and drop on the side containing only solvent. This will go on until the activities on both sides are equal. This creates a pressure difference across the membrane, which is used by plants to drive its fluids. It also causes difficulties to fresh-water fish when they are put into the sea. Suppose that in region (1) we have fresh water, and in region (2) we have water. $W$, and salt, $C$. The pressure is given by

$$PV = N k_B \Theta \log Z = k_B \Theta \log\left(1 + a_W + a_C\right). \qquad (4.63)$$

In (1), there is no salt, so $a_C = 0$ there. At equilibrium, the chemical potentials of water are equal on both sides, to $a_W$ say. Thus the difference between the potentials is

$$P(2) - P(1) = V^{-1} N k_B \Theta \log \frac{1 + a_W + a_C}{1 + a_W} \approx \frac{R^{-1}}{v} \frac{n_C}{n_W}.$$

Here, $R$ is the gas constant, $N_L k_B$, where $N_L$ is the Loschmidt number, the number of molecules per mole; $n_C$ and $n_W$ are the molar fractions of salt and water in the region (2), and $v$ is the volume of one mole of the salt and water. This is an example of the relation of van t'Hoff [van t'Hoff (1884)].

### 4.3.5   *Exchange Diffusion*

When there is more than one type of particle, diffusion can occur by random exchanges, say of nearest neighbours. Take the case when a chemical $A$ is dissolved in solute $W$. The sample space at $x$ can be modelled by $\{\emptyset, A, W\}$. Let $\mathcal{M}$ be the information manifold defined by the mean fields $A(x), W(x)$. Suppose that in a time $\Delta t$, the exchange of an adjacent pair at $x$ and $x+e$, say $(A, W)$ to give $(W, A)$, have probability $\kappa$. This defines a map $T(x, e)$. Follow this map by the projection onto $\mathcal{M}$. Then, as above, the means obey the equation of motion

$$A'(x) = A(x) + \kappa\left(W(x)A(x+e) - A(x)W(x+e)\right) \tag{4.64}$$

$$W'(x) = W(x) - \kappa\left(W(x)A(x+e) - A(x)W(x+e)\right). \tag{4.65}$$

The hole density is unchanged by this map, which can be denoted by $T^{AW}(x, e)$. This looks like (4.59), but now the quadratic terms do not cancel. We can again take the product of such maps over the even lattice, and then form the convex sum over the six directions of $e$, to define the exchange diffusion map involving $A$ and $W$.

Let us add the similar exchange-hops $T^{A}(x, e)$ and $T^{W}(x, e)$ at $x$ between $A$ and $\emptyset$ and between $W$ and $\emptyset$, with hopping parameters $\kappa_A$ and $\kappa_W$ respectively. We divide by 3 to maintain normalisation. To find the fixed points of the composite map thus obtained, we must put $A'(x) = A(x)$, $W'(x) = W(x)$ for all $x$. We can ignore the dynamics of the hole density $p_x(\emptyset)$ as the three fields sum to 1. The easy way to solve this system for the fixed points is to write the mean fields in terms of the canonical fields $\xi_A, \xi_W$, and the activities $a_A(x)$, $a_W(x)$, and $a_\emptyset(x) := 1$:

$$N_A(x) = Z_x^{-1} e^{-\xi_A(x)} \tag{4.66}$$

$$N_W(x) = Z_x^{-1} e^{-\xi_W(x)} \tag{4.67}$$

$$Z_x := 1 + a_A(x) + a_W(x) \tag{4.68}$$

$$a_A(x) := e^{-\xi_A(x)} \tag{4.69}$$

$$a_W(x) := e^{-\xi_W}. \tag{4.70}$$

Then the dynamics generated by the three exchange diffusions are given respectively by

$$N'_A(x) - N_A(x) = \kappa Z_x^{-1} Z_{x+}^{-1} \{a_W(x)a_A(x_+) - a_A(x)a_W(x_+)\}$$
$$N'_W(x) - N_W(x) = -\kappa Z_x^{-1} Z_{x+e}^{-1} \{a_W(x)a_A(x_+) - a_A(x)a_W(x_+)\}$$

$$N'_A(x) - N_A(x) = \kappa_A Z_x^{-1} Z_{x+e}^{-1} \{a_A(x+e) - a_A(x)\}$$
$$N'_W(x) - N_W(x) = 0$$

$$N'_W(x) - N_W(x) = \kappa_W Z_x^{-1} Z_{x+e}^{-1} \{a_W(x+e) - a_W(x)\}$$
$$N'_A(x) - N_A(x) = 0.$$

In the first two equations, we have used the notation $x_+ = x + e$. By the theorem on detailed balance, any fixed point of a convex sum of these maps, for all $x$, must be a fixed point of each map. Clearly, the fixed point of the last two maps must have $a_A(x)$ and $a_W(x)$ independent of $x$, and this is also a fixed point of the exchange diffusion between $A$ and $W$. This implies obviously that the canonical fields $\xi_A$ and $\xi_W$ must be independent of $x$, and so the chemical potential (linearly related to $\xi$) is constant in space; this is the correct thermodynamic equilibrium.

It would be wrong to interpret these rate equations as being 'activity-led'. Indeed, out of equilibrium, the partition functions depend on time, and the exchange diffusion remains density-led. The non-linearity of the diffusion in exchange with holes is hidden in the partition functions; only in a different model (with no hard cores) have we been able to show that the dynamics is activity-led; this theory is presented in Section 4.6. For liquids however, $N_W \gg N_A + p_\emptyset$ is quite common; we then get the *dilute limit*, in which the partition function is constant in time up to zeroth order in $N_A/N_W$. Then the equations are activity-led and also density-led up to first order.

### 4.3.6  *General Diffusions*

When a liquid such as water, $W$, contains chemicals $A_1, \ldots, A_n$, we may expect that all the processes of exchange between $A_j$ and $\emptyset$, $A_j$ and $W$, and between the $A_j$ and $A_k$ are present with various diffusion coefficients. We can set up the rate equations by forming the convex mixture of all of them. The point of equilibrium of such a system must be a point of equilibrium of each process separately, by the theorem of detailed balance. At equilibrium, therefore, the activities $a_W, a_1, \ldots a_n$ are independent of $x$.

The values of the $a_i$ at equilibrium are determined by the initial amounts of the substances, by the system of equations

$$N_W = Z^{-1}a_W; \quad N_i = Z^{-1}a_i; \quad Z = 1 + a_W + \sum_i a_i. \qquad (4.71)$$

The solution to this is $a_i = N_i/N_\emptyset$.

## 4.4 Chemical Reactions

### 4.4.1 *Unimolecular Reactions*

The simplest chemical reaction is when one molecule of $A$ can make a transition to one molecule of $B$, a different molecule. By the first law, they must have the same energy (to within the accuracy of the model). For example, $A$ and $B$ could be two isomers of a molecule, with the same energy. The inverse process must then be possible, with the same rate. The reaction can take place at an individual site. We can put $\Omega_x = \{\emptyset, W, A, B\}$. The transition matrix for the site $x$ is then

$$T_x := \begin{array}{c} \emptyset \\ W \\ A \\ B \end{array} \begin{pmatrix} 1 & 0 & 0 & 0 \\ 0 & 1 & 0 & 0 \\ 0 & 0 & 1-\kappa & \kappa \\ 0 & 0 & \kappa & 1-\kappa \end{pmatrix}. \qquad (4.72)$$

This gives for the dynamics of one time-step, at the site $x$:

$$p'(A) = (1-\kappa)p(A) + \kappa p(B) \qquad (4.73)$$
$$p'(B) = \kappa p(A) + (1-\kappa)p(B) \qquad (4.74)$$
$$p'(W) = p(W) \qquad (4.75)$$
$$p'(\emptyset) = p(\emptyset). \qquad (4.76)$$

Follow this map by the projection $Q$ onto the information manifold defined by the fields $N_x(W)$, $N_x(A)$, $N_x(B)$, and we get the density-led reactions at each $x$:

$$N'(A) - N(A) = \kappa(p(B) - p(A)) = \kappa(N(B) - N(A)) \qquad (4.77)$$
$$N'(B) - N(B) = -\kappa(p(B) - p(A)) = -\kappa(N(B) - N(A)). \qquad (4.78)$$

Clearly, a fixed point of this map must have equal mean numbers, and therefore chemical potentials, of $A$ and $B$, at the site $x$.

### 4.4.2 Balanced Reactions

In the section on cotransport, we considered the reaction

$$A + B \rightleftharpoons C + D, \tag{4.79}$$

in a space of two points, so we took $\Lambda = \{x, x+e\}$, and $\Omega_x = \{A, B, C, D\}$. Since the number of molecules in the final state is the same as in the initial state, this is called a 'balanced' reaction. This is easily generalised to reactions in which $n$ particles $\{A_i\}_{i=1..n}$ sitting on $n$ close-by sites $\Lambda_0 = \{x_1, \ldots, x_n\}$ can make a transition to $\{B_j\}_{j=1,\ldots,n}$ distributed over the same sites. We can also include a hole, so as to allow diffusion. Some of the chemicals $A_i$ might be the same as some of the $B_j$; in that case, these chemicals are called *catalysts*. We can take the sample space at a site to be $\Omega_x = \{\emptyset, \ldots A_i, \ldots B_j, \ldots\}$ where no repetitions of catalysts are needed. The region of the reaction is the set $\Lambda$, so the sample space is

$$\Omega(\Lambda) = \prod_{x \in \Lambda} \Omega_x.$$

The information manifold is the set of states independent at each point of $\Lambda$. So we start with a probability defined in terms of its marginals, of the form

$$p(\{\omega_x\}) = \prod_{x \in \Lambda_0} p_x(\omega_x). \tag{4.80}$$

There are many ways to define a transition Markov matrix; for example, we may assume that any permutation among the ingoing is equally likely to lead to any permutation among the outgoing particles. Calling this probability $\kappa$, there are $n!$ ingoing configurations at the $n$ sites, and the probability of one of these is (4.80). For a given resultant $B_j$ at site $x_k$, there are $(n-1)!$ outgoing configurations, so the rate equations are

$$p'_k(B_j) = (1 - n!(n-1)!\kappa)p_k(B_j)$$

$$+ \kappa(n-1)! \left\{ \sum_{p_{i_1}} (A_1) \ldots p_{i_n}(A_n) - p_{j_1}(B_1) \ldots p_{j_n}(B_n) \right\}. \tag{4.81}$$

There are similar equations for the update of the other concentrations. If some of the $A_i$ and the $B_j$ are the same chemical, and the catalysts are called $C_k$, the reaction takes the form

$$m_1 A_1 + \ldots + m_r A_r + M_1 C_1 + \ldots + M_s C_s$$

$$\rightleftharpoons l_1 B_1 + \ldots + l_t B_t + L_1 C_1 + \ldots L_s C_s. \tag{4.82}$$

The coefficients $m_1, \ldots, L_s$ are the numbers of chemicals in the reaction, and define its *stochiometry*. We see at once that the reaction stops if the concentration of a catalyst is zero, and that the rate of progress of the reaction is proportional to some power of the concentration of the catalyst. This type of law will not hold of catalysts that are not in solution. For example, Atkins [Atkins (2001)], p. 245 mentions that powdered tungsten acts as a catalyst inducing the decomposition of phosphene, $PH_3$; the rate is constant, independent of the concentration of phosphene, until all the phosphene is decomposed. In this case, the rate of decomposition is proportional to the surface area of the tungsten, rather than the quantity by weight. It can be modelled by the reaction

$$\emptyset + PH_3 \rightleftharpoons P + 3H$$

where the hole is on the surface of a fixed particle of tungsten. It is thought that many catalysts work simply by bringing their thermal energy into contact with the reacting chemicals, thus supplying the needed excitation energy.

We can write the probabilities $p_A, \ldots$ occurring in (4.81) in terms of the activities:

$$p_A = Z^{-1} a_A = Z^{-1} e^{-\xi_A} = Z^{-1} e^{\beta \mu_A};$$

at equilibrium the products must be equal on each side of the reaction. As the reaction is balanced, the products of the partition function at all points in $\Lambda_0$ is the same in both factors, and we are left with the equality of the sums of the chemical potentials, weighted with the stoichiometry.

We can imagine that a compound process actually proceeds in stages through three-body reactions, with perhaps very short-lived intermediate ions. Thus $A + B \rightleftharpoons C + D$ might be modelled by two simpler reactions

$$A + B \rightleftharpoons X \tag{4.83}$$

$$X \rightleftharpoons C + D, \tag{4.84}$$

where the rate of the decay of $X$ is very rapid. There would hardly be any difference in the dynamics of the observable ions, $A, B, C, D$ between this 'theory of the transition state', and the direct four-body reaction. Both would show an exponential approach to the same equilibrium. So we could limit the study to convex mixtures of three-body reactions, and still get a viable class of models. However, our main theorem, the fundamental theorem of chemistry, is nearly as easy to prove with general stoichiometry as in the case of three bodies, so we need to consider the theory with the transition state only if there is some evidence for the existence of some possibly short-lived $X$; as $X$ needs its own equation of motion, the amount of numerical work is increased whenever an intermediate state $X$ is introduced.

If it is known that, in the ions occurring in a reaction, the atoms line up in a particular geometric manner, then we could model this by using several sites of $\Lambda_0$ for each ion. Then reactions might not occur unless the reactants lay in a certain position or orientation. The simple choice made here is enough for our purposes, which is to show that the thermodynamics of chemical reactions can be modelled, and that driven systems can show periodic behaviour.

To incorporate energy as well as atomic number as a conserved quantity, we use a heat particle, located on each bond. Take for example the reaction

$$A + B + \gamma \rightleftharpoons C + D. \tag{4.85}$$

Here, $\gamma$ is the activation energy for the forward reaction. We recognise $-\gamma$ as the heat of reaction of the forward process. Since it is negative, we call the forward process *endothermic* and the backward process *exothermic*. This can proceed when $A$ is at site $x$, and $B$ is at site $x + e$, or *vice versa* where $e$ is one of the lattice basis vectors. The pair need to pick up a heat-particle of energy $\gamma$ from the bond $b$ between them. Thus $\Omega_b$ must contain the classical Fock space of quanta of energy $\gamma$. The reaction can proceed if the state of the bond has one or more quanta in it. It can be argued [Einstein (1917)] that as each quantum had an equal chance of being absorbed, the rate of reaction must be proportional to the number of quanta, $j$ present. The rate, for a thermal state of the bond, thus has a factor

$$\sum_{j=0}^{\infty} j \left(1 - e^{-\beta \epsilon}\right) = \left(1 - e^{-\beta \epsilon}\right)^{-1} e^{-\beta \epsilon} = p_\gamma \cdot \mathcal{N}_\gamma.$$

For very large $\beta$, that is, low enough temperatures, for which the activation energy is large compared to the thermal energy, we can ignore $e^{-\beta \epsilon}$ compared to 1, and the forward rate depends on temperature according to the Arrhenius formula. At larger temperatures, $\beta$ is small, and we can expand $e^{-\beta \epsilon} \approx 1 - \beta \epsilon + \dots$. Then the rate acquires a factor

$$e^{-\beta \epsilon} k_B \Theta / \epsilon$$

which is one of the popular corrections to the Arrhenius formula. Einstein used microscopic reversibility to get the backward rate of such a reaction. Suppose the sites $x$ and $x + e$ are occupied by $C$ and $D$, and $b$ contains $j$

quanta; then the final state will be $A + B$ and $j + 1$ quanta. So the forward rate would have been proportional to $j + 1$, and so by reversibility, the backward rate has the factor $j + 1$ too; summing over $j$ gives the factor

$$(\mathcal{N}_\gamma + 1) \cdot p_\gamma = (1 - e^{-\beta \epsilon})^{-1}$$

in the backward rate. Einstein called the term proportional to $n_\gamma$ the *stimulated emission*, proportional to the energy density. The extra 1 he called *spontaneous emission*. We see that the ratio of the forward to the backward rate is thus $e^{-\beta \epsilon}$ exactly, without any corrections. Results similar to these are usually obtained (in thermal equilibrium) by postulating an intermediate reaction, in which the initial state moves to a *transition state* in thermal equilibrium with both the reactants and the products [Atkins (2001); Eyring, Lin and Lin (1980)]. We do not need this kind of argument.

If the process and the reverse process both have an activation energy, then the reaction is

$$A + B + \gamma_1 \rightleftharpoons C + D + \gamma_2,$$

and the sample space of the bond $b$ is the tensor product of two Fock spaces, of quanta with energies $\epsilon_1$ and $\epsilon_2$. The above argument then shows that the forward rate will have factors

$$e^{-\beta \epsilon_1} \left(1 - e^{-\beta \epsilon_1}\right)^{-1} \left(1 - e^{-\beta \epsilon_2}\right)^{-1}.$$

This gives, for high temperatures, a correction to Arrhenius proportional to $\Theta^2$. Thus our model predicts that when a reaction, and its reverse reaction, both show an activation energy, then at high temperatures, the Arrhenius prefactor should be $\Theta^2$ rather than $\Theta$.

We now show how to deal with unbalanced reactions, such as the reaction $A + B \rightleftharpoons C$; to get the dynamics of such a reaction, we can add a hole to the right-hand side, and then use the same construction as for balanced reactions, with the hole playing the role of a chemical type, having zero chemical potential. We arrive at (4.6). This process reaches equilibrium when the densities are uniform in space, and the products are equal:

$$p(A)p(C) = p(\emptyset)p(B),$$

which leads to the equality of the chemical potentials on both sides of the reaction. The 'density-led' equations advocated in [Atkins (2001); Atkins and de Paula (2006); Laidler (1987)] leads to the incorrect fixed point, unless we introduce the artificial device of the transition state at equilibrium with both reactants and products.

## 4.5  Energy of Solvation

When salt is dissolved in water, heat is emitted and salt splits into chlorine and sodium ions. This is due to a chemical reaction between water and the salt. As we assume that the solvent, water, is in abundance, we can approximately model this by assuming a constant heat of reaction, equal to the potential energy when each sodium and chlorine ion is surrounded by water molecules. Since like charges repel, and unlike charges attract, the realistic dynamics of ions involves potentials between particles. This can be formulated, but it leads to a model outside our present simple theory; first, even to find the energy-shell for a many-body system is a big computing job. So an allowed Markov chain, $T$, is difficult to write down. Secondly, the map $Q$, from the state $pT$ onto its product state, does not preserve the energy of the system; we cannot use this simplifying map to write the state in terms of the density-field, without violating the first law. The consideration of ion dynamics is thus outside the scope of this book.

## 4.6  Activity-led Reactions

When the density is not high, the hard-core in the hard-core model should not show up. It has been suggested, for example [Keizer (1987)], that the rate equation for $A + B \rightleftharpoons C + D$ should be

$$\frac{dN_A}{dt} = K_- a_C a_D - K_+ a_A a_B, \qquad (4.86)$$

rather than being density-led, as in (4.36). Here, $N$ is the mean number, $N = \langle \mathcal{N} \rangle$, and $a$ is the activity. For a free particle with energy $\mathcal{E} = 2\pi\hbar\nu\mathcal{N} = E\mathcal{N}$ the grand canonical state is the probability on classical Fock space given by

$$\text{Prob}\{\mathcal{N} = j\} = Z(\beta, \mu)^{-1} \exp\{j\beta(\mu - E)\} \qquad (4.87)$$

and the activity is:

$$a = N/(1 + N) = \exp\{\beta(\mu - E)\}. \qquad (4.88)$$

Then rather than (4.37) the equilibrium condition becomes,

$$a_A a_B / a_C a_D - K_- / K_+ . \qquad (4.89)$$

Let us first consider the case when $K_+ = K_-$, so the heat of reaction is zero. Then (4.89) becomes

$$\exp\{-\beta(E_A - \mu_A)\} \exp\{-\beta(E_B - \mu_B)\}$$
$$= \exp\{-\beta(E_C - \mu_C)\} \exp\{-\beta(E_D - \mu_D)\}$$

which is satisfied if $\mu_A + \mu_B = \mu_C + \mu_D$ since by energy conservation we already have

$$E_A + E_B = E_C + E_D \ .$$

Similarly, activity-led equations for a general reaction $A_1 + A_2 + \ldots + A_m \rightleftharpoons B_1 + B_2 + \ldots + B_n$ can be written down (4.129). You are asked to check that if $q = 0$ then a grand canonical state is a fixed point if and only if the sum of the chemical potentials on each side of the reaction is the same.

We conclude that a grand canonical state with balanced chemical potentials is a fixed point of the activity-led dynamics, provided that the heat of reaction is zero. It is easy to modify the dynamical law to cover the cases when some of the chemicals are the same kind of molecule as others. The number of molecules entering the reaction is called the 'stoichiometry' of the molecule. If a type of molecule appears on both sides of the reaction, say $A_1 = B_1$, then this chemical is called a catalyst. The general conclusion, that equilibrium occurs when the chemical potentials on both sides are equal, is still true if there are catalysts present, whatever the stoichiometry. Now do (4.131).

So far, we have taken the heat of reaction to be zero. To include the case where heat is emitted, say in the forward reaction

$$A_1 + A_2 + \ldots + A_m \rightleftharpoons B_1 + B_2 + \ldots + B_n + q$$

we simply introduce $B_{n+1} = \gamma$, a heat-particle with quantum of energy equal to $q$. The activity of $\gamma$ is $a_\gamma = e^{-\beta q}$, and the chemical energies on each side of the reaction no longer sum up to the same total, but differ by $q$:

$$E_{A_1} + \ldots + E_{A_m} = E_{B_1} + \ldots + E_{B_n} + q \ .$$

As a result, the grand canonical state with balanced chemical potentials is a fixed point. The backward equation contains the factor $a_\gamma = e^{-\beta q}$, so the relation (4.38) is predicted to hold. This is even true when heat-particles appear on both sides of the equation, in which case heat is a catalyst that allows the reaction to start. This is what happens in the theory of the transition state. In that theory [Eyring, Lin and Lin (1980)], it is postulated that the reaction does not proceed directly, but first moves to a 'transition state', requiring an energy $E_a$, the activation energy. This is supplied by the kinetic energy, or in statistical dynamics, by a heat-particle. The energy of the heat-particle on the other side of the reaction must then be $q + E_a$; or we could have two or more heat-particles whose total energy is $q + E_a$. In any case, the forward-to-backward rates are in the ratio of $e^{-\beta q}$, as predicted by (4.38). So there is good reason to study the activity-led equations. The question arises, can they be derived by statistical dynamics? If so, then they obey the laws of thermodynamics. We now show that we can indeed find a $T$ conserving energy so that these equations arise by the application of $T$ followed by the $LTE$-map $Q_c$. We follow [Koseki and Streater (1993)] closely. We start with a detailed study of the simplest case, that of a unimolecular reaction.

A unimolecular reaction is a reaction

$$A \rightleftharpoons B . \tag{4.90}$$

We assume no heat flows, so that the states $A$, $B$ must have the same energy, number of atoms, etc. The possible configurations of $A$ are the numbers $\{0, 1, 2, ...\} = \Omega_A$ of particles in the cell, and the configuration space for $B$ is $\Omega_B = \{0, 1, 2, ...\}$, and for the system as a whole we choose $\Omega = \Omega_A \times \Omega_B$. A typical point of $\Omega$ will be denoted $\omega = (j, k)$. The numbers of $A$- and $B$-particles, $\mathcal{N}_A$, $\mathcal{N}_B$, are the random variables

$$\mathcal{N}_A(\omega) = j, \quad \mathcal{N}_B(\omega) = k \tag{4.91}$$

and the reaction (4.90) conserves $\mathcal{N} = \mathcal{N}_A + \mathcal{N}_B$. A probability $p$ on $\Omega$ is described by numbers $p_{jk} \geq 0$, $\sum_{j,k} p_{jk} = 1$.

We start at some time with $p_A \in \Sigma(\Omega_A)$ and $p_B \in \Sigma(\Omega_B)$, and form the independent product $p_A \otimes p_B \in \Sigma(\Omega)$. The reaction in one time-step is then described by a bistochastic map $T$:

$$T : \Sigma(\Omega) \longrightarrow \Sigma(\Omega) \tag{4.92}$$

conserving $\mathcal{N}$. The new distributions $p'_A$, $p'_B$ after one time-step are then obtained as the marginal distributions of $(p_A \otimes p_B)T$. Explicitly, $\mathcal{M}_A$, $\mathcal{M}_B$, the marginal operators, are

$$(p\mathcal{M}_A)_j = \sum_k p_{jk} \quad , \quad (p\mathcal{M}_B)_k = \sum_j p_{jk} \tag{4.93}$$

for any $p \in \Sigma(\Omega)$.

The operators $\mathcal{M}_A$, $\mathcal{M}_B$ do not alter the mean values of $\mathcal{N}_A$, $\mathcal{N}_B$ respectively. So far, the dynamics in one time-step is the map

$$(p_A, p_B) \mapsto p_A \otimes p_B \mapsto (p_A \otimes p_B)T$$
$$\mapsto ((p_A \otimes p_B)T\mathcal{M}_A) \bigotimes ((p_A \otimes p_B)T\mathcal{M}_B)$$
$$= p'_A \otimes p'_B . \tag{4.94}$$

Even if $p_A$ and $p_B$ are local equilibrium states of the form (4.87), in general $p'_A$ and $p'_B$ given by (4.94) will not be. The average value of $\mathcal{N}_A$ in (4.87) is

$$N_A = \sum_{j \geq 0} j p_{Aj} = Z_A^{-1} \sum_{j \geq 0} j a_A^j = (1 - a_A) \frac{a_A}{(1 - a_A)^2} = \frac{a_A}{1 - a_A} \ ;$$

thus

$$N_A = \frac{a_A}{1 - a_A}, \quad a_A = \frac{N_A}{1 + N_A}, \quad Z_A = 1 + N_A = (1 - a_A)^{-1}. \quad (4.95)$$

The local equilibrium state (4.87) can be expressed in terms of $N$, thus:

$$\mathrm{Prob}\{\mathcal{N}_A = j\} = \frac{1}{1 + N_A} \left( \frac{N_A}{1 + N_A} \right)^j. \quad (4.96)$$

Given any state $p \in \Sigma(\Omega)$, we have defined $pQ_c$ to be the local equilibrium state, (4.96), with the values of $N_A$ and $N_B$ determined by

$$N_A = \sum_1^\infty j p_A(j) \quad (4.97)$$

and a similar formula for $N_B$. By definition, $Q_c$ does not alter the mean value of $\mathcal{N}_A$ or $\mathcal{N}_B$. The dynamics of the model is defined by (4.94) followed by $Q_c$

$$p\tau = (p(\tau)_A, (p\tau)_B) = (p'_A Q_c, \ p'_B) Q_c \ . \quad (4.98)$$

By construction, $\tau$ takes a pair of local equilibrium states on $\Omega_A$, $\Omega_B$ to another pair; we can parametrize this either by the map $(N_A, N_B) \mapsto (N'_A, N'_B)$ or $(a_A, a_B) \mapsto (a'_A, a'_B)$.

It remains to specify the bistochastic map $T$. We shall give it as a symmetric Markov matrix on $\ell^2(\Omega)$, the Hilbert space of square-summable random variables with scalar product

$$\langle f, g \rangle = \sum_{j,k=0}^\infty f(j, k) g(j, k) \ .$$

An orthonormal basis for $\ell^2(\Omega)$ is given by the projection operators onto $(j, k)$, for each $(j, k) \in \Omega$:

$$\left. \begin{array}{ll} P(j, k) = 1 & \text{if} \quad \omega = (j, k) \\ \qquad\quad = 0 & \text{otherwise.} \end{array} \right\} \quad (4.99)$$

Then

$$\langle P(j, k), P(j', k') \rangle = \delta_{jj'} \delta_{kk'} \ . \quad (4.100)$$

Any random variable on $\Omega$ is the sum of multiples of the $P(j,k)$; for example,

$$\mathcal{N}_A = \sum_{j,k=0}^{\infty} j P(j,k) \quad , \quad \mathcal{N}_B = \sum_{j,k=0}^{\infty} k P(j,k) . \qquad (4.101)$$

A probability $p \in \Sigma(\Omega)$ can also be written in this way; thus the local equilibrium state $(a_A, a_B)$ can be written

$$(a_A, a_B) = \frac{1}{Z_A Z_B} \sum_{j,k=0}^{\infty} a_A^j a_B^k P(j,k). \qquad (4.102)$$

Then the expectation of an observable $f$ in the state $p$ is given by

$$\sum_{j,k=0}^{\infty} f(j,k)p(j,k) = \sum_{j,k,j',k'} \langle f(j,k)P(j,k), p(j',k')P(j',k') \rangle$$
$$= p \cdot f . \qquad (4.103)$$

Our operator $T$ is defined once it has been specified on each $P(j,k)$. For some $K \in (0, 1/2)$ let us put

$$
\begin{aligned}
TP(0,0) &= P(0,0) \\
TP(0,k) &= (1-K)P(0,k) + KP(1,k-1); & k \geq 1 \\
TP(j,0) &= (1-K)P(j,0) + KP(j-1,1); & j \geq 1 \\
TP(j,k) &= (1-2K)P(j,k) + KP(j-1,k+1) \\
&\quad + KP(j+1,k-1); & j,k \geq 1 . \quad (4.104)
\end{aligned}
$$

The first says that if there are no particles, then this remains true after one time step. The second says that if $A$ is absent and there are $k$ $B$'s, then one $B$ is converted to an $A$ with probability $K$, independent of the number $k \geq 1$. The third is the same with the roles of $A$ and $B$ reversed. The fourth says that if both $A$ and $B$ are present then each makes a transition to the other with rate $K$ depleting the original state with probability $1 - 2K$. (This is why we need $0 \leq K \leq 1/2$; we choose $0 < K < 1/2$ to ensure the ergodicity of $T$.) It is by choosing these transition rates to be independent of $j, k \geq 1$ that we get activity-driven reactions.

One easily checks from (4.104) that $T$ is a symmetric operator and $T1 = 1$: this means that $T$ is a symmetric Markov matrix. We now compute the change in the local equilibrium state induced by the map $\tau$. Since the state is determined by $N_A$, $N_B$ it is enough to compute the change in these averages. We also note that the marginal maps $\mathcal{M}_A$ and $\mathcal{M}_B$ and the equilibrating map $Q_c$ do not change the averages $N_A$, $N_B$, whose changes are thus exactly those caused by $T$:

$$N'_A = p\tau \cdot \mathcal{N}_A = p \cdot Q_c \mathcal{M}_A T \mathcal{N}_A = pT \cdot \mathcal{N}_A \ . \tag{4.105}$$

Now insert (4.101), (4.102) in (4.105); we get

$$N'_A = \frac{1}{Z_A Z_B} \sum_{j,k=0}^{\infty} \sum_{j',k'=0}^{\infty} a_A^j a_B^k \langle TP(j,k), j'P(j',k') \rangle \ . \tag{4.106}$$

Looking at (4.104), $T$ consists of a term proportional to $K$, and a constant term. Thus

$$TP(j,k) = P(j,k) + O(K) \qquad (K \to 0) \ . \tag{4.107}$$

The sum of terms in (4.106) independent of $K$ must sum to $N_A$, since $N'_A \to N_A$ as $K \to 0$.

To evaluate the term $O(K)$ in (4.106) it is best to separate the terms in (4.104) that contribute to the forward reaction, $A \to B$, from those that contribute to the reverse, $A \leftarrow B$. This is done for mathematical convenience; only the net flow can be actually measured. The forward terms must have $j \geq 1$ since otherwise $A \to B$ is not possible. Thus the first two lines of (4.104) do not contribute to the forward process. The third line gives the contribution of $j \geq 1$, $k = 0$, and is

$$\frac{1}{Z_A Z_B} \sum_{j \geq 0} a_A^j \sum_{j' \geq 0, k' \geq 0} \langle (1-K)P(j,0) + KP(j-1,1), j'P(j',k') \rangle \ ,$$

of which the contribution to $O(K)$ is

$$K \frac{1}{Z_A Z_B} \sum_{j \geq 1} a_A^j (-j + (j-1)) = -K \frac{1}{Z_A Z_B} \sum_{j \geq 1} a_A^j \tag{4.108}$$

using (4.100). The last line of (4.104) gives the contribution to (4.106) equal to

$$\frac{1}{Z_A Z_B} \sum_{j,k \geq 1} \sum_{j',k' \geq 0} a_A^j a_B^k \langle (1-2K)P(j,k) + KP(j-1,k+1)$$

$$+ KP(j+1,k-1), j'P(j',k') \rangle$$

$$= \frac{1}{Z_A Z_B} \sum_{j,k \geq 1} a_A^j a_B^k [(1-2K)j + K(j-1) + K(j+1)] \ ,$$

using (4.100). The contribution linear in $K$ vanishes, but one can recognise

$$\frac{1}{Z_A Z_B} \sum_{j,k \geq 1} a_A^j a_B^k [-Kj + K(j-1)] = \frac{-K}{Z_A Z_B} \sum_{j,k \geq 1} a_A^j a_B^k \tag{4.109}$$

as being the forward contribution, the reverse contribution being equal and opposite. The total forward contribution is thus the sum of (4.108) and (4.109):

$$\frac{-K}{Z_A Z_B} \left( \sum_{j \geq 1} a_A^j + \sum_{j \geq 1, k \geq 1} a_A^j a_B^k \right) = \frac{-K}{Z_A Z_B} \sum_{j \geq 1, k \geq 0} a_A^j a_B^k = -K a_A \ .$$

This is exactly the sum over all states of $B$, and only those states of $A$ that allow the forward reaction. Similarly, the sum of term in (4.106) linear in $K$ contributing to the reverse reaction $A \leftarrow B$ is

$$\frac{K}{Z_A Z_B} \left( \sum_{k \geq 1} a_B^k + \sum_{j, k \geq 1} a_A^j a_B^k \right) = \frac{K}{Z_A Z_B} \sum_{j \geq 0, k \geq 1} a_A^j a_B^k = K a_B \ ,$$

and it comes from all states of $A$ and only those states of $B$ that allow the reverse reaction. This proves the dynamics desired:

$$N'_A = N_A + K(-a_A + a_B) \ . \tag{4.110}$$

Obviously, at a fixed point, $a_A = a_B$ and $N_A = N_B$. Since $E_A = E_B$ (energy conservation in $A \rightleftharpoons B$) we must also have $\mu_A = \mu_B$ at equilibrium, which is a very special case of (4.129).

We now construct a stochastic model for the reaction

$$m_1 A_1 + \cdots + m_r A_r + M_1 C_1 + \cdots + M_s C_s$$
$$\rightleftharpoons l_1 B_1 + \cdots + l_t B_t + L_1 C_1 + \cdots + L_s C_s \tag{4.111}$$

such that the dynamics is activity-led, i.e. is given by (4.133). We first do the case of an isolated system, i.e. $K_+ = K_-$. We take the sample space to be

$$\Omega = \Omega_{A_1} \times \cdots \times \Omega_{A_r} \times \Omega_{B_1} \times \cdots \times \Omega_{B_t} \times \Omega_{C_1} \times \cdots \times \Omega_{C_s} \ . \tag{4.112}$$

We shall denote the numbers of $A_1, A_2, \ldots$ in the sample point by $(j_1, j_2, \ldots) = j$ and the numbers of $B_1, B_2, \ldots$ by $(k_1, k_2, \ldots) = k$, and the numbers of $C_1, C_2, \ldots$ by $(i_1, i_2, \ldots) = i$. Thus the typical sample point in $\Omega$ is

$$\omega = (j_1, j_2, \ldots; k_1, k_2, \ldots; i_1, i_2, \ldots) = (j, k, i) \ .$$

Let $P(j, k, i)$ denote the projection onto $\omega$; it is the function that takes the value 1 at $\omega$ and zero elsewhere. Then the local equilibrium state is

$$p = Z_{A_1}^{-1} \cdots Z_{C_s}^{-1} \sum_{j, k, i} a_{A_1}^{j_1} \cdots a_{B_1}^{k_1} \cdots a_{C_1}^{i_1} \cdots P(j, k, i) \tag{4.113}$$

where $a$'s are the activities defined in (4.88). The expectation of the random variable

$$\mathcal{N}_{A_\alpha} = \sum_{j,k,i} j_\alpha P(j_1, \ldots, j_\alpha, \ldots, j_r; k_1, \ldots; i_1, \ldots) \tag{4.114}$$

is equal to $p \cdot \mathcal{N}_{A_\alpha}$. To define the dynamics we must specify the bistochastic map $T$ corresponding to (4.111). It is enough to specify $T P(j, k, i)$ for all $j, k, i$. If any $j_\alpha$ is less than $m_\alpha$ or any $i_\alpha$ less than $M_\alpha$, then the forward reaction cannot go. Also if any $k_\beta$ is less than $l_\beta$ or any $i_\beta$ is less than $L_\beta$, then the backward reaction cannot go. Let

$$\Omega_+ = \{\omega : j_\alpha \geq m_\alpha, \alpha = 1, \ldots, r \text{ and } i_\alpha \geq M_\alpha, \alpha = 1, \ldots, s\} \tag{4.115}$$

$$\Omega_- = \{\omega : k_\alpha \geq l_\alpha, \alpha = 1, \ldots, t \text{ and } i_\alpha \geq L_\alpha, \alpha = 1, \ldots, s\} . \tag{4.116}$$

These are the configurations for which the forward and backward reactions can occur. We define $T$=identity on $P(\omega)$, $\omega \notin \Omega_+ \cup \Omega_-$. In a way similar to the unimolecular case, it will be convenient to define the forward and backward parts of $T$, $T_+$ and $T_-$, separately (although these cancel out on $\Omega_+ \cap \Omega_-$):

$$T_+ P(j, k, i) = (1/2 - K)P(j, k, i) + KP(j_1 - m_1, \ldots, j_r - m_r;$$
$$k_1 + l_1, \ldots, k_t + l_t; i_1 + L_1 - M_1, \ldots, i_s + L_s - M_s) \tag{4.117}$$

for $\omega \in \Omega_+$ and

$$T_- P(j, k, i) = (1/2 - K)P(j, k, i) + KP(j_1 + m_1, \ldots, j_r + m_r;$$
$$k_1 - l_1, \ldots, k_t - l_t; i_1 + M_1 - L_1, \ldots, i_s + M_s - L_s) \tag{4.118}$$

for $\omega \in \Omega_-$. Then $T = T_+ + T_-$ is symmetric and stochastic.

The change in $N_{A_1}$ due to the map $\tau = Q_c \mathcal{M}_{A_1} T$ in the forward direction is the $K$-dependent part of $p \cdot T_+ \mathcal{N}_{A_1}$ namely

$$Z_{A_1}^{-1} \cdots Z_{C_s}^{-1} \sum_{\omega \in \Omega_+} a_{A_1}^{j_1} \cdots a_{B_1}^{k_1} \cdots a_{C_1}^{i_1} \cdots a_{C_s}^{i_s}$$

$$\times \sum_{\omega' \in \Omega} j_1' \langle \{ -KP(j_1, \ldots, k_1, \ldots, i_1, \ldots)$$

$$+ KP(j_1 - m_1, \ldots, k_1 + l_1, \ldots, i_1 - M_1 + L_1, \ldots) \}, P(j_1', \ldots) \rangle$$

$$= K Z_{A_1}^{-1} \cdots Z_{C_s}^{-1} \sum_{\Omega_+} A_{A_1}^{j_1} \cdots A_{B_1}^{k_1} \cdots a_{C_1}^{i_1} \cdots (-j_1 + (j_1 - m_1))$$

$$= -m_1 K a_{A_1}^{m_1} \cdots a_{A_r}^{m_r} a_{C_1}^{M_1} \cdots a_{C_s}^{M_s} .$$

In this, we have used that $Z_{A_\alpha}^{-1} \sum_{j_\alpha \geq m_\alpha} a_{A_\alpha}^{j_\alpha}$ equals $a_{A_\alpha}^{m_\alpha}$ etc. Similarly the backward rate is

$$m_1 K a_{B_1}^{l_1} \cdots a_{B_t}^{l_t} a_{C_1}^{L_1} \cdots a_{C_s}^{L_s},$$

and we have proved (4.133) for discrete time. The equation for $N'_{B_\beta}$ is entirely similar. For the autocatalytic ions $C_1$, we get for $p \cdot T_+ \mathcal{N}_{C_1}$ instead of the line before (4.119) the expression

$$KZ_{A_1}^{-1} \cdots Z_{C_s}^{-1} \sum_{\omega \in \Omega_+} a_{A_1}^{j_1} \cdots a_{B_1}^{k_1} \cdots a_{C_1}^{i_1} \cdots (-i_1 + (i_1 - M_1 + L_1))$$

which, together with the backward reaction, and putting $N_1 = L_1 - M_1$, leads to the activity-led equation

$$N_1^{-1} \frac{dN_{C_1}}{dt} = K(a_{A_1}^{m_1} \cdots a_{A_r}^{m_r} a_{C_1}^{M_1} \cdots a_{C_s}^{M_s} - a_{B_1}^{l_1} \cdots a_{B_t}^{l_t} a_{C_1}^{L_1} \cdots a_{C_s}^{L_s}) \ .$$

This completes the proof of (4.133).

It follows from this analysis that we can apply the general results of statistical dynamics to these equations. The orbit of an initial point lies in the compact space of states with given mean numbers $N_j$ of various conserved atomic numbers, and entropy is a Lyapunov function. It can be shown that there is a unique fixed point for each possible set of $N_j$, provided that they lie in a closed time-invariant set not intersecting the boundary of state-space. The system therefore converges to a fixed point, which is the grand canonical state with the same mean values of all the conserved quantities as the initial state. The details of the proof are omitted, as they are very similar to the isothermal case, given in the next chapter.

In order to cope with reactions which produce heat $q$, we interpret one of the chemicals as a heat-particle with $q$ as its quantum. To cover the cases where there is a non-zero activation energy, we take one of the catalysts to be a heat-particle. In this way the factor (4.38) is obtained, where the beta of the factor is the instantaneous local beta of the thermalised heat-particle. The example of cotransport shows that a fixed point does not need to have the same chemical potential for all chemicals called by the same name, but in different beakers.

## 4.7 Exercises

**Exercise 4.119.** Show that the product state

$$p(\omega) = \prod_{x \in \Lambda} p_x(\omega_x)$$

is the state of maximum entropy among all states with the given marginals $\{p_x\}_{x \in \Lambda}$.

Exercise 4.120. Show that the equation of motion for the temperature in the isomer model is

$$E\frac{d\beta}{dt} - (1 - e^{-\beta E})(p_2 e^{\beta E} - p_1).$$                         (4.121)

Exercise 4.122. Show that in Example 4.1

$$p_1'(B) = p_1(B) - \kappa\,(p_1(B) - p_2(B))$$

$$- \lambda\left(2p_1(B)p_2(B) - p_1(A)p_2(C) - p_1(C)p_2(A)\right).$$

Exercise 4.123. Consider $\Lambda$ with two points, and the space $\{\emptyset, A, B, C\}$ at each site. Set up a model describing the process $A + C \rightleftharpoons B$.

Exercise 4.124. In three dimensions, when we quantise a free Schrödinger particle in a box of side $\ell$ with periodic boundary conditions, the wave-number must be quantised, and the number of states with kinetic energy between $E$ and $E + dE$ is approximately const. $E^{1/2}dE$. Find the constant. Modify the model in Example 4.3, with the assumption that each momentum state can hop to only one momentum state at the neighbouring point, with the same total energy as itself. Assume that all non-zero transition probabilities are the same. Find the dependence of density with height at equilibrium.

Exercise 4.125. Show that in Example 4.3 the density at height $x$ is given by

$$\rho_x = d^{-1}\frac{\exp\{-\beta(mdgx - \mu)\}}{1 + \exp\{-\beta(mgdx - \mu)\} - \exp\{-\beta mdg\}}$$                         (4.126)

and that in terms of the density at $x = 0$ we get

$$\rho_x = \frac{\rho_0 \exp\{-\beta mdgx\}}{1 - \rho_0 d\,(1 - \exp\{-\beta mdgx\})}.$$                         (4.127)

Note that a factor $d$ was omitted in [Streater (1994b)].

Exercise 4.128. We are given any $\beta \neq 0$, and an energy function $\mathcal{E}$ on a two-point set $\Omega$, which is zero on $\omega = \emptyset$, and $E$ on $\omega = 1$; show that any probability $(q = 1 - p,\, p)$ on $\Omega$ can be written as a grand canonical state $q = Z^{-1},\, p = Z^{-1}e^{-\beta(E-\mu)}$ for a uniquely determined $\mu$. Hence show that the fixed point of the model (4.4) corresponds to a constant chemical potential and beta. Show that the density as a function of height at equilibrium is the same as in Example 4.3.

Exercise 4.129. Show that the system of activity-led equations

$$\frac{dN_j}{dt} = -K(a_1 a_2 \ldots a_m - b_1 b_2 \ldots b_n) \tag{4.130}$$

where $a_j$ is the activity of $A_j$ and $b_k$ is the activity of $B_k$, $1 \leq j \leq m$, $1 \leq k \leq n$, has many fixed points, for example any grand canonical state for which the sum of chemical potentials on each side is the same.

Exercise 4.131. Consider the reaction

$$m_1 A_1 + \ldots + m_r A_r + M_1 C_1 + \ldots + M_s C_s$$
$$\rightleftharpoons l_1 B_1 + \ldots l_t B_t + L_1 C_1 + \ldots + L_s C_s. \tag{4.132}$$

Here, $m_1 \ldots m_r$ are the stoichiometries of the $r$ $A$-particles, $l_1 \ldots l_t$ are the stoichiometries of the $t$ $B$-particles. The $C$-particles are catalysts, which appear on the left and right of the reaction, with stoichiometries $M_1 \ldots$ and $L_1 \ldots$ respectively. The activity-led dynamics for this reaction is

$$\frac{1}{m_\alpha} \frac{dN_{A_\alpha}}{dt} = -K a_{A_1}^{m_1} \ldots a_{A_r}^{m_r} a_{C_1}^{M_1} \ldots a_{C_s}^{M_s} + K a_{B_1}^{l_1} \ldots a_{B_t}^{l_t} a_{C_1}^{L_1} \ldots a_{C_s}^{L_s}$$
$$= K\Delta$$
$$\frac{1}{L_\alpha - M_\alpha} \frac{dN_{C_\alpha}}{dt} = -K\Delta. \tag{4.133}$$

Show that a product of grand canonical states all with the same beta is a fixed point if and only if the sum of chemical potentials, including stoichiometry, is the same on both sides of the reaction.

Exercise 4.134. A jar containing the states $\{\emptyset, W, A, B\}$ undergoes the transmutation $A \rightleftharpoons B$, followed by the map $Q$ (with no change to $S$) giving up-date equations

$$p'(A) = (1 - \kappa)p(A) + \kappa p(B)$$
$$p'(B) = \kappa p(A) + (1 - \kappa)p(B).$$

Show that $S(p') - S(p) \geq 0$.
HINT: use the concavity of $-p \log p$.

Ostwald devised and made a thermostat, which enabled hir
temperature of the apparatus in which a chemical reaction w
When the temperature fell, the gas in a jar contracted an
tap which turned up the gas flame; when the temperatur
expanded and this turned down the flame. So to within the
able at that time, Ostwald was able to study reactions u
conditions. The change from isolated conditions, in which
is constant, to isothermal conditions, in which beta is con
a *Legendre transform*, and $\beta$ is called the Legendre varial
We note that energy is an extensive variable, proportional
whereas beta is an intensive variable, constant throughout
thermostatics the equilibrium state with a given mean ene
as the state maximising the entropy subject to the given m
the equilibrium state with a given temperature is obtained
not the entropy $S$, but the Legendre transform of $S$, denote
to the Helmholtz free energy, subject to the beta being ke
the next section we introduce the related transform of the w
and not just of the equilibrium state. Thus the isolated d
verted into the isothermal dynamics; we shall see that, unl
dynamics, the isothermal dynamics is linear, and so defin
Markov chain. The second law, that entropy increases alo
converted to the free-energy theorem, that the free energy
the orbit. This idea can be generalised to any of the other c
sive quantities, such as the atomic number. If we set up phy
that ensure that there are exactly the right flows of parti
chemical potential of a particular atom is constant, we are d
transform to a variable dual to that atomic number; there i

123

ing thermodynamic potential, the Gibbs free energy, that decreases along the orbit. The resulting dynamics will be called *isopotential*. In the second section, we prove the fundamental theorem of chemical kinetics, that any activity-led isothermal, isopotential dynamics converges to equilibrium. There can be no chaos or permanent oscillations. In the third section, we prove that any Markov chain with a strictly positive fixed point can be regarded as the isothermal version of a suitable statistical dynamics. We show that the relative entropy, related to the free-energy-difference between the state and the fixed point, decreases to zero, which leads to convergence of the state in a useful norm.

## 5.1 Legendre Transforms

Suppose we are given a function, $S$ say, of some variables, $p_1, p_2, \ldots, p_n$ say, and we want to find the value of the $n$-vector $p$ that maximises $S(p)$, subject to a constraint, say $E(p) = \overline{E}$. The case we shall use is when $E$ is a linear function of $p$, namely the mean energy: $E(p) = p \cdot \mathcal{E}$. According to the method of Lagrange multipliers, the values of $p$ giving the turning points can be obtained by seeking the turning points of $\Psi = S(p) - \beta E(p)$ without constraints. We can then find the stationary point $\overline{p}$ in terms of $\beta$, the Lagrange multiplier; $\beta$ is found by fitting the solution $p$, to the constraint, so that $\beta$ and $\overline{E}$ become functionally related, and the maximum value of $S$ becomes a function of $\overline{E}$. $\Psi$ itself can be written as a function of $\beta$, by using the relation between $\overline{E}$ and $\beta$ to eliminate $\overline{E}$. When written as a function of $\beta$, $\Psi$ is called the Legendre transform of $S$.

It is elementary to justify the method of Lagrange multipliers when $S$ and $E(p) = p \cdot \mathcal{E}$ are smooth functions of the components of $p$, and $E$ is a non-constant function, in a neighbourhood of the point of maximum, or, more generally, the turning point. It is easy to generalise from one to many variables, so we illustrate the argument when $S = S(p, q)$ is a function of two variables, and we wish to find its turning points subject to the constraint $E(p, q) = \overline{E}$. Solve the constraint for $q$ as a function of $p, E$, so that $S = S(p, q(p, E))$. Then by the chain rule, a small change in $p$ gives the change $dS$:

$$dS = \left(\frac{\partial S}{\partial p}\right)_q dp + \left(\frac{\partial S}{\partial q}\right)_p \left(\left(\frac{\partial q}{\partial p}\right)_E dp + \left(\frac{\partial q}{\partial E}\right)_p dE\right) = 0$$

at a turning point. Since we move on the surface $E(p, q) = \overline{E}$, we put

Substitute in Eq. (5.1) to arrive at

$$\left(\frac{\partial E}{\partial q}\right)_p \left(\frac{\partial S}{\partial p}\right)_q - \left(\frac{\partial S}{\partial q}\right)_p \left(\frac{\partial E}{\partial p}\right)_q = 0.$$

This is exactly the condition that $(S, E)$ fail to be a valid se
for the $(p, q)$-plane; so let $(p_0, \ q_0)$ satisfy Eq. (5.2). Th
the level sets of $S$ and $E$ are parallel: the surface $E(p,$
touches the surface $S(p, q) = S(p_0, q_0)$. Such points $(p_0, q$
in the $(p, q)$-plane parametrised by $\overline{E}$, and $S$ is a function
line of turning points. This result can be got by the meth
multipliers without so many technical difficulties; in Lagr
we seek the turning points of

$$\Psi = S(p, q) - \beta E(p, q)$$

without constraints, where $\beta$ is kept constant. The turning
by

$$\frac{\partial S}{\partial p} = \beta \frac{\partial E}{\partial p} \qquad\qquad \frac{\partial S}{\partial q} = \beta \frac{\partial E}{\partial q}.$$

When we eliminate $\beta$ we get the same condition as before, t
points of $S$ along the manifold with fixed $E$ are the points
surfaces of $S$ and $E$ touch. Moreover, the Lagrange multipl

$$\beta = \frac{\partial S}{\partial p} \bigg/ \frac{\partial E}{\partial p} = \frac{\partial S}{\partial q} \bigg/ \frac{\partial E}{\partial q} = \left(\frac{\partial S}{\partial E}\right)_q = \left(\frac{\partial S}{\partial E}\right)$$

At a turning point, we get the same answer whether we cho
$q$, as here, or $p$. From this we see that

$$d\Psi = \left(\frac{\partial S}{\partial p}\right)_E dp + \left(\frac{\partial S}{\partial E}\right)_p dE - \beta dE - E d\beta = \left(\frac{\partial S}{\partial p}\right)$$

$$\Psi(p, \beta(p, E)) - \beta(-E) = S(p, E).$$

Thus $\Psi$ and $S$ are each other's Legendre transforms.
asked to verify all this for the entropy and its Legendre
$\Omega$ is the countable space $\{0, 1, \ldots\}$ with some energy fu
the partition function is finite.

A more versatile theory is obtained by considering a
$\mathcal{E}_j$ to be parameters; then $\beta$ arises as an overall factor
equal to 1, and the thermodynamic potential $\Psi$ becc
all the $\mathcal{E}_j$. This, the analytic theory of Legendre trans
the existence of derivatives and uses the implicit fun
statistical mechanics, after the thermodynamic limit ha
not necessarily differentiable as a function of $p$; but it
function. To cover this case, the Fenchel transform
turning-point condition is replaced by a condition that
should touch the convex set given by the level surfaces

$$\Psi(\mathcal{E}_1, \mathcal{E}_2, \ldots) = \sup_{p \in \Sigma}\{S(p) - p \cdot \mathcal{E}\}.$$

One shows that $\Psi$ is a concave function of $\mathcal{E}$ and that its
is $S$ as a function of $p$. See [Wightman (1979); Thir
(2005)].

In the next section we describe how the partial L
(over the variables of the heat-particle) can transform th
isolated dynamics into the orbit of the system under iso

## 5.2 The Free-energy Theorem

We have seen that the sample space for a system is of
$\Omega_\gamma$ with energy $\mathcal{E}_c + \mathcal{E}_\gamma$, and that the dynamics of an

determined by an energy-conserving bistochastic map $T$. The initial state $p \in \Sigma(\Omega)$ is changed in one time-step to

$$p' = pT\mathcal{M}_c \otimes pT\mathcal{M}_\gamma = p\tau.$$

Suppose now that the state at time $t = 0$ is at a uniform beta $\beta$; this is expressed by saying that $p = p_c \otimes s_\beta$, where $s_\beta$ is the canonical state of the heat-particle. Unless the state $p$ is invariant under the dynamics $\tau$, it is most unlikely that the part of $p\tau$ that concerns the heat-particle, namely $s' = pT\mathcal{M}_\gamma$, is itself a canonical state. We can, however, mathematically implement the thermostat by changing this state $s'$ back to $s_\beta$, so we define the *isothermal* dynamics to be

$$p \mapsto p' = p\tau_\beta := pT\mathcal{M}_c \otimes s_\beta. \tag{5.7}$$

That is, the thermostat ensures that each mode of $(p_c \otimes s_\beta) T\mathcal{M}_\gamma$ is supplied with just the right amount of heat to restore it to the canonical state $s_\beta$.

We see that the map $\tau_\beta$ can be regarded as acting on $\Sigma(\Omega_c)$, where it is a linear map, unlike the Boltzmann map of the isolated system. We have done a partial Legendre transform, replacing the dynamics at constant mean energy by one with constant beta (or temperature, $\Theta = 1/k_B\beta$) which is the same as one with constant mean heat $s_\beta \cdot \mathcal{E}_\gamma$. We shall regard $p_c\tau_\beta$ as a state in $\Sigma(\Omega_c)$ parametrised by $\beta$. Then the Legendre transform of the entropy, called $\Psi$, is a Lyapunov function for the dynamics:

**Theorem 5.1 (The Free-Energy Theorem).** *Let* $\Psi(p_c) = S(p_c) - \beta p_c \cdot \mathcal{E}_c$; *then*

$$\Psi(p_c\tau_\beta) \geq \Psi(p_c) + \|s' - s_\beta\|^2/2.$$

*Proof.* We know that the bistochastic map $T$ increases the entropy (or leaves it the same). Also the Boltzmann map, which replaces $pT$ by $p' = p\tau = pT\mathcal{M}_c \otimes pT\mathcal{M}_\gamma$, increases the entropy. Let $p = p_c \otimes s$ and $p' = p'_c \otimes s'$. Then we have

$$S(p') = S(p'_c) + S(s') \geq S(p) = S(p_c) + S(s).$$

In isothermal dynamics, we always start with the state of the heat-particle in equilibrium, so $s = s_\beta$. Then

$$S(s) = -\sum_{\omega_\gamma} s_\beta(\omega_\gamma)(-\beta\mathcal{E}(\omega_\gamma) - Z_\gamma(\beta)).$$

Let us write

$$- \sum_{\omega_\gamma} s'(\omega_\gamma) \log s'(\omega_\gamma) = - \sum_{\omega_\gamma} s'(\omega_\gamma) \log s_\beta(\omega_\gamma)$$

$$+ \sum_{\omega_\gamma} s'(\omega_\gamma) \log \left( s_\beta(\omega_\gamma)/s'(\omega_\gamma) \right).$$

By Kullback's inequality,

$$\sum_{\omega_\gamma} s'(\omega_\gamma) \log \left( s_\beta(\omega_\gamma)/s'(\omega_\gamma) \right) \leq - \| s_\beta - s' \|^2 / 2. \tag{5.8}$$

As we saw in Sec. 4.1, the mean energy is conserved by the isolated dynamics; thus

$$\sum_{\omega_c} p_c(\omega_c) \mathcal{E}_c(\omega_c) + \sum_{\omega_\gamma} s_\beta(\omega_\gamma) \mathcal{E}_\gamma(\omega_\gamma)$$

$$= \sum_{\omega_c} p_c'(\omega_c) \mathcal{E}_c(\omega_c) + \sum_{\omega_\gamma} s'(\omega_\gamma) \mathcal{E}_\gamma(\omega_\gamma). \tag{5.9}$$

We put these estimates together in the following chain of inequalities:

$$S(p_c') - \sum_{\omega_\gamma} s'(\omega_\gamma) \log s_\beta(\omega_\gamma) = S(p_c') - \sum_{\omega_\gamma} s'(\omega_\gamma) \log s'(\omega_\gamma)$$

$$- \sum_{\omega_\gamma} s'(\omega_\gamma) \log \left( s_\beta(\omega_\gamma)/s'(\omega_\gamma) \right)$$

$$\geq S(p_c') - \sum_{\omega_\gamma} s'(\omega_\gamma) \log s'(\omega_\gamma) + \| s' - s_\beta \|^2 / 2$$

$$= S(p') + \| s' - s_\beta \|^2 / 2$$

$$\geq S(p) + \| s' - s_\beta \|^2 / 2$$

$$= S(p_c) + S(s_\beta) + \| s' - s_\beta \|^2 / 2$$

$$= S(p_c) - \sum_{\omega_\gamma} s_\beta(\omega_\gamma) \log s_\beta(\omega_\gamma) + \| s' - s \|^2 / 2.$$

Hence

$$S(p_c') - \sum_{\omega_\gamma} s'(\omega_\gamma) \left( -\beta \mathcal{E}_\gamma(\omega_\gamma) - \log Z_\gamma(\beta) \right) \geq S(p_c)$$

$$- \sum_{\omega_\gamma} s_\beta(\omega_\gamma) \left( -\beta \mathcal{E}_\gamma(\omega_\gamma) - \log Z_\gamma(\beta) \right) + \| s' - s_\beta \|^2 / 2.$$

The $\log Z$ terms cancel, and using (5.9) we have

$$-dQ = \sum_{\omega_\gamma} \left( s'(\omega_\gamma) - s_\beta(\omega_\gamma) \right) \mathcal{E}_\gamma(\omega_\gamma)$$

$$= \sum_{\omega_c} \left( p_c(\omega_c) - p_c'(\omega_c) \right) \mathcal{E}_c(\omega_c) = U - U' \tag{5.10}$$

where $U$ is the chemical energy of the chemicals, and $dQ$ is the heat that has to be added to keep the temperature constant. Thus we get

$$S'_c - \beta U' \geq S_c - \beta U + \|s' - s_\beta\|^2/2. \qquad (5.11)$$

This is the free energy theorem [Streater (1992, 1993b)]. □

It is usually expressed in terms of the internal energy, including the kinetic energy, of the chemicals, and the total entropy, including that of the heat. However, the contributions of the heat-particle to the internal energy and to the entropy are unchanged in the isothermal case, and so our result is not in conflict with the usual claim that the *free energy*, $U - \Theta S$, decreases with time, where $\Theta$ is the temperature.

We see that in one time-step the system gains entropy at least $dQ/\Theta + \|s' - s_\beta\|^2/2$. We may therefore define the mysterious concept, a *quasi-static* process, as one for which $s' = s_\beta$, that is, one for which the dynamics $T$ does not alter the distribution of the heat-particle in any way. This is the same as saying that the change in entropy is $dQ/\Theta$, the smallest it can be.

There is nothing special about the energy, and the degrees of freedom of the heat-particle, compared with other conserved quantities; for each conserved quantity, say atomic number $\mathcal{N}$, we can define the corresponding *isopotential* dynamics by exchanging particles with the surroundings after each time-step in such a way that the chemical potential of this atom is returned to its initial value. Physically, this means that it is in contact with a bath at a specified chemical potential. In practice it would be difficult to prevent heat escaping with the particles, so isopotential dynamics is usually not considered except at a fixed beta. So the convention has arisen of including a factor $\beta$ along with $\mu$ in the grand canonical state; this gives $\mu$ the dimensions of energy. To implement this map mathematically, we need the sample space to be the product of the classical Fock space of the atom, and the rest, and for the dynamical law $T$ to map the number-shells into themselves (as well as the energy-shells into themselves). Then, a similar calculation to that in the free-energy theorem shows that the system gains entropy of at least $\mu dN$ when $dN$ atoms are absorbed in one time-step to keep the chemical potential constant at $\mu$. This applies to any number of atoms that are conserved in the dynamics and whose chemical potential is kept fixed. The dynamical laws are obtained from the isolated dynamics by simply keeping the beta and the chemical potentials fixed for the controlled atoms, and taking the same formulae. In this set-up, the flow of atoms in the isopotential dynamics is taken to be free atoms, not atoms bound in other molecules.

We can prove the law (4.38) for isothermal dynamics under the assumption that $T$ is a symmetric operator and that the dynamics is activity-led. This means that $T = T^d$ when $\Sigma(\Omega)$ is identified with the subset of $\ell^2$ random variables, using the privileged basis, namely, the points of $\Omega$. For, suppose we consider the general reaction (4.130) with some of the symbols on each side representing heat-particles of various energies, $\varepsilon_j$. Then in the activity-led model the forward rate will contain factors $a_{\gamma,j} = e^{-\beta\varepsilon_j}$ for the heat-particles absorbed, and the backward rate will contain factors of the same type for each heat-particle emitted in the reaction. The ratio of these rates will be $e^{\beta q}$, where $q = \sum \varepsilon_{\text{in}} - \sum \varepsilon_{\text{out}}$ is the heat emitted, the heat of reaction. In fact, in the activity-led equations, the forward rate itself is proportional to $e^{-\beta Q_{\text{in}}}$, where $Q_{\text{in}}$ is the activation energy, the heat needed to get the reaction started. This is the Ahrennius law, and some chemical reactions agree with this prediction. When it is found not to be true, chemists tend to look for reasons, such as the occurrence of more than one route for the reaction to occur; on the other hand, we would look for other models, not leading to the activity-led law; but the factor $e^{-Q_{\text{in}}}$ in the forward direction will always be present, coming from the canonical state of the heat-particle.

## 5.3   Chemical Kinetics

The question as to whether complex chemical rate-equations always converge to equilibrium was answered in a series of papers by chemists [Shear (1968); Horn and Jackson (1972); Horn (1972-73); Clarke (1980)]. These authors worked with density-led equations, for which an underlying stochastic model might not exist. Rondoni [Rondoni (1994)] has proved a similar result for a wide class of equations, which he calls 'X-led'. Here we give a proof for the activity-led case, and some particular density-led models which are derived from statistical models in the previous Chapter. We shall take the space $\Lambda$ to be a finite set, and each $\Omega_x$, with $x \in \Lambda$, will be the hard-core sample space of the chemicals (including the empty site) for the density-led case, and the set-product of the classical Fock spaces of chemical species for the activity-led case. We shall consider the dynamics to be a convex combination of a finite number, $R$, of independent reactions in which chemicals $A_1, \ldots, A_r$ react to form chemicals $B_1, \ldots, B_t$, with chemicals $C_1, \ldots, C_s$ acting as catalysts. We take it that the stoichiometries are $m_1 \ldots m_r$ for the $A$'s, $l_1 \ldots l_t$ for the $B$'s, and $M_1 \ldots M_s$ for the initial catalysts, and $L_1 \ldots L_s$ for the final catalysts. In symbols we have the 'simple' reaction

$$m_1 A_1 + \cdots + m_r A_r + M_1 C_1 + \cdots + M_s C_s$$
$$\rightleftharpoons l_1 B_1 + \cdots + l_t B_t + L_1 C_1 + \cdots + L_s C_s. \tag{5.12}$$

Consider first the density-led case. We include enough points in $\Lambda$ to allow all the reactants, that are needed for the forward and backward reactions, to be present. The choice of sample space, $\Omega = \prod \Omega_x$, ensures that the configuration with one $A$-particle at $x \in \Lambda$ and no other particle present, is distinct from the configuration where the single $A$-particle is at $y \neq x$, but the configuration with two $A$-particles at $x$ and $y$ does not distinguish the particles. Thus we are adopting the field point of view. We can then accommodate any stoichiometry by having a non-zero transition probability from the state with several chemicals of the same type at nearby points in $\Lambda$, to the state representing the products that they combine to form. The stoichiometry then enters in the definition of the transition matrix, but not in the choice of sample space, provided that $\Lambda$ is large enough: it must have enough sites to allow all the interacting particles to be present. Let $\nu = |\Lambda| \cdot |\Omega_x|$. This is the total number of components of all the marginal probabilities $p_x$ with $x \in \Lambda$. Let $\mathcal{K} = [0,1]^\nu$. A point $p$ in $\Sigma = \Sigma(\Omega)$ which is independent over $\Lambda$ (which is the case after the stoss map) defines a point in $\mathcal{K}$ once we fix on an ordering convention. In the dynamics from a given initial point of $\Sigma$, there will be some conserved quantities, and these will say that certain mean atomic numbers are constant. This implies that the motion inside $\mathcal{K}$ must lie in certain affine planes, whose position is fixed by the initial values of these atomic numbers. The conditions that the marginal probabilities add up to 1 also define affine planes. It is easy to see that a boundary point of $\mathcal{K}$ which comes from a point in $\Sigma$ must come from a boundary point of $\Sigma$.

The linear part of the dynamics will be given by a convex sum of primitive symmetric stochastic operators each of which links a certain initial configuration of particles with a final configuration of products. For example, we may wish to include a non-zero probability that the reaction $A + B \rightleftharpoons C$ occur when $A$ is at $x$, $B$ is at $x + 1$, and the product $C$ is created at $x + 1$, where $B$ was; then the inverse of this must be included. We may choose to omit the reaction in which the product $C$ is created at $x$. Then the model expresses that the particle $A$ accretes to $B$ to make $C$, rather than that $B$ accretes to $A$ to make $C$. In the hard-core model, with this choice of dynamics, the site $x$ is left empty after the reaction, and must be empty before the reaction for the inverse reaction to be possible. We assume that the inverse rate is the same as the forward rate, (the principle of microscopic reversibility). In artificial ecology, this model represents a hungry predator $A$ at $x$ leaping on a prey $B$ at $x + 1$ to end up with a fat predator $C$ at $x + 1$. In this case the probability of the inverse reaction is zero, and so the following analysis applies only to the chemical case. The initial state, taken to be the product over $\Lambda$ of the marginal states, is the product of the occupation probabilities of $A$ at $x$ and $B$ at $x+1$. The probability of the final state will be the product of the occupation probability of $C$ at $x+1$ and the empty probability at $x$. After the action of the linear part, we apply the Boltzmann map and replace the final state by the product over $\Lambda$ of the marginals. We obtain a version of *density-led* dynamics, but modified from the classical law of mass action by the presence of these empty-state probabilities, just as if the empty state were a particle. You have already met this in (4.123). Thi modification of the classical density-led law of mass action was suggested in [Streater (1992)]. The full dynamics $\tau$ of complex reactions will be a convex linear combination of such maps $\tau_k$, and by the principle of detailed balance, equilibrium for the complex reaction will necessarily be a fixed point of each of these $\tau_k$, by the strict concavity of the entropy.

The fundamental theorem of chemical kinetics says that this version of the law of mass action leads to convergence to equilibrium as time becomes large [Streater (1992)]. We already have most of the ingredients of the proof: the motion takes place within the compact set $\mathcal{K}$, and there is a strict Lyapunov function, the entropy. If there is a unique fixed point, then Lyapunov's theorem (3.23) will ensure convergence. However, typically there is not a unique fixed point in the whole of $\Sigma$. If an essential ingredient, say a catalyst or more generally one of the chemicals on each side of the reaction, has zero probability, then the reaction will not proceed. Thus there may be fixed points on the boundary of the state space, in addition to the equilibrium states in the interior. It was shown in [Rondoni (1991)] that the boundary fixed points are *repellers* at least for the usual law of mass action, which needs balanced reactions. The same proof works here, with the probability of an empty site replacing a chemical concentration from time to time. Thus, if $p \in \Sigma$ is an interior point, all its components are positive; then $p' = p\tau_k$ cannot be a boundary point; for if one component, say $p'_1$, of the marginal state onto the first factor $\Omega_1$ of $\Omega$, has a zero component, then we get an equation of the form

$$0 = \sum T_{1,\ldots} p_1 \cdots p_N$$

where all the $p$s are positive. This implies that all the components of $T$ involved in the sum are zero, which is impossible, as $T$ is a stochastic matrix. Hence if $p$ is an interior point, so is $p'$, and so no simple reaction $\tau_k$ can take a point from the interior to the boundary in one fell swoop. The complex reaction is a convex sum of these, and so cannot take an interior point to the boundary in one time-step either. We must also exclude the possibility that a sequence of points on an orbit under time-evolution can lead from an interior point to the boundary. This proviso is not needed if the time is continuous. We satisfy this easily by choosing the initial state to have entropy not less than at any point on the boundary. The set of such states, $\mathcal{K}_0$, is compact, and is mapped to itself by the dynamics, which increases the entropy. We consider the dynamical system got by iterating $\tau$ acting on $\mathcal{K}_0$. We denote a general element of Span $\mathcal{K} = \mathbf{R}^\nu$ by $p$. Now let $p(0)$ lie in the interior of $\mathcal{K}_0$. Our problem is to show that $\tau^n p(0)$ converges to a limit as $n \to \infty$. If $p(0)$ is a fixed point of $\tau$ we are done. So we may assume that $\tau p(0) \neq p(0)$. Our method, not dissimilar to the original methods [Shear (1968); Horn and Jackson (1972); Horn (1972-73); Clarke (1980)], is to show that the motion takes place on a certain hyperplane through $p(0)$, and that this hyperplane contains a unique fixed point. We treat the isolated system first. It is a neat remark of Rondoni [Rondoni (1991)] that any one of the $\tau_k$ has exactly $\nu - 1$ linear conserved quantities, including the $|\Lambda|$ quantities $\sum p_j$, and so essentially only one of the $\nu$ variables changes. The orbit therefore lies on the intersection of $\nu - 1$ hyperplanes, which is a line through the initial state called *the line of reaction*.

The simple reaction $\tau_k$ gives an orbit starting at $p(0)$ and lying on the line of reaction $L_k$ passing through $p(0)$, defined by $\nu - 1$ planes

$$L_k = \{p \in \mathcal{K} : N_j^{(k)}(p) = N_j^{(k)}(p(0)), \ j = 1, \ldots, \nu - 1\}.$$

Here $N_j^{(k)}$, $j = 1, \ldots, \nu - 1$ are the conserved quantities for $\tau_k$, including the conservation of probability of the marginals,

$$\sum_{\omega_x} p_x(\omega_x) = 1. \tag{5.13}$$

Now consider the complex reaction; suppose that $R$ of the lines $L_k$ are linearly independent, but no more. Then $\{L_k\}$ span an $R$-dimensional convex cone with vertex $p(0)$, and the lines can be made to pass through the origin by subtracting $p(0)$ from each point in the line. Thus, $\{L_k -$

$p(0): k = 1, \ldots, R\}$ form part of a linear basis in Span $\mathcal{K}$. The conserved quantities $N_j^{(k)}$ are elements of the dual space to Span $\mathcal{K}$; this is obvious for the atomic numbers, as they are random variables; but it is also true that the marginal probabilities in Eq. (5.13) are given by sums of certain components of $p$ and so are conserved linear functionals. The conserved functionals $N_j^{(k)}$ vanish on the corresponding axis $L_k - p(0)$; for, they are conserved, so $N_j^{(k)}(p) = N_j^{(k)}(p(0))$ for $p = \tau p(0) \neq p(0)$ on the line $L_k$. Then $N_j^{(k)}(p - p(0)) = 0$, so $N_j^{(k)}$ annihilates $p - p(0)$ as well as 0, and being linear, kills all points on the line. Conversely, any linear functional that annihilates all the points on the $R$ lines of reaction is conserved. Thus the conserved quantities, which are not all linearly independent, span a space of dimension exactly $\nu - R$ in the dual to Span $\mathcal{K}$, since they annihilate a space of dimension $R$. Let us take as linear coordinates in Span $\mathcal{K}$ a certain $\nu - R$ linearly independent conserved functionals $N_j$, $j = 1, \ldots, \nu - R$, together with $R$ parameters $t_1, \ldots, t_R$, one on each axis $L_k - p(0)$. The $\nu$ labels for the chemicals in the hard-core sample spaces $\Omega_x$ for $x \in \Lambda$ can be identified with $\nu$ of the corners of $\mathcal{K}$, in which $A_x$ is the point $p_x(A) = 1$ and all other components are zero. This is not a possible point in $\Sigma(\Omega)$, since it violates all except one of the constraints in (5.13). But nevertheless these form a basis for Span $\mathcal{K}$. Let $p \in \mathcal{K}$ be a fixed point of the map $\tau$ with the same values of the conserved quantities as $p(0)$, and let $f = \log p$, meaning that $f$ is a function on the set of labels: $f(A_x(j)) = \log p_x(j)$. Define $f$ to be zero at $0 \in \mathcal{K}$. This $f$ has a unique linear extension to Span $\mathcal{K}$, which we also denote by $f$. By the principle of detailed balance, $p$ is a fixed point of each $\tau_k$, which is the dynamics of the simple reaction

$$A_{k_1} + \cdots + A_{k_K} \rightleftharpoons A'_{k_1} + \cdots + A'_{k_K}$$

whose fixed point obeys

$$p_{k_1} p_{k_2} \cdots p_{k_K} = p'_{k_1} p'_{k_2} \cdots p'_{k_K}$$

which in terms of $f$ can be written

$$f(k_1) + \cdots + f(k_K) = f(k'_1) + \cdots + f(k'_K). \tag{5.14}$$

For each given $k$, each side of (5.14) is the value of $f$ evaluated at a point of Span $\mathcal{K}$, the left-hand side being the value of $f$ at $P = A_{k_1} + \cdots + A_{k_K}$ and the right-hand side being the value of $f$ at a different point, $P' = A'_{k_1} + \cdots + A'_{k_K}$. Both $P$ and $P'$ have the same value for each of the $\nu - 1$ conserved quantities $N_j^{(k)}$, $j = 1, 2, \ldots, \nu - 1$, so the difference $P - P'$ lies on the axis $L_k - p(0)$, which is the simultaneous null space of $N_j^{(k)}$; hence $f$, which is linear, vanishes at a non-zero point on this axis, and so must vanish along $L_k - p(0)$. Similarly, $f$ vanishes on all $R$ axes $L_k - p(0)$, $k = 1, \ldots, R$, and therefore is a linear function of the chosen $\nu - R$ linearly independent conserved quantities $N_j(p)$, $j = 1, \ldots, \nu - R$. So we get

$$\log p_i = \sum_{j=1}^{\nu-R} \lambda_j N_j(\mathbf{p}_i) \tag{5.15}$$

where $\mathbf{p}_i = (0, \ldots, 1, \ldots)$ with 1 in the $i$th place, and $\lambda_j$ are some real numbers. But this system of equations is identical to what we get by using the method of Lagrange multipliers to maximise $S(p)$ subject to the $\nu - R$ independent linear constraints $N_j(p) = $ constant. Because of the strict concavity of $S$, there is only one turning point in the interior of $\Sigma$, so the uniqueness of the fixed point in density-led dynamics is proved, and consequently, by Lyapunov's theorem, the convergence of the system to its unique fixed point is established.

We get a bonus, in that the convergence to equilibrium is established for the dynamics based on $T$ followed by the stoss map, in which we replace the state by the product of its marginals at the points in $\Lambda$, and we do not need the further randomisation which would be caused if we applied the $LTE$ map as well. The fixed point is a grand canonical state; for each $x \in \Lambda$ there is a conserved quantity $N_x$, being the total marginal probability, and we could have chosen these to be among the linearly independent conserved numbers. Then the local partition function $Z_x$ is related to the corresponding Lagrange multiplier $\lambda_x$ by $\lambda_x = \log Z_x$. The other $\lambda$'s are the various chemical potentials.

We get the fundamental theorem for isothermal and isopotential dynamics in the same way; the only difference is that the state of the heat-particle, and of the chemicals whose chemical potential is kept fixed, is fixed to be the canonical or grand canonical state with given parameters; it does not affect the uniqueness of the fixed point. In place of the entropy of the whole system, the free energy of the chemical subsystem undergoing the dynamics is a Lyapunov function, so we prove convergence by Lyapunov's theorem.

For the activity-led dynamics, the sample space is infinite, since at each $x$ we have the classical Fock space $\Omega_x = \{0, 1, 2, \ldots\}$. This leads to problems in showing that the limit state has the same energy as the mean energy during the dynamics [Streater (1984)]. However, the $LTE$-dynamics takes place through a subset of $\Sigma$ labelled by finitely many parameters; this enables us to find a compact set mapped to itself by the dynamics. We now sketch the argument.

Let us choose a symmetric operator $T$ to define the linear part of the dynamics as in Sec. 4.6 to give activity-led dynamics when we apply the $LTE$-map. A state at time $t$ is then a grand canonical state of each of the $r + s + t$ chemicals at each of the $|\Lambda|$ points $x$. This is determined by the mean number fields $n(x)$, one for each particle. The reduced state-space is thus finite-dimensional, and labelled by $|\Lambda|(r + s + t)$ non-negative real numbers. In any realistic chemical reaction at fixed temperature, there is a conserved quantity, the total mean number of atoms. If the initial state has a finite mean number of atoms, then the dynamics, which tells us the number of molecules of each type at each point, takes place in a bounded set inside a finite-dimensional space. The $R$ independent reactions give rise to $\nu - R$ conserved numbers, in the same way as in density-led reactions. The region of $\Sigma$ determined by the initial values of the conserved quantities can be parametrised by a subset of the space of the activities $a$ of each chemical at each point, and these parameters lie between 0 and 1. The boundary where some $a$'s are zero might be fixed points, as they correspond to the absence of a chemical that might be needed for the reaction to go. So if the grand canonical state we get from applying $LTE$ to $pT$ has zero activity, this means that $pT$ has zero particles of this type, and so has a zero entry at all points except $0 \in \Omega_x$. So $p$ is a boundary point of $\Sigma$. One can show [Rondoni (1991)] that boundary points are repellers. Since all the mean numbers remain bounded along the orbit, it follows that the dynamics takes place in a compact submanifold of $\Sigma$. There are $R$ independent lines of reaction, as above, giving $\nu - R$ independent conserved quantities. The equations for the fixed point say that the product of the initial activities is the product of the final activities, for each simple reaction. This leads to an equation similar to (5.14), except that the number of terms on each side might not be equal. Then the same argument shows that there is a unique fixed point for each set of initial values of the conserved quantities. Thus we can apply the Lyapunov direct method to prove convergence of the activity-led dynamics.

The upshot of this section is that every chemical system, whether isolated or isothermal, converges to its fixed point, and that permanent oscillations, or chaos, are not possible. Any really interesting chemistry must come from 'driven' systems.

## 5.4  Convergence in Norm

Let $T_c$ be a stochastic map on the algebra $\mathcal{A}_c$, having a strictly positive left fixed point $p_0$. We have seen that the relative entropy $S(p_c|p_0)$ is not increased by the isothermal dynamics. Since the relative entropy is bounded below by zero, it follows that $S(p_c|p_0)$ converges along the orbit. Under ergodic conditions, the relative entropy converges to zero. In this section we show that the relative entropy is greater than a certain metric distance of $p_c$ from $p_0$, given by a useful norm, first used by Fisher [Fisher (1925)], and now the subject of much study [Amari (1985)]. We thus prove that the states along the orbit converge in this metric when the ergodic conditions hold. We follow [Streater (1993b)].

We first note that the so-called condition of detailed balance, advocated by Glauber [Glauber (1963)] is not the most general condition occurring in isothermal dynamics. Let us choose the natural basis in $\mathrm{Span}\,\Omega_c$, namely, the points of $\Omega_c$ itself. Then $T_c$ and its adjoint become infinite matrices. The condition of detailed balance according to Glauber is

$$T_{cij} = e^{\beta(\mathcal{E}_i - \mathcal{E}_j)} T_{cji}. \tag{5.16}$$

This is just like the relation between the forward and backward chemical rates, and it has the same source; the principle of microscopic reversibility as used in the above theory of chemical kinetics is that the bistochastic map $T$ on $\mathcal{A}(\Omega)$ should be a symmetric matrix. In any reaction conserving energy, the difference between the energies of the initial and final states is made up by the energies of one or more heat-particles. In isothermal dynamics the heat-particles are in the canonical state at some beta; we denoted this state by $s_\beta$. Then we get $T_c$ from $T$ by taking the marginal distribution of the chemicals after one time-step:

$$p'_c = p_c T_c = ((p_c \otimes s_\beta) T)\, \mathcal{M}_c. \tag{5.17}$$

Then if $T$ is symmetric (in the natural basis) and energy-conserving, we see that the ratio of the forward to the backward rate is $s_i/s_j$, which gives (5.16). In this derivation, it does not matter whether there is one or several heat-particles, on one or on both sides of the equation. Now, the symmetry of $T$ is not at all a natural assumption in applications of statistical dynamics to neural nets, or if the model includes velocity, which is odd under time-reversal. To be able to do thermodynamics, we just need $T$ to be bistochastic. By Lemma 3.3, this will be true if $T$ is stochastic, and the model is invariant under time-reversal.

Returning to isothermal dynamics, how is this bistochasticity of $T$ reflected in the properties of $T_c$? The first thing to remark is that the canonical state $p_{c\beta}$ is a fixed point of $T_c$. In fact, we know already that $p_{c\beta} \otimes s_\beta$ is a fixed point of any energy-conserving map $T$, since this state is constant on each energy-shell. It will turn out, in the next section, that no further conditions than this, that $p_{c\beta}$ is a fixed point, can be derived from the bistochasticity of $T$. We therefore take this, or rather the following condition easily shown to be equivalent to it, as the definition of detailed balance. We first introduce a new metric topology on the algebra. To avoid confusion with the free energy, denoted $F$, in this section we shall use $f$, $g$ to denote random variables. For $f$ and $g$ in $\mathcal{A}_c$, define the scalar product for each $\beta > 0$ by

$$\langle f, g \rangle_\beta = Z_\beta^{-1} \sum_{j \in \Omega_c} e^{-\beta \mathcal{E}_c(j)} f_j g_j, \qquad (5.18)$$

where

$$Z_\beta = \sum_{j \in \Omega_c} e^{-\beta \mathcal{E}_c(j)}.$$

This idea is borrowed from the theory of quantum statistical mechanics [Haag, Hugenholtz and Winnick (1967)]. We denote the norm coming from this scalar product by

$$\|f\|_\beta = \left( Z_\beta^{-1} \sum_{j \in \Omega_c} e^{-\beta \mathcal{E}_c(j)} |f(j)|^2 \right)^{1/2}. \qquad (5.19)$$

Let $\mathcal{A}_\beta$ be the completion of $\mathcal{A}_c$ in the metric given by $\| \bullet \|_\beta$. This contains quite a few unbounded random variables, because of the strongly convergent factors $e^{-\beta \mathcal{E}_c}$. The dual space $\mathcal{A}_\beta^*$ of this Banach space is the completion of $\mathrm{Span}\,\Omega_c$, of real measures of finite support, in the metric defined by the dual norm

$$\|p\|_{-\beta} = \left( Z_\beta \sum_{j \in \Omega_c} e^{\beta \mathcal{E}_c(j)} |p_j|^2 \right)^{1/2}. \qquad (5.20)$$

We shall see that this is the natural norm governing the convergence of states evolving under isothermal conditions, rather than the $L^1$-norm usually used for measures. In this we follow [Streater (1993c)]. It is also the norm in which the *equivalence of ensembles* holds. The rising exponential $e^{\beta \mathcal{E}_c(j)}$ forces the high-energy tail of $p$ to converge much faster than if we use the $L^1$-norm, and leads to a larger norm and stronger topology.

Let $T_c$ be a bounded operator on $\mathcal{A}_\beta$, which is a Hilbert space with scalar product $\langle \bullet, \bullet \rangle_\beta$, and denote the adjoint of $T_c$ by $T_c^{(\beta)}$. Note that $T_c^{(\beta)}$ acts on $\mathcal{A}_\beta$.

**Definition 5.2.** We say that $T_c : \mathcal{A}_\beta \to \mathcal{A}_\beta$ obeys *detailed balance* (relative to $\beta \mathcal{E}_c$) if $T_c$ and $T_c^{(\beta)}$ are both stochastic maps.

Apart from being positive maps, if $T_c$ satisfies detailed balance, then $T_c 1 = T_c^{(\beta)} 1 = 1$. We note that the canonical measure $p_\beta$ lies comfortably in $\mathcal{A}_\beta^*$. We have an easy lemma.

**Lemma 5.3.** *A stochastic map $T_c$ obeys detailed balance if and only if $p_\beta$ is a left fixed point of $T_c$.*

**Proof.** Note that if $p_\beta T_c = p_\beta$, then for all $f \in \mathcal{A}_c$,

$$\langle l, f \rangle_\beta = p_\beta \cdot f = p_\beta T_c \cdot f = p_\beta \cdot T_c f = \langle 1, T_c f \rangle_\beta$$
$$= \langle T_c^{(\beta)} 1, f \rangle_\beta;$$

since $\mathcal{A}_c$ is dense in $\mathcal{A}_\beta$, we get $T_c^{(\beta)} 1 = 1$ as required. For the converse, if $T_c^{(\beta)} 1 = 1$,

$$p_\beta T_c \cdot f = p_\beta \cdot T_c f = \langle 1, T_c f \rangle_\beta = \langle T_c^{(\beta)} 1, f \rangle_\beta = \langle 1, f \rangle_\beta = p_\beta \cdot f.$$

Since this is true for all $f \in \mathcal{A}_c$, we have $p_\beta T_c = p_\beta$. $\square$

The Lemma says that our formulation of detailed balance is obeyed by the isothermal dynamics $\tau_\beta$ used in the first half of this book. Also, it is the most general formulation possible in any classical theory that converges to equilibrium, as in the classical case it imposes no further conditions other than that $T_c$ be stochastic with equilibrium $p_\beta$ as fixed point. For, the positivity of $T_c$ implies that its matrix elements are non-negative. The positivity of $T_c^{(\beta)}$ then follows from the simple relation

**Lemma 5.4.**

$$p_\beta(j) T_{cjk}^{(\beta)} p_\beta(k)^{-1} = T_{ckj}.$$

**Proof.**

$$p_\beta(j) T_{cjk}^{(\beta)} p(k) f(j) = p_\beta T_c^{(\beta)} p \cdot f = \langle T_c^{(\beta)} p, f \rangle_\beta = \langle p, T_c f \rangle_\beta$$
$$= p_\beta p \cdot T_c f = p_\beta(k) p(k) T_{ckj} f(j).$$

Since this is true for all $p$ and $f$, we have the result. $\square$

At $\beta = 0$, $p_\beta$ is the unit operator, and our version of detailed balance reduces to the statement that $T_c$ is bistochastic. This is enough to ensure that the entropy is monotonic increasing in time. We shall see that at $\beta > 0$, detailed balance is enough to ensure that the thermodynamic function $\Psi = S_c - \beta\langle\mathcal{E}_c\rangle$ is monotonic increasing.

We now show that the equilibrium state can be characterised as the state with the smallest $-\beta$-norm.

**Theorem 5.5.** *Suppose $\beta\mathcal{E}_c$ is such that $Z_\beta = \sum_{j\in\Omega_c} e^{-\beta\mathcal{E}_c(j)} < \infty$. Then the unique probability measure with the smallest $-\beta$-norm is the equilibrium state $p_\beta$.*

**Proof.** Minimise $\sum_{j\in\Omega_c} e^{-\beta\mathcal{E}_c(j)} p_c(j)^2$ over $p_c \in \Sigma_c$ by introducing a Lagrange multiplier $\lambda$:

$$\frac{\partial}{\partial p_k}\left(\sum_j e^{\beta\mathcal{E}_c(j)}p_j^2 + \lambda\sum_j p_j\right) = 2p_k e^{\beta\mathcal{E}_c(k)} + \lambda = 0$$

at a turning point, giving the unique answer

$$p_k = \text{const. } e^{-\beta\mathcal{E}_c(k)}.$$

The second derivative is

$$\frac{\partial}{\partial p_l}\left(2p_k e^{\beta\mathcal{E}_c(k)} + \lambda\right) = 2e^{\beta\mathcal{E}_c(k)}\delta_{kl}$$

which is a diagonal matrix with positive entries, and so is positive definite. The turning point is therefore a minimum.                    $\square$

We now come to the main effective result of this section, which is an adaptation of a result of Koseki [Koseki (1993)].

**Theorem 5.6.** *If $T_c$ obeys detailed balance, then $T_c$ is a contraction in the norm $\|\bullet\|_{-\beta}$. If in addition $T_c$ is irreducible, with positive diagonal elements, then $p_\beta$ is the only fixed point of $T_c$, and if $p \neq p_\beta$ then $\|pT_c\|_{-\beta} < \|p\|_{-\beta}$.*

Note: $T_c$ irreducible means that for any $i, j \in \Omega_c$, there is a chain of finite length $i_1 = i$, $i_2$, $i_3$, $\ldots, i_n = j$ such that $T_{i_k,i_{k+1}} > 0$, for $k = 1,\ldots,n-1$. Whereas in statistical dynamics the bistochastic map $T$ is not irreducible on $\Omega$ as it maps each energy-shell to itself, the isothermal map $T_c$ is typically irreducible, unless there are conserved quantities such as atomic number in addition to the energy.

**Proof.** Obviously for a sum over any finite subset $\Omega_0$ of $\Omega_c$ we have for any $p, q \in \Sigma(\Omega_c)$:

$$\sum_{i,j} Z_\beta T_{cij} e^{-\beta \mathcal{E}_i} \left( e^{\beta \mathcal{E}_j} q_j - e^{\beta \mathcal{E}_i} p_i \right)^2 e^{-\beta \mathcal{E}_i} \geq 0 \qquad (5.21)$$

and this inequality persists in the limit as $\Omega_0$ increases to $\Omega_c$ if the sums converge. Let $q = pT_c$. The first part is the sum of non-negative terms

$$Z_\beta \sum_{i,j} T_{cij} e^{-\beta(\mathcal{E}_i - \mathcal{E}_j)} e^{\beta \mathcal{E}_j} q_j^2 = Z_\beta \sum_{i,j} T_{ji}^{(\beta)} e^{\beta \mathcal{E}_j} q_j^2.$$

By detailed balance the sum of $T_{ji}^{(\beta)}$ over $i$ converges to 1 as $\Omega_0$ increases to $\Omega_c$, so the first part reduces to

$$Z_\beta \sum_j e^{\beta \mathcal{E}_j} q_j^2 = \|pT_c\|_{-\beta}^2.$$

The cross terms give

$$-2Z_\beta \sum_{i,j} T_{cij} e^{\beta \mathcal{E}_j} p_i q_j$$

and the sum over $i$ is

$$\sum_{i \in \Omega_c} T_{cij} p_i = q_j = (pT_c)_j.$$

The cross-terms thus give us

$$-2\|pT_c\|_{-\beta}^2.$$

The last term is exactly $\|p\|_{-\beta}^2$, since $T_c$ is stochastic. We conclude that

$$-\|pT_c\|_{-\beta}^2 + \|p\|_{-\beta}^2 \geq 0$$

as desired.

Let $p \in \Sigma(\Omega_c)$ be such that $T_c$ is not a proper contraction at $p$. We shall show, using the assumptions of the theorem, that $p - q = p_\beta$. Since we get a proper contraction unless every term is zero, we get

$$e^{\beta \mathcal{E}_j} q_j = e^{\beta \mathcal{E}_i} p_i \qquad \text{for all } i, j \text{ with } T_{cij} > 0.$$

Since $T_{cii} > 0$ for all $i \in \Omega_c$, we have $p = q$. Starting with a component of $p$, say $i = 0$, with $p_0 \neq 0$, and changing the origin of the energy so that $\mathcal{E}_0 = 0$, we get

$$p_j = e^{-\beta \mathcal{E}_j} p_0$$

for all $j$ linked to 0 by one step. Then we get

$$p_k = e^{-\beta(\mathcal{E}_k - \mathcal{E}_j)} p_j = e^{-\beta \mathcal{E}_k} p_0$$

for all $k$ linked to 0 by two steps. In this way we get

$$p_k = e^{-\beta \mathcal{E}_k} p_0 \qquad \text{for all } k \in \Omega_c.$$

Since $p$ is normalised we see that $p = p_\beta$. $\qquad \square$

We can obtain a result similar to the uniqueness part of the Perron-Frobenius theorem, that 1 is a simple eigenvalue. For, if $p'$ is another fixed point of $T_c$, it can be taken to be real, since its real and imaginary parts in the privileged basis are also eigenvectors. Then the equation $pT = p$ leads, as in the theorem, to $p = p_\beta$.

We now seek a theorem on the convergence of the isothermal dynamics which will cover the case when the sample space is infinite, such as in the important case when the classical Fock space is present. It is true that the state space $\Sigma(\Omega_c)$ is compact in the weak* topology even when $\Omega_c$ has infinitely many points, and that this can lead to a convergence theorem [Streater (1984)]. However, we cannot be sure that the limit state has the same mean energy as those along the sequence, unless some condition is placed on the initial state. The problem is that energy is an unbounded observable, and that some energy might be lost 'along the energy ladder' if the notion of convergence is too weak. The following rather strong condition on the initial state leads to convergence in the norm $\| \bullet \|_{-\beta}$, which is good enough for most people.

**Definition 5.7.** Let us say that a state $p \in \Sigma(\Omega_c)$ is close to equilibrium $(C, \beta)$ if there exists a number $C$ such that

$$p(j) \leq C Z_\beta^{-1} e^{-\beta \mathcal{E}_j} \qquad \text{for all } j \in \Omega_c.$$

This includes all equilibrium states with beta greater than $\beta$, all states with a maximum energy, and all states of the form $f p_\beta$ with $f \in \mathcal{A}_c$. The concept is stable under the dynamics:

**Lemma 5.8.** *If $p$ is close to equilibrium $(C, \beta)$, and $T$ obeys detailed balance $(\beta)$, then $pT$ is close to equilibrium $(C, \beta)$.*

**Proof.** Note that as $T$ is positively preserving,

$$\left( C Z_\beta^{-1} e^{-\beta \mathcal{E}} - p \right) T(j) \geq 0.$$

Since $T$ leaves $p_\beta$ fixed, we get

$$C Z_\beta^{-1} e^{-\beta \mathcal{E}_j} - (pT)_j \geq 0$$

which proves the lemma.                                                    $\square$

Now comes the good estimate:

**Theorem 5.9.** *Let $\Omega$ be a sample space and $p$ and $q$ be non-negative functions in $\ell^1(\Omega)$, and let $S(q|p) = \sum_{j \in \Omega} q(\omega) (\log q(\omega) - \log p(\omega))$. Suppose that there exists $C > 0$ such that*

$$p(\omega) \leq C q(\omega) \qquad \text{for all } \omega \in \Omega.$$

*Then*

$$S(q|p) \geq \frac{1}{2C'^2} \sum_\omega q(\omega)^{-1}|p(\omega) - q(\omega)|^2 - \|p\|_1 + \|q\|_1$$

*where* $C' = \max\{C, 1\}$.

Remark. If $\omega$ is such that $q(\omega) = 0$ then $p(\omega) = 0$ and we interpret the contribution as zero.

**Proof.** The Taylor series with Lagrange remainder gives

$$S(q|p) = \sum q(\omega) (\log q(\omega) - \log p(\omega))$$

$$= \sum q(\omega) \left( \log q(\omega) - \left[ \log q(\omega) + \frac{p(\omega) - q(\omega)}{q(\omega)} \right. \right.$$

$$\left. \left. - \frac{(q(\omega) - p(\omega))^2}{2\xi(\omega)^2} \right] \right)$$

$$= \sum (q(\omega) - p(\omega))^2 q(\omega)/(2\xi(\omega)^2) - \|p\|_1 + \|q\|_1.$$

Here $\xi(\omega)$ is the intermediate value, obeying

$$q(\omega) < \xi(\omega) < p(\omega) \text{ or } p(\omega) < \xi(\omega) < q(\omega)$$

except if $p(\omega) = q(\omega)$. If $\omega$ is such that $\xi(\omega) < q(\omega)$ we get for that term in the sum a contribution greater than

$$\frac{1}{2}|q(\omega) - p(\omega)|^2 \frac{q(\omega)}{q(\omega)^2}.$$

If $\omega$ is such that $\xi(\omega) < p(\omega) < C\,q(\omega)$, we get a term greater than

$$\frac{1}{2}|q(\omega) - p(\omega)|^2 \frac{q(\omega)}{C^2 q(\omega)^2}.$$

In either case, summing gives the result. $\qquad\square$

So if $\Omega$ is the space of chemicals $\Omega_c$ and $q = p_\beta$, and if $\tau_\beta$ defines a dynamics such that the relative entropy $S(p_\beta|p\tau_\beta^n)$ converges to zero, then the dynamics also converges in the norm $\| \bullet \|_{-\beta}$, if the initial state is close to equilibrium. This is not such a convenient result, since the free energy is related to the other relative entropy $S(p|p_\beta)$. For this we have the following

**Theorem 5.10.** *Suppose that $T_c$ obeys detailed balance $(\beta)$ and let*

$$F(p) = p \cdot \mathcal{E}_c - \beta^{-1} S(p)$$

*denote the free energy, defined on a subset of $\Omega_c$. Then*

$$F(pT_c) \leq F(p).$$

*If further $p$ is close to equilibrium $(C, \beta)$, then*

$$F(pT_c) - F(p) \leq -(2\beta C)^{-1} \left( \|p\|_{-\beta}^2 - \|pT_c\|_{-\beta} \right).$$

Our proof uses the

**Lemma 5.11.** *Let* $s(x) = -x \log x$ *and* $x = \sum \lambda_i x_i$, $\lambda_i \geq 0$, $\sum \lambda_i = 1$. *Then if* $x \neq x_i$, $i = 1, 2, \ldots, n$ *we have*

$$\sum \lambda_i s(x_i) = s(x) - \frac{1}{2} \sum \lambda_i (x_i - x)^2 / \xi_i$$

*where* $x_i < \xi_i < x$ *or* $x < \xi_i < x_i$.

***Proof.***    [Proof of lemma] By Taylor's theorem with Lagrange's remainder,

$$s(x_i) = s(x) + (x_i - x)[-1 - \log x] + (x_i - x)^2 \left(-1/(2\xi_i)\right).$$

Multiply by $\lambda_i$ and sum; we get the lemma, since $\sum \lambda_i (x_i - x) = 0$.    □

***Proof.***    [Proof of theorem] Choose $j$ and try this lemma with

$$\lambda_i = T_{cij} e^{-\beta(\mathcal{E}_i - \mathcal{E}_j)} = T_{ji}^{(\beta)}$$

and with $x_i = e^{\beta \mathcal{E}_i} p_i$. Then

$$x = \sum_i T_{cij} e^{-\beta \mathcal{E}_i + \beta \mathcal{E}_j} e^{\beta \mathcal{E}_i} p_i = e^{\beta \mathcal{E}_j} q_j$$

where $q = pT_c$. So for each $j$

$$-\sum_i T_{cij} e^{-\beta(\mathcal{E}_i - \mathcal{E}_j)} \left(e^{\beta \mathcal{E}_i} p_i (\log p_i + \beta \mathcal{E}_i)\right) = -e^{\beta \mathcal{E}_j} q_j (\log q_j + \beta \mathcal{E}_j)$$

$$-\sum_i T_{cij} e^{-\beta(\mathcal{E}_i - \mathcal{E}_j)} \left(e^{\beta \mathcal{E}_i} p_i - e^{\beta \mathcal{E}_j}\right)^2 / (2\xi_{ij}).$$

Cancel $e^{\beta \mathcal{E}_j}$ to get

$$\sum_i T_{cij} \left(-p_i \log p_i - \beta p_i \mathcal{E}_i\right) = -q_j \log q_j - \beta q_j \mathcal{E}_j$$

$$-\sum_i T_{cij} e^{-\beta \mathcal{E}_i} \left(e^{\beta \mathcal{E}_i} p_i - e^{\beta \mathcal{E}_j} q_j\right)^2 / (2\xi_{ij}).$$

Now for non-zero terms, either

$$e^{\beta \mathcal{E}_i} p_i < \xi_{ij} < e^{\beta \mathcal{E}_j} q_j \leq Z_\beta^{-1} C$$

or

$$e^{\beta \mathcal{E}_j} q_j < \xi_{ij} < e^{\beta \mathcal{E}_i} p_i \leq Z_\beta^{-1} C.$$

In either case, $1/\xi_{ij} > Z_\beta C^{-1}$. Summing over $j$ we get

$$S(p) - \beta p \cdot \mathcal{E} \leq S(q) - \beta q \cdot \mathcal{E}$$

$$-\frac{Z_\beta}{2C} \sum_{ij} T_{cij} e^{-\beta \mathcal{E}_i} \left(e^{\beta \mathcal{E}_i} p_i - e^{\beta \mathcal{E}_j} q_j\right)^2,$$

which gives the result in view of the proof of Theorem 5.6.    □

This theorem has a useful corollary:

**Theorem 5.12.** *If $T_c$ has positive diagonals and is irreducible, and $p$ is close to equilibrium, then $pT_c^n \to p_\beta$ as $n \to \infty$, the topology of the convergence being given by the metric $\| \bullet \|_{-\beta}$.*

This is a good result, since it does not depend on the finiteness of $\Omega_c$ or the reduction of the number of variables to a finite number by the $LTE$ map.

We now state without proof our version of the theorem on the equivalence of ensembles; the proof is given in detail in [Streater (1994c)]. Let $\Omega_x = \{0, 1, 2, \ldots\}$ and let $|\Lambda|$ be a finite set, and take the sample space to be $\Omega = \Omega(\Lambda) = \prod_x \Omega_x$. Let $\mathcal{E}(x) = \varepsilon \omega_x$ and

$$\mathcal{E}(\Lambda_0) = \sum_{x \in \Lambda_0} \mathcal{E}(x).$$

The energy in $\Lambda$, the random variable $\mathcal{E}(\Lambda)$, divides $\Omega$ into the union of energy-shells $\Omega_E$. Define the microcanonical state for $\Lambda$ with energy $E$ to be

$$p_\Lambda(\omega) = \begin{cases} |\Omega_E|^{-1} & \text{if } \omega \in \Omega_E \\ 0 & \text{otherwise.} \end{cases}$$

Define the marginal distribution of a probability $p_\Lambda \in \Sigma(\Lambda)$ onto $\Lambda_0 \subseteq \Lambda$ to be

$$p(\Lambda, \Lambda_0, \omega_0) = \sum_{\omega \in \Omega : \omega | \Lambda_0 = \omega_0} p_\Lambda(\omega). \tag{5.22}$$

Let us take $\Lambda$ to infinity in such a way that the energy per site, $E/|\Lambda|$ remains constant. Then for each fixed $\Lambda_0$, the marginal state in $\Lambda_0$ of the microcanonical state for $\Lambda$ converges to the canonical state for $\Lambda_0$. Thus

**Theorem 5.13 (Equivalence of Ensembles).** *As $\Lambda \to \infty$,*

$$p(\Lambda, \Lambda_0, \bullet) \to p_{\Lambda_0, \beta}(\bullet) \quad \text{in the norm } \| \bullet \|_{-\beta}. \tag{5.23}$$

*Here,*

$$p_{\Lambda_0, \beta}(\omega_0) = Z(\Lambda_0)^{-1} \exp\{-\beta \mathcal{E}(\Lambda_0)(\omega_0)\}$$

*and $\beta$ is fixed by*

$$\sum_{\omega_0} p_{\Lambda_0, \beta} \mathcal{E}(\Lambda_0)(\omega_0) = E|\Lambda_0|/|\Lambda|.$$

There is a similar theorem for an assembly of free fermions.

This theorem might be used to justify the step between random linear dynamics, which converges to the microcanonical state, and the convergence to the canonical state; the latter needs non-linear dynamics which worries some people (not me).

## 5.5  Dilation of Markov Chains

We now show that any Markov chain with a finite sample space $\Omega_c$ and
a strictly positive fixed point can be regarded as the isothermal dynamics
of some bistochastic map on a suitable sample space, conserving a certain energy-function. We follow [Streater (1995)]. The construction is not
unique, but $\Omega$ can be taken to be as small as $\Omega_c \times \Omega_c$.

We first note that if $p^{(0)} \in \Omega_c$ is strictly positive, then there exists a
stochastic map $T_c$ with $p^{(0)}$ as fixed point. Indeed,

$$T_c = \begin{pmatrix} p_1^{(0)} & p_2^{(0)} & \cdots & p_n^{(0)} \\ p_1^{(0)} & p_2^{(0)} & \cdots & p_n^{(0)} \\ \cdot & \cdot & \cdots & \cdot \\ p_1^{(0)} & p_2^{(0)} & \cdots & p_n^{(0)} \end{pmatrix} \tag{5.24}$$

has the much stronger property that $p_c T_c = p^{(0)}$ for all $p_c$; that is, it takes
any state to equilibrium in one step. The stochastic matrix $T_c$ is the limit,
as $n \to \infty$, of $T_1^n$, where $T_1$ is any ergodic stochastic matrix with $p^{(0)}$
as fixed point. We can also find a family $\{T_\lambda\}$ of stochastic matrices, all
having $p^{(0)}$ as fixed point, with $T_\lambda$ arbitrarily close to the identity, namely
$T_\lambda = (1 - \lambda)1 + \lambda T_c$. These remarks mean that the following structure
theorem is not empty.

**Theorem 5.14.** *Let $\Omega_c = \{1, 2, \ldots, n\}$. Let $p^{(0)} \in \Sigma(\Omega_c)$ have positive
components: $p_j^{(0)} > 0 \quad j = 1, 2, \ldots, n$. Then there exists a neighbourhood
$N$ of the unit $n \times n$ matrix and a sample space $\Omega_\gamma$, with $|\Omega_\gamma| = n$, and an
energy function $\mathcal{E} = \mathcal{E}_c + \mathcal{E}_\gamma$ on $\Omega = \Omega_c \times \Omega_\gamma$, such that for any stochastic
matrix $T_c \in N$ with $p^{(0)}$ as fixed point, there exists a bistochastic map $T$ on
$\mathcal{A}(\Omega)$ conserving $\mathcal{E}$ such that*

$$p_c T_c = (p_c \otimes s) T \mathcal{M}_c$$

*for all $p_c \in \Sigma(\Omega_c)$; here, $s$ is the canonical state of $\mathcal{A}(\Omega_\gamma)$ with energy $\mathcal{E}_\gamma$
and $\beta = 1$. Moreover, if $T_c$ obeys the usual form of detailed balance, then
$T$ is symmetric.*

**Proof.**    This is by construction. Consider the set $\{0\} \cup \Omega_c$, the given
sample space with one point adjoined. Define the *fan* of $\Omega_c$, denoted Fan $\Omega_c$,
to be the graph whose vertices are the points of $\{0\} \cup \Omega_c$, and whose edges
are the pairs $(\omega_c, 0)$. We can regard the edges as bonds between the points
of the system and a common heat-bath, $\{0\}$. We define the space $\Omega_\gamma$ of the

heat-particle to be the set of edges of Fan $\Omega_c$. Then $|\Omega_\gamma| = n$, as promised. Define the energy function $\mathcal{E}_c$ on $\Omega_c$ by

$$\mathcal{E}_c(\omega_c) = -\log p^{(0)}(\omega).$$

This is well defined and positive, since $0 < p^{(0)}(\omega_c) < 1$. We have adjusted the lowest energy so that the partition function is 1. Since $p^{(0)}(\omega) = e^{-\mathcal{E}_c(\omega)}$ we see that $\sum e^{-\mathcal{E}_{cj}} = 1$. We define $\mathcal{E}_\gamma(\omega_c, 0) = -\mathcal{E}_c(\omega_c)$, and the energy function on $\Omega = \Omega_c \times \Omega_\gamma$ to be

$$\mathcal{E}(\omega_c, (\omega_c', 0)) = \mathcal{E}_c(\omega_c) - \mathcal{E}_c(\omega').$$

We now construct a bistochastic matrix $T$ on $\mathcal{A}(\Omega)$ which causes hopping from $i \in \Omega_c$ to $j \in \Omega_c$ by picking up a heat-particle from the bond $b_i = (i, 0)$, hopping to $j$, and releasing a heat-particle to the bond $b_j = (j, 0)$. That is, we have the process

$$i + b_i \rightleftharpoons j + b_j.$$

The other bonds and vertices do not interact. Since the transitions occur only on the energy-shell $\Omega_0 \subset \Omega$ of zero energy, our reaction conserves energy, as claimed. We shall now define an $n \times n$ bistochastic matrix $T_0$ on $\mathcal{A}(\Omega_0)$. The action of $T$ will coincide with that of $T_0$ on the subspace, and will be the unit operator (of size $(n^2 - n) \times (n^2 - n)$) on the remaining points. $T$ is bistochastic and conserves energy, as it has the form

$$T = \begin{pmatrix} T_0 & 0 \\ 0 & 1 \end{pmatrix}. \tag{5.25}$$

Let

$$\lambda = \sum_{j=1}^{n} e^{\mathcal{E}(j)} \tag{5.26}$$

and put

$$\operatorname{diag} p^{(0)} = \begin{pmatrix} p^{(0)}(1) & 0 & \cdots & 0 \\ 0 & p^{(0)}(2) & \cdots & 0 \\ \cdots & \cdots & \cdots & \cdots \\ 0 & 0 & \cdots & p^{(0)}(n) \end{pmatrix}. \tag{5.27}$$

We define $T_0$ in terms of $T_c$ by

$$T_0 = 1 + \lambda \operatorname{diag} p^{(0)}(T_c - 1). \tag{5.28}$$

Let $p^{(1)}$ denote $(1, 1, \ldots, 1)$. Then $p^{(1)} \operatorname{diag} p^{(0)} = p^{(0)}$. A matrix $A$ is stochastic if and only if its elements are non-negative, and $p^{(1)} A = p^{(1)}$. We now check the second property for $T_0$ and $T_0^d$. We use that $T_c$ is stochastic, so that $T_c p(1)d = p^{(1)d}$, and that $p^{(0)}T_c = p^{(0)}$.

$$
\begin{aligned}
T_0 p^{(1)d} &= p^{(1)d} + \lambda \operatorname{diag} p^{(0)} T_c p^{(1)d} - \lambda \operatorname{diag} p^{(0)} p^{(1)d} \\
&= p^{(1)d} + \lambda \operatorname{diag} p^{(0)} p^{(1)d} - \lambda \operatorname{diag} p^{(0)} p^{(1)d} = p^{(1)d}
\end{aligned}
$$

since $T_c$ is stochastic. Hence the rows of $T_0$ add up to 1. Also,

$$
p^{(1)} T_0 = p^{(1)} + \lambda p^{(1)} \operatorname{diag} p^{(0)} (T_c - 1) = p^{(1)} + \lambda p^{(0)} (T_c - 1) = p^{(1)} \quad (5.29)
$$

since $p^{(0)}$ is a fixed point of $T_c$. Hence the columns of $T_0$ add up to 1. We now check the positivity. The off-diagonal elements of $T_0$ are non-negative. The $k$th diagonal element is

$$
1 + \lambda p_k^{(0)} (T_{ckk} - 1)
$$

which is positive if for each $k$, $T_{ckk}$ is close enough to 1. This defines the required neighbourhood $N$ of the unit matrix, in which $T_0$ is bistochastic.

It remains to show that we recover the Markov matrix $T_c$ as the isothermal dynamics given by the bistochastic matrix $T$. Recall that $T$ consists of the matrix $T_0$ in the block labelled by rows and columns $(j, b_j)$, $T$ being the unit matrix on the rest of the space, as in (5.25). This may be expressed in terms of matrix elements as:

$$
T_{i,b_j;k,b_\ell} = \delta_{i,j}\delta_{k,\ell}\left(\delta_{i,k} + p^{(0)}(i)(T_{ci,k} - \delta_{i,k})\right) + (1 - \delta_{i,j}\delta_{k,\ell})\delta_{i,k}\delta_{j,\ell}. \quad (5.30)
$$

One time-step of the isothermal dynamics defined by $T$ is

$$
p_c \mapsto p_c' = \sum_{\omega_\gamma}(p_c \otimes s\, T)(\omega_c, \omega_\gamma). \quad (5.31)
$$

Here, $\omega_c$ runs over $k \in \{1, 2, \ldots, n\}$ and $\omega_\gamma$ runs over $\ell = b_\ell = (\omega_\ell, 0)$. The heat particle is in the canonical state

$$
s(\ell) = \frac{e^{\mathcal{E}(\ell)}}{\sum_j e^{\mathcal{E}(j)}}. \quad (5.32)
$$

Then $\sum_j s(j) = 1$ and

$$
p^{(0)}(i)s(i) = e^{-\mathcal{E}(i)}e^{\mathcal{E}(i)}\left(\sum_j e^{\mathcal{E}(j)}\right)^{-1} = \lambda^{-1}. \quad (5.33)
$$

Both these are used in the following.

$$p'_c(k) = \sum_{i,j,\ell} p_c(i)s(j)T_{i,j;k,\ell}$$

$$= \sum_{i,j,\ell} p_c(i)s(j)\delta_{i,j}\delta_{k,\ell}T_{0,i,k} + \sum_{i,j,\ell} p_c(i)s(j)[\delta_{i,k}\delta_{j,\ell}$$

$$- \delta_{i,j}\delta_{k,l}\delta_{i,k}\delta_{j,\ell}]$$

$$= \sum_{i} p_c(i)s(i)T_{0,i,k} + p_c(k) - p_c(k)s(k)$$

$$= \sum_{i} p_c(i)s(i)[\delta_{i,k} + \lambda p^{(0)}(i)(T_{c,i,k} - \delta_{i,k})]$$

$$+ p_c(k) - p_c(k)s(k)$$

$$= p_c(k)s(k) + \lambda \sum_{i} p_c(i)p^{(0)}(i)s(i)T_{c,i,k}$$

$$- \lambda p^{(0)}(k)p_c(k)s(k) + p_c(k) - p_c(k)s(k)$$

$$= \sum_{i} p_c(i)T_{c,i,k}.$$

Thus the isothermal dynamics coincides with the given Markov chain.

As for detailed balance, we see that

$$T_0^d = 1 + \lambda(T_c^d - 1)\text{diag}\,p^{(0)}$$

which is $T_0$ if $\text{diag}\,p^{(0)}T_c = T^d\text{diag}\,p^{(0)}$, namely Glauber detailed balance holds, (5.16). □

The relative entropy of the Markov chain coincides with the free energy of the isothermal dynamics. By Theorem 3.11 and Exercise 3.27 we canv decompose the bistochastic map $T$ into permutations conserving energy. This gives us a natural decomposition of the convex set of stochastic maps, all with the same fixed point $p^{(0)}$, into extremal points. This result seems to be new.

The doubling up of the space from $\Omega_c$ to $\Omega_c \times \Omega_\gamma$ used here will be a key method in the quantum case.

## 5.6 Exercises

Exercise 5.34. Consider the sample space $\Omega = \{0, 1, \ldots\}$ with some energy-function $\mathcal{E}$ with a finite partition function. Define, for any state $p \in \Sigma(\Omega)$ the Legendre transform $\Psi$ of the entropy by $\Psi = S(p) - \beta p \cdot \mathcal{E}$, in which the variable $\mathcal{E}$ is singled out. Show that $\beta$ is the dual variable, and that $S$ is the inverse Legendre transform of $\Psi$.

Exercise 5.35. Show that for $2 \times 2$ matrices, the neighbourhood $N$ for which the construction of Sec. 5.5 works does not include all stochastic matrices with $p^{(0)}$ as fixed point in general.

# Chapter 6

# Driven Systems

The fundamental theorem on chemical kinetics shows that to get interesting behaviour such as chaos or oscillations, a chemical model needs to be *driven*; more than one of the intensive parameters (beta, chemical potential) must be kept fixed at non-equilibrium values by outside intervention. These parameters might then be different at different points of $\Lambda$, even at a stationary point. They are then not, strictly speaking, intensive variables any more; the term *canonical* variable remains valid.

## 6.1 Sources and Sinks

A source is a point in the space $\Lambda$ at which particles, or heat, is fed into the system from outside; a sink is a point from which particles or heat are withdrawn from the system. The rates are determined by the experimenter, and so these rates are *controlled variables*. Mathematically we can implement the control within statistical dynamics by forcing the state $p' = pT$ after one time-step to conform to the constraints. Here we follow [Streater (1994a)]. Experimentally, it is often a good description of the constraint to say that certain points $x \in \Lambda_0 \subseteq \Lambda$ are maintained at prescribed betas $\beta(x)$ or chemical potentials $\mu_j(x)$ for one or more chemicals $j$. To be able to express this, the sample space at $x$ should contain the sample space of the constrained chemical as a factor. Thus we must assume that $\Omega_x = \Omega_j \times \Omega_{\neq j}$. For consistency of notation, we shall speak of beta as a chemical potential (of the heat-particle). Let $\mathcal{M}_j$ be the marginal projection onto the algebra of random variables of the $j$-th chemical. We can presume that the state $p'\mathcal{M}_x\mathcal{M}_j$ at a controlled point is independent of the rest of the system, since the flow of heat and particles into the point $x$ from the outside destroys the correlations. The state can therefore be assumed to be a product

151

state, thus

$$p(\omega) = p_{\mu_j}(\omega_{x,j}) \otimes p(\omega_{\neq x,j}).$$

Here, $\mu_j$ is the chemical potential of the $j$-th chemical at which we maintain the point $x$. This property of factorisation of the state is not preserved by the dynamics $T$, and the external control must be used to restore the constraint after one time-step. As promised, we implement this control mathematically by just doing it; the first step is

$$p \mapsto pT\mathcal{M}_j \otimes pT\mathcal{M}_{\neq j}. \tag{6.1}$$

This is then replaced in the first factor by flows of heat and chemicals from the outside, to enforce the controlled state $p_{\mu_j}$, so that one time-step of the controlled dynamics is

$$p\tau = p_{\mu_j} \otimes pT\mathcal{M}_{\neq j}. \tag{6.2}$$

We can then repeat the operation, once for each controlled variable, to arrive at the desired model of the driven system.

There are alternative controls that can be modelled. For example, if a site can be occupied by any one of several chemicals, $A$, $B$, ... but not more than one (as in the hard-core models), we may control the site by insisting that it always contain only a certain chemical, say $A$. Then after a time-step, we sweep out the product of the reaction from the site, and replace it by the chemical $A$. More generally, we may replace the state at the controlled site by a chosen statistical mixture of all possibilities. The grand canonical state $p_{\mu_j}$ is just one of these. These more general controls are perhaps not worth studying unless we have in mind some experimental set-up that can implement them. We work out some important cases to illustrate the method.

## 6.2   A Poor Conductor

In the following model, heat is transmitted along a chain of atoms by local excitations from one to the next. In a good conductor such as copper the bulk of the heat is carried by electrons in the conduction band and very little by atomic excitations. We therefore describe the present model as a poor conductor. We take [Streater (1993a)] as usual $\Lambda$ to be a finite lattice, $\{0, 1, \ldots, N\}$ and $\Omega_x = \{0, 1\}$. The configuration space is then $\Omega = \prod \Omega_x$. We interpret $\omega_x = 0$ as saying that the atom at $x$ is in its ground state, and $\omega_x = 1$ as saying that the atom at $x$ is in its excited state, with energy $\varepsilon$ say. The energy of a configuration $\omega$ is thus taken to be

$$\mathcal{E}(\omega) = \varepsilon \sum_x \omega_x. \tag{6.3}$$

We assume that $\Lambda$ is provided with a set of edges, making it into a graph. These will be taken to define the nearest neighbours of the model. We take it that there are no multiple edges, and denote the set of edges, or bonds $b = (x, y)$ by $\mathcal{B}$. For each bond $b \in \mathcal{B}$ we define the transition matrix $T_x$, which only alters the coordinates $(x, y)$ of $b$, as

$$
T_x = \begin{array}{c} \\ 00 \\ 01 \\ 10 \\ 11 \end{array}
\begin{array}{cccc} 00 & 01 & 10 & 11 \end{array} \\
\left( \begin{array}{cccc} 1 & & & \\ & 1-\lambda & \lambda & \\ & \lambda & 1-\lambda & \\ & & & 1 \end{array} \right). \tag{6.4}
$$

In this, the rows and columns are labelled by $\omega_x, \omega_{x+1}$, which can take the values 0 and 1. Any state in $\Sigma(\Omega_x)$ is a pair of non-negative numbers whose sum is 1, and is thus determined by one parameter. It can therefore quite generally be written as a thermal state

$$p_x(\omega_x) = (1 + e^{-\beta\varepsilon})^{-1} e^{-\beta\mathcal{E}(\omega_x)} \tag{6.5}$$

for some parameter $\beta$, which is uniquely determined. Given any state $p \in \Sigma(\Omega)$, its marginal state $p\mathcal{M}_x$ can be written as in (6.5), which defines the local beta of the state $p$. In this case, where $|\Omega_x| = 2$, the stoss map is the same as the *LTE* map. We take as the dynamics of the system, due to $T_x$, to be the combined map

$$p \mapsto pT_x \mapsto \otimes_x pT_x \mathcal{M}_x. \tag{6.6}$$

Since the state is, after each time-step, forced to be a product state, it is determined by $(p_0, \ldots, p_N)$, which in turn is determined by the probabilities that the upper level is occupied at the various points. Put $n(x) = p_x(t)$, the density at time $t$. This is the mean number of occupation of the upper level at $x$, and so the state at time $t$ is described by this *density-field*. We now take the change in state in one time step to be a convex mixture of these $T_x$, as $x$ runs from 0 to $N - 1$. We again call the new rate constant $\lambda$. Let us denote by $n'(x)$ the density field at time $t + 1$. We have already seen that this leads to the discrete heat equation

$$
n'(x) - n(x) = \begin{cases} \lambda \left( n(x+1) - 2n(x) + n(x-1) \right), & \text{for } 0 < x < N \\ \lambda \left( n(0) - n(1) \right), & \text{for } x = 0 \\ \lambda \left( n(N-1) - n(N) \right), & \text{for } x = N \end{cases}.
$$

This, then, describes the dynamics of an isolated system. In the driven system, we maintain the ends, $x = 0$ and $x = N$, at fixed betas; this just means that $n(0)$ and $n(N)$ are returned to their initial values after each time-step. Let us denote these initial values by $n_0$ and $n_N$ and suppose that $n_0 > n_N$. A stationary state is one which is unchanged in one time-step. Such a state is said to represent *stasis*. It is not in equilibrium unless $n_0 = n_N$. Indeed, a stationary solution can easily be found:

$$n(x) = n_0 - \frac{x}{N} (n_0 - n_N) . \tag{6.7}$$

In this solution, the energy-density falls linearly as $x$ increases. This is in contrast to the commonly studied model, where it is the temperature that obeys the heat equation, rather than the energy-density, as here. Although at stasis heat enters at $x = 0$ and leaves at $x = N$, this might not be true at all times. If the initial state is hotter at $x = 1$ than $T_0$, the temperature of the external control at $x = 0$, then heat will leave the hot end at the beginning of the motion. The entropy of the system may thus either decrease or increase; not even the free energy is a Lyapunov function. This is typical of driven systems: some have Lyapunov functions, and so converge to stasis; others may converge without an obvious Lyapunov function. Yet others might oscillate or diverge to infinity. It is clear that our present model converges to stasis.

We can find the rate of entropy production by the system at stasis. We may treat the ends, $x = 0$, $x = 1$ and $x = N - 1$, $x = N$ separately. Thus, if $(n_0, n_1)$ is the density of the first two points at stasis, from above $n_1 = n_0 - N^{-1} (n_0 - n_N)$. After one time-step in which $T$ has acted, but no compensating heat flow has yet flowed, we have a new density,

$$n'_0 = n_0 + \lambda (n_1 - n_0) = n_0 - \lambda N^{-1} (n_0 - n_N) ;$$

it is clear that we must add an amount of heat equal to

$$Q = \varepsilon \lambda N^{-1} (n_0 - n_N)$$

to restore the mean value of the local energy at $x = 0$, and thus the beta, to its original value. The value of $n_1$ is unchanged (we are at stasis). The entropy at $x = 0$ is

$$S_0 = -n_0 \log n_0 - (1 - n_0) \log (1 - n_0) .$$

Passing to the limit of continuous time, we see that

$$\dot{S} = -\dot{n}_0 \log n_0 - \dot{n}_0 - (1 - \dot{n}_0) \log(1 - n_0) - (-\dot{n}_0)$$

$$= \lambda N^{-1} (n_N - n_0) \log \left( \frac{1 - n_0}{n_0} \right)$$

$$= N^{-1} (n_N - n_0) \varepsilon \beta = Q / k_B \Theta$$

in terms of $Q$ and $\Theta = (k_B \beta)^{-1}$, the temperature. Thus, in the limit of continuous time the entropy production (cancelled by the feeding in of hot heat) is quasi-static: the entropy loss of the environment is exactly $Q/(k_B \Theta)$ as given by elementary thermodynamics.

The stationary state of this model defines a stochastic process known as a Markov population process [Kingman (1969)]. In 1978, a justification of the construction was given by Davies [Davies (1978)]. He starts with a Hamiltonian system, in which each site of a chain is linearly coupled to an oscillator with coupling strength $\lambda$. Thus each site has its own private heat-bath, except that the ends of the chain are coupled to reservoirs whose state is maintained at fixed temperatures. The time-evolution of the system is then obtained by tracing over the private heat-baths. At the end, the weak-coupling limit is taken in the sense of van Hove. The density-matrix of the system then obeys equations very similar to those of the model given here. In particular, the correlations between different points vanishes in the limit, and the *LTE* hypothesis holds. It can be argued that this justifies the model. This is, however, very much a matter of style rather than substance. It is certainly nice to have at least one Hamiltonian leading to the process of the model in a certain limit. Unfortunately there are many other Hamiltonian schemes whose weak-coupling limit have unphysical features, thus defeating one of the main points of the Hamiltonian method. In particular, energy can be lost to the private heat-baths which are inside the chain, contradicting the first law. This is not the same as losing heat to the reservoirs at the ends, which represents heat flowing out of the system. It is even possible for a limit of Hamiltonian motion to violate the conservation of probability. In Davies's model, the internal heat-loss does not occur, and probability is conserved, because of the clever construction. Thus, the Hamiltonian method must be supplemented by an intuition that avoids these unwanted features. Statistical dynamics is the mathematics behind this intuition.

We now turn to a driven system which leads to oscillations.

## 6.3 A Driven Chemical System

The following is a simplified version of an oscillating system inspired by the Brussellator [Rondoni and Streater (1994, 1993); Streater (1995)]. It involves four chemicals, $A$, $B$, $X$ and $Y$, which undergo the unidirectional reactions

$$B + X \xrightarrow{k_1} Y, \tag{6.8}$$

$$X \xrightarrow{k_2} 0, \tag{6.9}$$

$$2X + Y \xrightarrow{k_3} 3X, \tag{6.10}$$

$$A \xrightarrow{k_4} X. \tag{6.11}$$

The meaning of (6.9) is that $X$ makes a transition to an inert chemical with rate $k_2$, or else to reactants that are swept away after each time-step. The reaction (6.10) is *autocatalytic*: two molecules of $X$ must be present before the molecule $Y$ can be persuaded to make the transition to $X$.

The unidirectional reactions violate the laws of chemistry and in particular, the Boltzmann ratio. They can be understood by postulating that a fifth particle is produced on the right in each reaction, and that this particle is swept away by the experimenter as fast as it is produced. This prevents the back reaction if the new particle is essential for the inverse reaction to proceed. A natural choice for the new particle is the heat-particle. This is tantamount to saying that all reactions are exothermic, and that the heat produced is swept away by keeping the system at zero temperature.

The system is in a single cell, a *well-stirred reactor* and the sample space will be taken to be

$$\Omega = \Omega_A \times \Omega_B \times \Omega_X \times \Omega_Y.$$

Here, $\Omega_A$ etc. are all the same, the classical Fock space

$$\Omega_A = \{0,\ 1,\ 2,\ldots\} = \Omega_B = \Omega_X = \Omega_Y.$$

Let $p = p(\omega_A, \omega_B, \omega_X, \omega_Y)$ be a probability on $\Omega$, and denote as usual the marginals on the factors by $p_A$ and so on. For example,

$$p_A(\omega_A) = \sum_{\omega_B=0}^{\infty} \sum_{\omega_X=0}^{\infty} \sum_{\omega_Y=0}^{\infty} p(\omega_A, \omega_B, \omega_X, \omega_Y).$$

Denote by $N_A$ etc. the mean values of the random variables $\mathcal{N}_A(\omega) = \omega_A$, etc. Thus,

$$N_A = \sum_{\omega_A=0}^{\infty} \omega_A p_A(\omega_A).$$

This is the mean number of $A$-particles when the state is $p$. The activity is defined in terms of the mean number by $a = N/(N+1)$. We use the activity-led equations of Sec. 4.6. The equations with the heat-particle swept away are thus

$$\frac{dN_A}{dt} = -k_4 a_A \, , \tag{6.12}$$

$$\frac{dN_B}{dt} = -k_1 a_B a_X \, , \tag{6.13}$$

$$\frac{dN_X}{dt} = -k_1 a_B a_X - k_2 a_X + k_3 a_X^2 a_Y + k_4 a_A \, , \tag{6.14}$$

$$\frac{dN_Y}{dt} = k_1 a_B a_X - k_3 a_X^2 a_Y \, . \tag{6.15}$$

We want to consider the driven system in which the activities $a_A$ and $a_B$ are maintained at constant values by feeding the system as they are used up. It is very easy to do this: we omit the first two equations, and regard $a_A$ and $a_B$ as constants in the last two equations. We then get the autonomous equations for the two variables $a_X$ and $a_Y$ (after eliminating $N$ in favour of $a$):

$$\frac{da}{dt} = \mathbf{F}(\mathbf{a}) \tag{6.16}$$

where $\mathbf{a} = (a_X, a_Y)$ and $\mathbf{F} = (F_1, F_2)$, with

$$F_1 = \left( -k_1 a_B a_X - k_2 a_X + k_3 a_X^2 a_Y + k_4 a_A \right) (1 - a_X)^2 \, , \tag{6.17}$$

$$F_2 = \left( k_1 a_B a_X - k_3 a_X^2 a_Y \right) (1 - a_Y)^2 \, . \tag{6.18}$$

Such a system can be studied by Lyapunov's first method: the fixed point is given by $\mathbf{F} = 0$, which gives two equations

$$-k_1 a_B a_X - k_2 a_X + k_3 a_X^2 a_Y + k_4 a_A = 0 \, , \tag{6.19}$$

$$k_1 a_B a_X - k_3 a_X^2 a_Y = 0 \, , \tag{6.20}$$

with the unique solution

$$\overset{\circ}{a}_X = \frac{k_4}{k_2} a_A \, , \qquad\qquad \overset{\circ}{a}_Y = \frac{k_1 k_2 a_B}{k_3 k_4 a_A} \, . \tag{6.21}$$

For this fixed point to lie in the allowed region $0 \leq a < 1$, there are limits on the values that the parameters can take. The next step is to study the stability of the fixed point; we expand $F$ in a Taylor series about $(\overset{\circ}{a}_X, \overset{\circ}{a}_Y)$, and we keep only the linear terms. The term of zeroth order vanishes, by definition of the fixed point. So, write $a_X = \overset{\circ}{a}_X + X$ and $a_Y = \overset{\circ}{a}_Y + Y$, and rewrite the equations for the new variables $\mathbf{Z} = (X, Y)$; we get the linear equation

$$\frac{d\mathbf{Z}}{dt} = A\mathbf{Z} \tag{6.22}$$

where $A$ is the $2 \times 2$ matrix

$$A = \begin{pmatrix} (k_1 a_B - k_2)(1 - \overset{\circ}{a}_X)^2 & k_3 \left(\frac{k_4 a_A}{k_2}\right)^2 (1 - \overset{\circ}{a}_X)^2 \\ -k_1 a_B (1 - \overset{\circ}{a}_Y)^2 & -k_3 \left(\frac{k_4 a_A}{k_2}\right)^2 (1 - \overset{\circ}{a}_Y)^2 \end{pmatrix}. \qquad (6.23)$$

Linear equations have exponential solutions, which are sinusoidal with an exponential growing or damping factor. Indeed, if $A$ has eigenvalues $\lambda_1$ and $\lambda_2$, which are complex in general, with eigenvectors $\phi_1$ and $\phi_2$, then the solution is

$$\mathbf{Z} = e^{\lambda_1 t}\phi_1 + e^{\lambda_2 t}\phi_2. \qquad (6.24)$$

The number, $\lambda = \max\{Re\,\lambda_1,\ Re\,\lambda_2\}$ is called the Lyapunov index of the fixed point $\mathbf{Z} = 0$. If $\lambda < 0$, then all solutions starting in the neighbourhood of the fixed point converge exponentially to it, at the rate $e^{\lambda t}$. We then say that it is a local attractor. If $\lambda > 0$ then the solutions move away from the fixed point, (except possibly on one orbit) as long as the linear approximation is a good one; then $\mathbf{Z} = 0$ is called a repeller. If both eigenvalues are purely imaginary, then in the linear approximation the solution is periodic about $\mathbf{Z} = 0$, and we see oscillatory solutions of the non-linear equations as well, provided that we start close enough to the fixed point. The present model exhibits these features, for suitable choices of the parameters. For example we may choose $a_A = a_B = 1/2$, and $k_1 = 5$, $k_2 = 11/8 = k_4$, $k_3 = 8$. Then the fixed points become $\overset{\circ}{a}_X = 1/2$, $\overset{\circ}{a}_Y = 5/8$, and the matrix $A$ becomes

$$A = \begin{pmatrix} 9/32 & 1/2 \\ -45/128 & -9/32 \end{pmatrix}, \qquad (6.25)$$

whose eigenvalues are purely imaginary. Figure (1) shows the plot of $a_X$ and $a_Y$ against time, and also the orbit, in which time is eliminated and $a_X$ is plotted against $a_Y$. The initial point, the pair of activities of $X$ and $Y$, has been chosen very close to the fixed point $(1/2, 5/8)$. It is clear that oscillations occur for these values of the parameters. Examination of the pictures shows that the period of oscillation, about 20, agrees with the prediction you will make if you do Exercise 6.34.

When the eigenvalues are purely imaginary, and there are two variables as here, then there is a periodic solution to the fully non-linear system; this follows from the Poincaré-Bendixon theorem [Platt (1971)]. Then we say that there is a limit-cycle; Fig. (2) shows that the limit-cycle attracts the orbit even when the initial state is rather close to the boundary. For a range of values of the parameters near these critical values, there will continue to be oscillatory behaviour. This is in contrast to isolated or isothermal systems. For these, we proved in the fundamental theorem of chemical kinetics, that there is no such possibility. Not all driven systems exhibit oscillations, as you will see in Exercise 6.36.

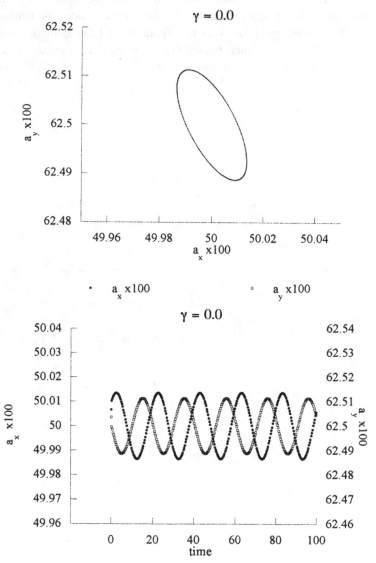

Fig. (1) If the initial point is very close to the critical point, the pair of activities $a_X, a_Y$ executes simple harmonic motion; the upper picture shows the orbit, $a_X$ versus $a_Y$, and the lower picture shows the time-development of both activities, with the values of $a_X$ on the left ordinate, and the values of $a_Y$ on the right.

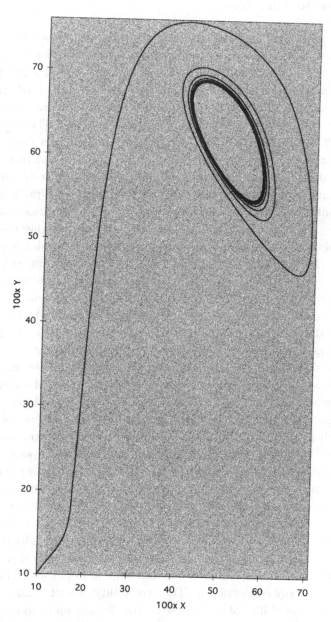

Fig. (2) If the initial point is far from the critical point then the orbit rapidly homes in to the limit cycle.

## 6.4   How to Add Noise

When a real system is subject to random interference we say that it suffers from *noise*. A mathematical model of the system without noise must be modified if it is to describe the noisy situation. In a classical theory of particles obeying Newton's laws, it is possible to add a random force to the deterministic force of a particle, whose motion is then governed by a stochastic differential equation, the Langevin equation. The nature of the random interference is modelled by choosing the random force appropriately. Various choices are white noise, coloured noise and thermal noise. One can also try 'multiplicative' noise, in which one of the dynamical variables is multiplied by a random term, or the adding of a random term to the Hamiltonian, with its consequent affect on the equations of motion. One of the interesting questions in the theory is then whether the addition of the noise has changed the salient features of the dynamics [Freidlin and Wentzell (1984); Kifer (1988)]. For example, if the system without noise exhibits a bifurcation to periodicity or chaos, is the value (of the control parameter) at which bifurcation occurs altered in the presence of noise? In statistical dynamics we have replaced the particle picture by the field picture, and the Newton's laws by a motion through the state space $\Sigma(\Omega)$, and the traditional description of noisy systems does not appear to be possible. In this section, we present a natural way to introduce noise into statistical dynamics, in a way that is consistent with the first and second laws of thermodynamics. Our construction is not obviously related to a stochastic differential equation; it is, however, similar to the theory of noisy neural nets [Amit (1989)], and this might justify our use of the term 'noise'. Suppose that the system is described by the sample space $\Omega_c$ and energy $\mathcal{E}_c$, and a dynamics $\tau$. Our procedure is to extend the system by introducing a heat-particle, and corresponding space $\Omega_\gamma$, with energy-function $\mathcal{E}_\gamma$, and an energy-conserving bistochastic map $T$ which gives the dynamics on the product space $\Omega = \Omega_c \times \Omega_\gamma$. Let $\tau_\beta$ be the corresponding isothermal dynamics of the combined system. If $\tau_\infty$ coincides with the original dynamics, $\tau$, then we say that $\tau_\beta$ is (one model of) the system with thermal noise. The noise allows transitions to occur that cannot go in the original model, because of energy conservation. The probability of such transitions is related to the probability of the inverse by the Boltzmann ratio $e^{-\beta W}$, where $W$ is the energy gap.

We now illustrate the idea with a study of the oscillating chemical reaction of the previous section. The equations for the activity-led isothermal reactions

$$B + X \overset{k_1}{\rightleftharpoons} Y + \gamma , \tag{6.26}$$

$$X \overset{k_2}{\rightleftharpoons} \gamma , \tag{6.27}$$

$$2X + Y \overset{k_3}{\rightleftharpoons} 3X + \gamma , \tag{6.28}$$

$$A \overset{k_4}{\rightleftharpoons} X + \gamma , \tag{6.29}$$

are

$$\frac{da_X}{dt} = \{k_1(a_Y a_\gamma - a_B a_X) - k_2(a_X - a_\gamma) + k_3(a_X^2 a_Y - a_X^3 a_\gamma)$$
$$+ k_4(a_A - a_X a_\gamma)\}(1 - a_X)^2 , \tag{6.30}$$

$$\frac{da_Y}{dt} = \{k_1(a_B a_X - a_\gamma a_Y) - k_3(a_X^2 a_Y - a_X^3 a_\gamma)\}(1 - a_Y)^2 , \tag{6.31}$$

which reduce to (6.16) when $a_\gamma = 0$. We take the same values of the parameters $k_i$ and $a_A$, $a_B$ as we had in the previous section, so that with $a_\gamma = 0$ we get a limit cycle. We also denote by $\overset{\circ}{a}_X$ and $\overset{\circ}{a}_Y$ the same values, $1/2$ and $5/8$ as before. Note that these are no longer the co-ordinates of a fixed point of the dynamics if $a_\gamma > 0$. The new fixed point will be close to this if $a_\gamma$ is small, by continuity. The correction to $(\overset{\circ}{a}_X, \overset{\circ}{a}_Y)$ will be a power series in $a_\gamma$ beginning with the first power. The matrix $A$ will also be modified, this time exactly by a matrix linear in $a_\gamma$, thus:

$$A(a_\gamma) = A + a_\gamma B.$$

This is because the equations of motion are linear in $a_\gamma$. We are thus left with problem of finding the new Lyapunov index for the new fixed point. I now show that we can find the Lyapunov index without finding the new fixed point, even up to order $a_\gamma$. Indeed, write as before,

$$a_X = \overset{\circ}{a}_X + X \qquad\qquad a_Y = \overset{\circ}{a}_Y + Y$$

and $\mathbf{Z} = (X, Y)$; then to first order in $\mathbf{Z}$, but exact in $a_\gamma$, the equations of motion take the form

$$\dot{\mathbf{Z}} = (A + a_\gamma B)\mathbf{Z} + a_\gamma \mathbf{b}.$$

Clearly, the fixed point has been displaced from $\mathbf{Z} = 0$ to

$$\mathbf{Z} = -a_\gamma A(a_\gamma)^{-1}\mathbf{b} = \mathbf{Z}_0$$

say. Then make a shift of dependent variable from $\mathbf{Z}$ to $\mathbf{Z}' = \mathbf{Z} - \mathbf{Z}_0$ and the resulting equation for $\mathbf{Z}'$ is homogeneous with the same matrix

$A(a_\gamma) = A + a_\gamma B$. If $a_\gamma$ is small, the eigenvalues of $A(a_\gamma)$ will be close to those of $A$, so they are complex conjugates of each other with the same real part. The Lyapunov index is $Tr\, A(a_\gamma)/2$, as can be seen by finding the solutions of

$$\det \begin{pmatrix} a - \lambda & b \\ c & d - \lambda \end{pmatrix} = 0 \tag{6.32}$$

in the case when there are complex roots. In our case, $Tr A = 0$, (the Lyapunov index for the system without noise was zero), so $\mathrm{Tr}\, (A + a_\gamma B) = a_\gamma Tr\, B$. It therefore suffices to find the diagonal elements of $B$, which means finding, respectively, the coefficient of $a_\gamma X$ on the right-hand side of (6.30) and the coefficient of $a_\gamma Y$ on the right-hand side of (6.31). Recall that $a_X = \mathring{a}_X + X$, $a_Y = \mathring{a}_Y + Y$. Exercise 6.37 shows that the sought terms are

$$a_\gamma X : (1 - \mathring{a}_X)^2 (-3k_3 \mathring{a}_X^2 - k_4)$$
$$-2(1 - \mathring{a}_X)(k_1 \mathring{a}_Y + k_2 - k_3 \mathring{a}_X^3 - k_4 \mathring{a}_X)$$
$$a_\gamma Y : (1 - \mathring{a}_Y)[-k_1(1 - \mathring{a}_Y) + 2k_1 \mathring{a}_Y - 2k_3 \mathring{a}_X^3]$$

which gives for the matrix $B$, using the values of the model:

$$B = \begin{pmatrix} -149a_\gamma/32 & \bullet \\ \bullet & 57a_\gamma/64 \end{pmatrix}. \tag{6.33}$$

Thus the Lyapunov index for the noisy system is $\lambda = (-149/32 + 57/64)a_\gamma/2$, which is negative. Thus the new fixed point has been converted into an attractor by the noise. This is illustrated in Fig. (3); we see that the orbit spirals in, and that as functions of time the activities of $X$ and $Y$ decay to their values at the fixed point.

This model shows that we can get a dynamic phase transition: if the parameters were not quite those of the model, but were such that when $a_\gamma = 0$ the Lyapunov index was small and positive, then at zero temperature the fixed point would be a repeller, and the system would spiral out and away for large times. As we increase the temperature, the Lyapunov index would decrease until it reached zero, beyond which it is an attractor. This defines a critical temperature separating two different asymptotic behaviour of the system, at least for motions beginning near the fixed point.

## 6.5  Exercises

Exercise 6.34. In the driven chemical activity-led reactions given by (6.16), show that the condition for a limit-cycle is

$$k_3 k_4^2 a_A^2 k_2^{-2}(1 - \overset{\circ}{a}_Y)^2 = (k_1 a_B - k_2)(1 - \overset{\circ}{a}_X)^2 \qquad (6.35)$$

and that the eigenvalues are then $\pm(1 - \overset{\circ}{a}_X)(1 - \overset{\circ}{a}_Y)k_4 a_A (k_3/k_2)^{1/2} i$.

Exercise 6.36. Consider the model obtained from that in Sec. 6.3 by replacing $2X + Y \rightleftharpoons 3X + \gamma$ by $X + Y \rightleftharpoons 2X$. Show that the two eigenvalues of the linearised equation are real.

Exercise 6.37. For the noisy model of Sec. 6.4, show that the matrix $B$ has diagonal elements $-149 a_\gamma/32$ and $57 a_\gamma/64$.

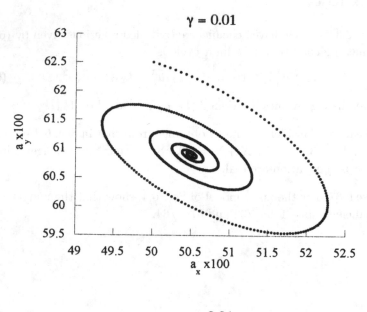

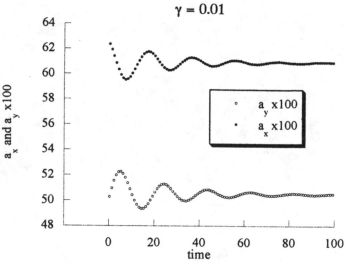

Fig. (3) In the presence of noise, the limit cycle has disappeared and the neutral fixed point has become an attractor. The upper picture shows the orbit, and the lower picture shows the time development. The value of $a_\gamma$ is 0.01.

# Chapter 7

# Fluid Dynamics

In this new chapter we shall derive the equations of fluid dynamics. The dynamics of the Brownian particle are sometimes obtained as a limit of the full dynamics of a compressible fluid, but this is not very convincing. It is not enough to argue that this is because the Smoluchowski equation of motion is a poor approximation to such complicated dynamics. Indeed, we shall find that the Smoluchowski equations come from a reasonable limit of the dynamics of the anterior theory, the hard-core gas. The diffusion term would arise as a kind of Ito correction to the continuity equation for the density. This correction is small and is omitted from the approximations leading to the Navier-Stokes equations. It is then not easy to reinstate the term; this explains the convoluted arguments of some derivations of the Smoluchowski equations. In our theory, we get the compressible Navier-Stokes equations with temperature; this leads to the Smoluchowski equation by putting the velocity to zero. The heat equation is also obtained in the same way.

The usual theory has two problems; the first is that the continuity equation has no diffusion term, and the second is that the fluid is incompressible, as expressed by the assumption that $\partial_j u_j = 0$. With these assumptions, Leray showed that there exist global weak solutions [Leray (1934)]; here, 'global' means valid for all positive time, and 'weak' means that the solution holds after smearing a version of the equation by a smooth test-function. If the initial data are small enough, there exist global strong solutions, which are unique [Fujita and Kato (1964)]. Uniqueness possibly fails if one has only a weak solution. It was also proved that there exist some strong solutions with large initial data, namely in a neighbourhood of data that are close to symmetrical [Ponce, Racke, Titi and Sideris (1994)]. A prize has been proposed for a proof of strong solutions for general initial data, or for a counter-example.

We suggest that our system, Eqs. (7.55)–(7.57), should have global strong solutions for any smooth initial conditions of compact support. This is in contrast with the usual Navier-Stokes system with temperature, which may have weak solutions but possibly not strong solutions if the initial data are large and not near a solution with symmetry.

## 7.1 Hydrostatics of a Gas of Hard Spheres

We take space to be $\Lambda \subseteq (a\mathbf{Z})^3$, and suppose the length $a$, representing the diameter of a molecule, to be so small compared with the variation of the macroscopic fields that we can replace all sums over $\Lambda$ by integrals. The possible configurations of the fluid are the points in the product sample space

$$\Omega = \prod_{x \in \Lambda} \Omega_x,$$

so a configuration is specified by the collection $\{\omega_x\}_{x \in \Lambda}$. For each $x$,

$$\Omega_x = \left\{ \emptyset, (\epsilon\mathbf{Z})^3 \right\}.$$

Here, $\epsilon$ is a small parameter having the dimension of momentum. If the system is in a configuration $\omega$, such that $\omega_x = \emptyset$, then we say that the site $x$ is empty. If $\omega_x = k$, we say that the site $x$ is occupied, by a particle of momentum $k$. This simple exclusion of more than one particle on each site incorporates the hard-core repulsion between the particles, which are thus hard spheres sitting at some of the points of $\Lambda$.

The *state* of the system is a probability on $\Omega$, denoted by $\mu$. We denote the set of states by $\Sigma$. The 'slow variables' of our model are the 5 extensive conserved random fields

$$\mathcal{N}_x(\omega) = \begin{cases} 0 & \text{if } \omega_x = \emptyset \\ 1 & \text{if } \omega_x = k \end{cases}$$

$$\mathcal{E}_x(\omega) = \begin{cases} 0 & \text{if } \omega_x = \emptyset \\ k \cdot k/2m + \Phi(x) & \text{if } \omega_x = k \end{cases}$$

$$\mathcal{P}_x(\omega) = \begin{cases} 0 & \text{if } \omega_x = \emptyset \\ k & \text{if } \omega_x = k \end{cases}$$

Here, $\Phi(x)$ is the external potential energy per particle. The slow variables appearing in hydrodynamics are the $n = 5|\Lambda|$ means in the state $\mu$:

$$N_x = \mathbf{E}_\mu[\mathcal{N}_x]; \qquad E_x = \mathbf{E}_\mu[\mathcal{E}_x]; \qquad \mathbf{\Pi}_x = \mathbf{E}_\mu[\mathcal{P}_x].$$

The (von Neumann) entropy of any state $\mu$ is

$$S(\mu) := -k_B \sum_\omega \mu(\omega) \log \mu(\omega). \tag{7.1}$$

In information geometry, the choice of slow variables $\{X_1, X_2 \ldots, X_n\}$ defines the *information manifold* $\mathcal{M}$, which consists of states of maximum entropy among all states with given means, say

$$\mathbf{E}_\mu[X_j] := \mu \cdot X_j = \eta_j, \qquad\qquad j = 1, \ldots, n.$$

By the Gibbs-Jaynes principle, such a state has the form

$$\mu(\omega) = \exp \left\{ -\sum_{j=1}^n \xi_j X_j(\omega) - \log \Xi \right\}$$

where the dual, or *canonical*, variables $\xi_j$ are Lagrange multipliers determined uniquely by the means $\eta_i$. In our case, for each $\boldsymbol{x}$, the duals to energy, number, and momentum are, respectively, $\beta_{\boldsymbol{x}}, \xi_{\boldsymbol{x}}, \zeta_{\boldsymbol{x}}$, and so the state has the form

$$\mu(\omega) = \prod_{\boldsymbol{x}} \Xi_{\boldsymbol{x}}^{-1} \exp \left\{ -\xi_{\boldsymbol{x}} \mathcal{N}_{\boldsymbol{x}}(\omega) - \beta_{\boldsymbol{x}} \mathcal{E}_{\boldsymbol{x}}(\omega) - \zeta_{\boldsymbol{x}} \cdot \mathcal{P}_{\boldsymbol{x}}(\omega) \right\}. \tag{7.2}$$

Such a state is said to be in *local thermodynamic equilibrium, LTE*. Equilibrium holds when $\beta$... are independent of $\boldsymbol{x}$. In [Streater (2000)] we found the (grand) partition function at point $\boldsymbol{x}$,

$$\Xi_{\boldsymbol{x}} = 1 + \left( \frac{2\pi m}{\epsilon^2 \beta_{\boldsymbol{x}}} \right)^{3/2} \exp \left\{ -\xi_{\boldsymbol{x}} - \beta_{\boldsymbol{x}} \Phi(\boldsymbol{x}) + m \, \zeta_{\boldsymbol{x}} \cdot \zeta_{\boldsymbol{x}} / 2\beta_{\boldsymbol{x}} \right\}$$

by replacing the sum over the momentum lattice of size $\epsilon$ by a Gaussian integral. The product structure of an *LTE* state means that an observable at a point of $\Lambda$ is independent of an observable at any other. Note that $\mu_{\boldsymbol{x}}\{\emptyset\} = (1 - N_{\boldsymbol{x}})$. It then follows from (7.2) that

$$1 - N_{\boldsymbol{x}} = \Xi^{-1}. \tag{7.3}$$

If $\omega_{\boldsymbol{x}} \neq \emptyset$, the state $\mu$ can be written in Maxwell form

$$\mu(\boldsymbol{x}, \boldsymbol{k}) = N_{\boldsymbol{x}} p(\boldsymbol{x}, \boldsymbol{k})$$
$$= N_{\boldsymbol{x}} Z^{-1} \exp\{ -\beta_{\boldsymbol{x}} \Phi(\boldsymbol{x}) - \beta_{\boldsymbol{x}} \boldsymbol{k} \cdot \boldsymbol{k} / (2m) - \zeta_{\boldsymbol{x}} \cdot \boldsymbol{k} \}$$

where $Z$ is the canonical partition function:

$$Z_x = \left( \frac{2\pi m}{\epsilon^2 \beta_{\boldsymbol{x}}} \right)^{3/2} \exp \left\{ -\beta_{\boldsymbol{x}} \Phi(\boldsymbol{x}) + \frac{m \zeta_{\boldsymbol{x}} \cdot \zeta_{\boldsymbol{x}}}{2\beta_{\boldsymbol{x}}} \right\}. \tag{7.4}$$

We note the identity for each $\boldsymbol{x}$

$$\Xi = 1 + e^{-\xi}Z = 1 + e^{-\xi-\beta\Phi}Z_0,$$

where $Z_0$ is the canonical partition function when $\Phi = 0$. The external potential does not influence the local velocity distribution $p$, as it is cancelled out by the partition function. The mean fields are related to the canonical fields by

$$E_{\boldsymbol{x}} = -\frac{\partial}{\partial\beta_{\boldsymbol{x}}}\log\Xi_{\boldsymbol{x}} = N(\boldsymbol{x})\left(\Phi(\boldsymbol{x}) + \frac{3}{2\beta_{\boldsymbol{x}}} + \frac{m\zeta_{\boldsymbol{x}}\cdot\zeta_{\boldsymbol{x}}}{2\beta_{\boldsymbol{x}}^2}\right)$$

$$N_{\boldsymbol{x}} = -\frac{\partial}{\partial\xi_{\boldsymbol{x}}}\log\Xi_{\boldsymbol{x}} = \frac{\Xi_{\boldsymbol{x}} - 1}{\Xi_{\boldsymbol{x}}} = \frac{Ze^{-\xi\boldsymbol{x}}}{1 + Ze^{-\xi\boldsymbol{x}}}$$

$$\Pi_{\boldsymbol{x}}^i = -\frac{\partial}{\partial\zeta_i}\log\Xi_{\boldsymbol{x}} = -\frac{mN_{\boldsymbol{x}}\zeta_{\boldsymbol{x}}^i}{\beta_{\boldsymbol{x}}}.$$

The hydrodynamic variables are the mass-density $\rho = mNa^{-3}$, the velocity field $\boldsymbol{u} = -\boldsymbol{\zeta}/\beta$, the temperature $\Theta = (k_B\beta)^{-1}$, and the potential energy per unit mass, $\phi = \Phi/m$. We shall therefore eliminate $\xi$, $\beta$, $\zeta^j$ in favour of $\rho$, $\Theta$, $u^j$. Only $\xi$ remains to be found:

$$e^{-\xi\boldsymbol{x}} = Z_{\boldsymbol{x}}^{-1}N_{\boldsymbol{x}}/(1 - N_{\boldsymbol{x}}),\tag{7.5}$$

or

$$\xi = -\beta\Phi - \log N + \log(1 - N) + \log Z_0\tag{7.6}$$

where $Z_0$ is the canonical partition function when $\Phi = 0$. We need its gradient:

$$\partial_j\xi = \frac{3\partial_j\Theta}{2\Theta} - \frac{m}{2k_B}\frac{u^iu^i}{\Theta^2}\partial_j\Theta + \frac{mu^i}{k_B\Theta}\partial_ju^i$$
$$-\frac{\partial_jN}{N} - \frac{\partial_jN}{1-N} - \frac{\partial_j\Phi}{k_B\Theta} + \frac{\Phi}{k_B}\frac{\partial_j\Theta}{\Theta^2}.\tag{7.7}$$

We denote by $E, N$ and $\boldsymbol{\Pi}$ the total values of the mean energy, number and momentum; then (7.1) gives for the entropy at equilibrium

$$\Theta S(\mu) = E + k_B\Theta\xi N - \boldsymbol{u}\cdot\boldsymbol{\Pi} + k_B\Theta\log\Xi.\tag{7.8}$$

Compare this with the thermostatic formula

$$\Theta S = E + k_B\Theta\xi N - \boldsymbol{u}\cdot\boldsymbol{\Pi} + PV\tag{7.9}$$

where $P$ is the pressure and $V$ is the volume; we see that

$$P = k_B\frac{|\Lambda|}{V}\Theta\log\Xi = -k_B\Theta a^{-3}\log(1 - N).\tag{7.10}$$

## 7.2 The Fundamental Equation

Our model for the dynamics is a path $\{\mu(t)\}_{t\geq 0}$ through $\Sigma(\Omega)$ determined by the Liouville motion, interrupted by a thermalisation at random points $\boldsymbol{x}$, occurring at random times. After a thermalisation at $\boldsymbol{x}$, the state changes to one in which observables at $\boldsymbol{x}$ are statistically independent of those at any $\boldsymbol{y} \neq \boldsymbol{x}$, and, restricted to $\Omega_{\boldsymbol{x}}$, the state is in thermodynamic equilibrium. The law for the random time between collisions is of exponential form, but the rate of the process depends on the local density of the gas, and its temperature, and so on the state itself. This means that the dynamics falls outside the usual theory of Markov chains, in which the updated state is linear in the current state, and the transition matrix is the same for all states; we are in the business of non-linear Markov chains. Non-linearity itself is not the problem; the ultimate goal of this work is to put an external potential $\Phi$ into our version [Streater (2000)] of the Navier-Stokes equations, which are non-linear. However, we are going to use the concept of conditional probability to derive the master equation, and care is needed if we stray from a path in $\Omega$ to a path in $\Sigma(\Omega)$ which does not come from a path in $\Omega$. We shall adopt the following way out, which makes use of the assumption that the density is low, and so the collision probability is small. In calculating the probability per unit time that a configuration $\omega$ at time $t \in (t_0 - t, t_0)$ move to another point of $\Omega$, we shall assume that the hydrodynamic parameters satisfy the Euler equations, (7.28)-(7.30) in the small time interval $(t_0 - T, t_0)$. The initial values of the hydrodynamic parameters in these equations are taken to be those of the true state at time $t_0 - T$. Here we only need to consider $T < 20t_\ell$, since the survival probability falls exponentially. We then assume that the Markov process between $t_0 - T$ and $t_0$ is linear as usual, and takes place in an ambient Euler fluid which determines the rate of thermalisation. We show that, neglecting collisions, the means of the slow variables do indeed satisfy the Euler equations, showing self-consistency. We can then show that there is no production of entropy in this case.

The Liouville dynamics is determined as follows. If at time $t = 0$ a configuration $\omega \in \Omega$ has a particle at $\boldsymbol{x}$, then it follows Newton's laws

$$\frac{d\boldsymbol{x}}{dt} = \frac{\boldsymbol{k}}{m}, \qquad \frac{d\boldsymbol{k}}{dt} = -\nabla\Phi := m\boldsymbol{f} \qquad (7.11)$$

for a time. Before we take the continuum limit, space is $\Lambda$, which is discrete; a particle following (7.11) will nearly always leave the lattice. In that case, after any given random time $t$ after which a thermalization occurs, we place the resulting thermalized particle at the lattice site nearest to $x(t)$, ties being decided by tossing a coin. In the limit $a \to 0$ we expect this to introduce a negligible error in the location being assigned to the particle.

Let $\mu \in \Sigma(\Omega)$ and define $N_x = \mu\{\omega_x \neq \emptyset\}$, the probability that $x$ is occupied in the state $\mu$. Then $N_x = \mu \cdot \mathcal{N}_x$. Let $p_x(k)$ be the conditional probability that $\omega_x = k$, given that $\mathcal{N}_x = 1$:

$$p_x(k) = \mu\{\omega : \omega_x = k | \omega_x \neq \emptyset\} = N_x^{-1}\mu\{\omega : \omega_x = k\}. \qquad (7.12)$$

Thus,

$$\mu\{\omega : \omega_x = k\} = N_x p(x, k). \qquad (7.13)$$

This does not mean of course that $\mathcal{N}_x$ and $\mathcal{P}_x$ are independent in the state $\mu$.

Given that no collisions occur, the particles obey (7.11), and this induces the Liouville motion on the states, namely, Boltzmann's equation with no collision term:

$$\frac{\partial \mu(x, k)}{\partial t} + \frac{k}{m} \cdot \partial \mu(x, k) + f \cdot \nabla_k \mu(x, k, t) = 0. \qquad (7.14)$$

These dynamics of the probabilities do not correspond to an underlying motion in $\Omega$ for all time. We can find initial conditions for which two particles both arrive within a distance $a/2$ from the same point at the same time. Such a configuration does not lie in $\Omega$. Our assumptions of no collisions is true with high probability for a few time steps, but collisions are almost sure to occur eventually. We will replace the problem of tracking which collisions actually occur if we follow (7.11) by Boltzmann's idea of introducing a probability for collisions. Instead of giving, as Boltzmann did, the probability that a pair of particles with given position and momentum produce another specified pair, it will be enough to assume that the collision is 100% efficient (in the terminology of [Boon and Rivet (2001)]). This means that after a collision at $x$, of a particle with momentum $k$, we replace this configuration by a particle at $x$ with momentum $k'$, with probability determined by the Maxwell distribution of mean momentum $k$ and mean energy $k \cdot k/(2m) + \Phi$. By construction, this does not alter the means of the slow variables. We shall refer to this event as a thermalisation, rather than as a collision. We shall assume that the mean free path is large on the molecular scale, and neglect the possibility that a snapshot of the lattice catch a particle in the process of thermalising: with high probability it will be in Newtonian motion between collisions. From any initial configuration $\omega$, we can follow the dynamics as time progresses, following this dynamics.

It is a random path through $\Omega$, that is, a process, defined for $t_0 - T \leq s \leq t_0$, and described by the family of points $\{\mu(s)\}_{t_0 - T \leq s \leq t_0}$.

Let $t$ denote the random time between collisions, whose law is determined by the conditional probability $w(\boldsymbol{x}, \boldsymbol{k}, t_0; t)dt$ that a particle, certainly at $\boldsymbol{x}$ at time $t_0$ with momentum $\boldsymbol{k}$, travel under Newton's laws a free time $t$ and thermalise in the interval $(t_0 + t, t_0 + t + dt)$. Since it will thermalise eventually,

$$\int_0^\infty w(\boldsymbol{x}, \boldsymbol{k}, t_0; t)dt = 1. \tag{7.15}$$

The important small parameter is the mean free time, also called the relaxation time, $t_\ell$:

$$t_\ell(\boldsymbol{x}, \boldsymbol{k}, t_0) := \int_0^\infty t\, w(\boldsymbol{x}, \boldsymbol{k}, t_0; t)\, dt. \tag{7.16}$$

This is invariant under the Galilean group, say $\mathcal{G}$, unlike the mean free path, which is related by $\ell = \boldsymbol{k} t_\ell / m$. For free particles, $\Phi = 0$, these were found in [Streater (2000)]. They are not affected by the potential, so we use the same values here. We also introduced

$$C(\boldsymbol{y}, \boldsymbol{k}, t)dt = \text{Prob}\{\text{particle at } (\boldsymbol{y}, \boldsymbol{k}, t) \text{ thermalize in } (t, t + dt)\}$$

$$W((\boldsymbol{x}, \boldsymbol{k}, t_0, t) = \text{Prob}\{\text{particle at } (\boldsymbol{x}, \boldsymbol{k}, t_0) \text{ survive until } t_0 + t\}.$$

Put $z(t) := (\boldsymbol{x}(t), \boldsymbol{k}(t)) := \tau_t(z)$, the solution to Newton's equations with initial point $z := (\boldsymbol{x}, \boldsymbol{k})$. Then we have the relation

$$w(\boldsymbol{x}, \boldsymbol{k}, t_0; t) = W(\boldsymbol{x}, \boldsymbol{k}, t_0; t)\, C(z(t), t + t_0) \tag{7.17}$$

and just as in [Streater (2000)] we get from (7.15),

$$W(\boldsymbol{x}, \boldsymbol{k}, t_0; t) = \exp\left\{ -\int_0^t C(z(t_1), t_0 + t_1)\, dt_1 \right\}. \tag{7.18}$$

As in [Streater (2000)], Eq. (36), $C$ is proportional to the density $\rho$. Thus if $\rho$ is bounded below by a positive number, we get an exponential decrease for $W$, the survival probability, along an orbit.

The fundamental equation will relate the probability distribution $\mu(z, t_0)$ at an arbitrary point $z$ in phase space at the time $t_0$ to a Maxwell distribution $\bar{\mu}$ at the same point, with small correction terms. Let $z(s)$ denote (for $s > 0$) the point in phase space on the backward Newtonian orbit of the point $z$ at time $t_0 - s$. For each sample path $\omega(\cdot)$, in which $z$ is occupied at time $t_0$, there is a unique time $t_0 - t$ at which the last thermalisation occurred, in that no further thermalisations took place on the orbit $\{z(s)\}_{0<s<t}$. It follows that the free time of the particle thermalised at $t_0 - t$ must have a free time $t'$ say, with $t' > t$.

In the following, 'event' will mean an event in the sample space (the path space of $\Omega$) of the process constructed above in the time interval $(t_0 - T, t_0)$. Let $E(s)$ denote the event, that there is a particle at $z(s)$ at time $t_0 - s$. We shall consider the conditional probabilities of collision and free motion along the orbit, given $E(0)$. We shall then recover a formula for $\mu(z(0))$ using Bayes's theorem. Liouville's theorem, that $d^3x\, d^3k$ is invariant under Newton's laws, enables us to deduce equations relating the density of probability from equations relating probabilities. For example we shall write $\Pr\{E(0)\} = \mu(\boldsymbol{x}, \boldsymbol{k}, t_0)$, referring to the densities.

Let $F(t)$ be the event, that $E(0)$ occurred, and the last collision occurred at time $t_0 - t$ producing a particle at the phase point $z(t)$. Then $F(t) \subseteq E(0)$, since the event produced exactly the right state, $z(t)$, which evolves to our point $z(0)$, as it undergoes no further collisions. Let $F(a, b)$ denote the event, that the last collision was at some $s$, $a \leq s \leq b$. If $(a, b) \cap (c, d) = \emptyset$, then $F(a, b)$ and $F(c, d)$ are disjoint. So, assuming smoothness, $F(t)$ has a probability density on $\mathbf{R}^+$, say $f(t)$. Let $G(t, t')$ be the event that $z(t)$ is occupied at time $t_0 - t$, and that the particle's free time is $t'$. Let $H(t_1)$ be the event that $F(s)$ occurred for some $s < t_1$. We shall need the formula

$$\Pr\{F(t)\}/\Pr\{H(t')\} = f(t) \Big/ \int_0^{t'} f(s)\,ds$$

$$= \frac{1}{t'}\left(1 + \frac{f'(0)}{f(0)}(t - t'/2)\right) + O(t'), \tag{7.19}$$

which is true if $f$ is smooth and $f(0)$ is not zero. We see that $G(t, t') \cap H(t') \subseteq E(0)$. Hence certainly

$$G(t, t') \cap F(t) \cap H(t') \subseteq E(0).$$

Hence

$$\Pr\{G(t, t') \cap F(t) \cap H(t') \cap E(0)\} = \Pr\{G(t, t') \cap F(t) \cap H(t')\}.$$

Thus

$$\Pr\Big\{G(t, t') \cap F(t) \cap H(t') \Big| E(0)\Big\}$$
$$:= \Pr\{E(0)\}^{-1}\Pr\{G(t, t') \cap F(t) \cap H(t') \cap E(0)\}$$
$$= \mu(z(0), t_0)^{-1}\Pr\{G(t, t') \cap F(t) \cap H(t')\}.$$

Let $\Pr_{H'}\{.\}$ denote the conditional probability of an event, given $H(t')$. Then from what we have just seen,

$$\Pr_{H'}\left\{G(t,t') \cap F(t)\middle| E(0)\right\}$$

$$:= \Pr\left\{G(t,t') \cap F(t) \cap H(t')\middle| E(0)\right\} \Pr\{H(t')\}^{-1}$$

$$= \Pr\{G(t,t') \cap F(t) \cap H(t')\}\mu^{-1} \Pr\{H(t')\}^{-1}.$$

Now,

$$\int_0^\infty dt' \int_0^{t'} dt \Pr_{H'}\left\{F(t) \cap G(t,t')\middle| E(0)\right\} = 1$$

as the last collision must have happened for some $t$ and some $t' > t$. Hence

$$\int_0^\infty dt' \int_0^{t'} dt \Pr\{G(t,t') \cap F(t) \cap H(t')\}\mu^{-1} \Pr\{H(t')\}^{-1} = 1.$$

As $\mu(z(0), t_0)$ does not depend on $t$ or $t'$, we get

$$\mu(z(0), t_0) = \int_0^\infty dt' \int_0^{t'} dt \frac{\Pr\{G(t,t') \cap F(t) \cap H(t')\}}{\Pr\{H(t')\}}.$$

We can omit $H$ from the numerator, as $t < t'$ is enforced by the region of integration. So

$$\mu = \int_0^\infty dt' \int_0^{t'} dt \frac{\Pr\{G(t,t') \cap F(t)\}}{\Pr\{H(t')\}}$$

$$= \int_0^\infty dt' \int_0^{t'} dt \frac{\Pr\{G(t,t')| F(t)\} \Pr\{F(t)\}}{\Pr\{H(t')\}}. \tag{7.20}$$

Also, if $t < t'$, then

$$\Pr\left\{G(t,t')\middle| F(t)\right\} = \overline{N}\overline{p}(z(t), t_0 - t)w(z(t), t_0 - t; t'), \tag{7.21}$$

(what we have been calling a collision is a thermalization). In applying (7.19), we note that the case $f(0) = 0$ corresponds to no collisions, so we may assume that $f(0) \neq 0$. We expand (7.21) up to first order in $t$, as $t \leq t'$, and $t' \leq T = 20t_\ell = O(t_\ell)$; this gives

$$\Pr\left\{G(t,t')\middle| F(t)\right\} = \overline{\mu}(z, t_0)w(z, t_0; t')$$

$$- t\left[\partial_i k_i/m + \partial_0 - (\partial_j \Phi)\frac{\partial}{\partial k_j}\right]\overline{\mu}(z, t_0)w(z, t_0; t').$$

Put this, and use (7.19), in (7.20) to get

$$\mu(z, t_0) = \int_0^\infty dt' \left[ \overline{\mu}(z, t_0) w(z, t_0, t') \right.$$

$$\left. - t'/2 \left( \frac{k_i \partial_i}{m} + \partial_0 - (\partial_j \Phi) \frac{\partial}{\partial k_j} \right) \overline{\mu}(z, t_0) w(z, t_0; t') \right]. \qquad (7.22)$$

The unknown term involving $f'/f$ does not appear, because $\int_0^{t'} (t - t'/2) dt$ vanishes, and $\int_0^{t'} (t - t'/2) t \, dt$ is of second order in $t'$, and so can be neglected. We can now do the integral over $t'$, using (7.15) and (7.16). This gives

$$\mu = \overline{\mu} - \frac{1}{2} \left( \frac{k_j \partial_j}{m} + \partial_0 - (\partial_j \Phi) \frac{\partial}{\partial k_j} \right) (\overline{\mu} \, t_\ell). \qquad (7.23)$$

This equation was derived in [Streater (2003)] using another method (aka guesswork), in the case when $\Phi = 0$.

It does not seem fruitful to seek accuracy up to and including $O(t_\ell^2)$. This would involve introducing unknown parameters, such as $f'/f$; worse, we would have to keep terms of second order in taking the continuum limit of the lattice; this introduces diffusion terms into the equations, which come from the Ito corrections to calculus. A similar extension of the work in [Chapman and Cowling (1970)], who start with the Boltzmann equation rather than a master equation, is generally agreed not to have been worth the effort. Keeping only terms of first order in $t_\ell$ leads to the surprising result that the equations of motion are determined, knowing only that $\overline{\mu}$ is in *LTE*. We do not need to assume, as is done in [Balescu (1997)], that the means of the slow variables in the approximating *LTE*-state $\overline{\mu}$ are the same as the true means, in the state $\mu$. Indeed, this turns out not to be the case; the reason is that $\overline{\mu}$ is conditioned by the fact that a thermalisation has occurred, and is not just the maximum entropy estimate of $\mu$. Then $\mu(\boldsymbol{x}, t_0)$ is the sum of such terms from various nearby points $(\boldsymbol{x} - \boldsymbol{k}t/m, t_0 - t)$ and for a given $\boldsymbol{k}$, all the contributions are from one side of $\boldsymbol{x}$, so the means should differ unless the state is constant in space and time. If we neglect the difference between $\mu$ and $\overline{\mu}$, which is the cause of dissipation in the Navier-Stokes equations, we arrive at the Euler equations, and show that they are free of dissipation, as expected.

## 7.3    The Euler Equations

The Euler equations follow from the approximation of zeroth order, in which the difference between $\mu$ and $\bar{\mu}$ is neglected. Mean fields for states in *LTE* are computable in terms of Gaussian integrals.

The velocity field of a particle is the random field $\Upsilon_{\boldsymbol{x}} := \boldsymbol{P_x}/m$, and the mean current of a random variable $\chi$, conserved or not, is, in the continuum, low-density limit

$$\mathbf{J}_\chi = \int d^3k\, \mu \boldsymbol{\Upsilon} \chi. \tag{7.24}$$

In the following equations, we use $u_i$, $i = 1,2,3$ to denote a three-vector, and $\partial_i$ to denote the $i$-th component of $\nabla$; repeated indices, such as $u_i u_i$, will imply that we sum the repeated index from 1 to 3. Our assumption is that on thermalisation there is no change in the means of $\mathcal{N}_{\boldsymbol{x}}$, $\boldsymbol{P_x}$ or $\mathcal{E}_{\boldsymbol{x}}$. Since the space integrals of $N_{\boldsymbol{x}}$ and $E_{\boldsymbol{x}}$ are to be conserved in time, their equations of motion are of the form

$$\dot{\rho} + \partial_j J_\rho^j = 0 \tag{7.25}$$

$$\dot{E} + \partial_j J_E^j = 0. \tag{7.26}$$

Momentum is not conserved; in a small volume at $\boldsymbol{x}$ the momentum $\boldsymbol{k}$ of a particle in time $dt$ is changed to $\boldsymbol{k} - \nabla\Phi\, dt$, so the momentum obeys

$$\frac{d\varpi^i}{dt} + \partial_j J_{\varpi^i}^j + \frac{\rho}{m}\partial_i\Phi = 0. \tag{7.27}$$

In [Streater (2000)] we showed that we can evaluate the currents (7.24) of hydrodynamics exactly if $\mu$ is in *LTE*, with parameters $\rho, u^i, \Theta$ say. By the same method, (7.25)-(7.27) can be written as:

$$\frac{d\rho}{dt} + \partial_j \left(\rho u_j\right) = 0 \tag{7.28}$$

$$\frac{d}{dt}\left(\rho u_i\right) + \partial_j\left(\rho u_i u_j\right) + \partial_i P + \rho\partial_i\phi = 0 \tag{7.29}$$

$$\frac{d}{dt}\left\{\rho\left(\frac{3k_B}{2m}\Theta + \frac{1}{2}u_j u_j + \phi\right)\right\}$$
$$+\partial_j\left\{\rho u_j\left(\frac{3k_B}{2m}\Theta + P/\rho + \frac{1}{2}u_i u_i + \phi\right)\right\} = 0. \tag{7.30}$$

Here, $\phi = \Phi/m$. These are the Euler equations. Note that the pressure $P$ appearing in these equations seems to that of an ideal gas, not the equilibrium pressure of the model; the two differ by a term which goes to zero as the density goes to zero. This reflects that the derivation of the fundamental equation depends on the assumption that the density is small. We shall see that it is better to replace the ideal pressure with the equilibrium pressure given by (7.10); this has the effect that the Euler equations do not change the entropy as time goes by.

## 7.4  Entropy Production

We find from von Neumann's formula,

$$S := -\sum_{\omega} \mu(\omega) \log \mu(\omega) \qquad (7.31)$$

that the following holds:

$$\dot{S} = -\sum_{\omega} \dot{\mu}(\omega) \log \mu(\omega) - \sum_{\omega} \dot{\mu}(\omega). \qquad (7.32)$$

The last term is zero, so we may regard $-\log \mu$ as a random variable $\mathcal{S}$ whose average rate of change is given in the Schrödinger picture by $\dot{\mu} \cdot \mathcal{S}$. The current of the entropy would then be, per site

$$a^3 \mathcal{J}_s^j(x) = \mathcal{S}(\omega)\Upsilon^j = \xi\mathcal{N}\Upsilon^j + \beta\mathcal{E}\Upsilon^j + \zeta^i\mathcal{P}_i\Upsilon^j + \log\Xi\,\Upsilon^j, \qquad (7.33)$$

when $\mu = Np$ and $p$ is in *LTE*.

We must clarify the definition of $\Upsilon^i(\omega)$ when $\omega = \emptyset$, as zero is not a good definition; the information in there being no particles at $x$ is carried along with the general fluid. Let us define $\Upsilon^i(\emptyset) := u^i$ when the state $\mu$ has velocity $u^i$. In that case, the mean of $\Upsilon^i$ in the state $\mu$ is

$$\mu(\emptyset)u^i + \sum_{k} Np \cdot \Upsilon^i = (1 - N)u^i + Nu^i = u^i.$$

The mean density of the entropy current, $J_s^j$, in the state $\mu$ then has three terms:

$$\xi a^{-3}\mu \cdot (\mathcal{N}\Upsilon^j) + a^{-3}\log\Xi\,\mu \cdot \Upsilon^j = \frac{\xi}{m}J_\rho^j - \log(1 - N)u^j$$

$$a^{-3}\zeta^i\mu \cdot (\mathcal{P}_i\Upsilon^j) = \zeta^i J_{\varpi_i}^j$$

$$a^{-3}\beta\mu \cdot (\mathcal{E}\Upsilon^j) = \beta J_{E}^j.$$

We expect that entropy will be conserved, because of Kossakowski's argument [Kossakowski (1969)]: in the absence of collisions, the projections onto the information manifold do not create any entropy if time is continuous. See also [Balian (2006a)]. Without collisions, Boltzmann's equation also gives zero entropy production, if the assumption of *LTE* is made. So we expect that the rate of change of entropy is balanced by the outflow through the boundary:

$$\dot{s} + \partial_j \left( \frac{\xi}{m} J_\rho^j + \zeta^i J_{\varpi_i}^j + \beta J_E^j - \log(1-N)u^j \right) = 0, \qquad (7.34)$$

where $s$ denotes the entropy density. On the other hand, we have

$$a^3 s = \xi N + \zeta^i \Pi_i + \beta E + \log \Xi,$$

giving

$$a^3 \frac{ds}{dt} = N \partial_t \xi + \Pi_i \partial_t \zeta^i + E \partial_t \beta + \frac{1}{\Xi} \left( \frac{\partial \Xi}{\partial \xi} \partial_t \xi + \frac{\partial \Xi}{\partial \zeta^i} \partial_t \zeta^i + \frac{\partial \Xi}{\partial \beta} \partial_t \beta \right)$$

$$+ \xi \partial_t N + \zeta^i \partial_t \Pi_i + \beta \partial_t E. \qquad (7.35)$$

Now,

$$\frac{1}{\Xi} \frac{\partial \Xi}{\partial \xi} = -N \text{ etc.}$$

so the first line of (7.35) vanishes, to leave

$$a^3 \partial_t s = \xi \partial_t N + \zeta^i \partial_t \Pi_i + \beta \partial_t E.$$

We divide by $a^3$ to get the densities $\rho$, $a^{-3}E$ and $\varpi_i$, and the relation

$$\int_\Lambda \partial_t s \, d^3 x = \int_\Lambda \left( \frac{\xi}{m} \partial_t \rho + \zeta^i \partial_t \varpi_i + \beta \partial_t E/a^3 \right) d^3 x \qquad (7.36)$$

$$= -\int_\Lambda \left( \frac{\xi}{m} \partial_j J_\rho^j + \zeta^i \left( \partial_j J_{\varpi_i}^j + \rho \delta_{ij} \partial_j \phi \right) + \beta \partial_j J_E^j \right)$$

$$= \int_\Lambda \left( \zeta^i \rho f_i + \partial_j \xi J_\rho^j + \partial_j \zeta^i J_{\varpi_i}^j + \partial_j \beta J_E^j \right) \qquad (7.37)$$

$$- \oint_{\partial \Lambda} \left( \xi J_\rho^n + \zeta^i J_{\varpi_i}^n + \beta J_E^n \right) d\sigma_n \qquad (7.38)$$

where $n$ denotes the normal direction to the boundary. The term in (7.37) involving $f_i = -\partial_i \Phi/m$ represents the free-energy change due to the work done by the external field on the particles; we shall see that this is cancelled out by another term, showing that this work done is not thermalised by the dynamics. In spite of the apparent Onsager form of (7.37), we cannot identify it as the entropy production and (7.38) as the flow through the boundary, because the surface term differs from the entropy current, as in (7.34) by $-\log(1-N)u^j$. Indeed, if we put the Euler currents in (7.37) we do not get zero, even when $\Phi = 0$. We must therefore add and subtract the missing term, to get

$$\partial_t s = \int_\Lambda \left( \zeta^i \rho f_i + \partial_j \xi J_\rho^j + \partial_j \zeta^i J_{\varpi_i}^j + \partial_j \beta J_E^j \right) d^3 x$$

$$\int_\Lambda -\partial_j (a^{-3} \log(1 - N) u_j) d^3 x \qquad\qquad (7.39)$$

$$-\oint_{\partial \Lambda} J_s^i d\sigma_i. \qquad\qquad (7.40)$$

If we now put in the Euler currents, (7.28)-(7.30), we still do not find exactly zero for the entropy production, (7.39). Indeed, the entropy production involves the logarithm, whereas the Euler equations are polynomial, so cancellation is not possible. We find that (7.37) vanishes up to terms $O(N^2)$, that is, in the limit of low density. This reflects the low-density assumption inherent in the axiom that the *BBGKY* currents, (7.24), are the actual currents of particles flowing out of the region $d^3 x$. While this is a reasonable model for point-like particles, it neglects the fact that if the lattice-site just beyond the boundary is occupied, then a particle moving out of the region will not be able to land, as no configuration in the sample space can have two particles at the same site. Our dynamics did not need to specify the rule as to what will happen, as (in the low density limit) the probability that the point is occupied is small.

It turns out that if we modify the pressure in the Euler equations, to be the equilibrium pressure of the gas with hard core, (7.10), rather than that of the ideal gas, then the entropy production is exactly zero, in conformity with Kossakowski's theorem:

**Theorem 7.1 (Entropy conservation in Euler's equations).**
*Take the pressure $P$ in (7.29) and (7.30) to be that of the hard-core gas at equilibrium, (7.10). Then the entropy production, (7.39), is zero.*

**Proof.** We first recall the relation between the canonical variables and the hydrodynamic variables. We saw that

$$\beta = \frac{1}{k_B \Theta} \qquad\qquad \zeta^j = -\beta u^j = -\frac{u^j}{k_B \Theta},$$

so from (7.7) we get for the entropy production (7.39):

$$\dot{s}_{\text{prod}} = \zeta^i \rho f_i$$

$$+ \frac{\rho u^j}{m}\left(\frac{3}{2}\frac{\partial_j \Theta}{\Theta} - \frac{m}{2k_B}\frac{u^i u^i}{\Theta^2}\partial_j\Theta + \frac{m}{k_B}\frac{u^i}{\Theta}\partial_j u^i\right.$$

$$\left. - \frac{\partial_j N}{N} - \frac{\partial_j N}{1-N} - \frac{m}{k_B}\frac{\partial_j \phi}{\Theta} + \frac{m}{k_B}\frac{\phi}{\Theta^2}\partial_j\Theta\right)$$

$$+ (\rho u_i u_j + \delta_{ij}P)\left(-\frac{\partial_j u^i}{k_B\Theta} + \frac{u^i}{k_B}\frac{\partial_j\Theta}{\Theta^2}\right)$$

$$+ \rho u_j\left(\frac{3k_B}{2m}\Theta + \frac{P}{m} + \frac{1}{2}u^i u^i + \phi\right)\frac{-\partial_j\Theta}{k_B\Theta^2}$$

$$+ a^{-3}\frac{\partial_j N}{1-N}u^j - a^{-3}\log(1-N)\partial_j u^j$$

$$= 0.$$

Thus as claimed, there is no production of entropy in the dynamics defined by the discrete Euler equations. □

The Euler equations have not been shown to possess smooth solutions for all time, unless the initial values are sufficiently small; they may be unstable, in the sense that small changes in the initial conditions might lead to solutions which differ by large amounts. It is better, then, to include dissipation in the equations, and this will be discussed below.

We now give the short Euler equations. For any function of $x$ and $t$, let

$$D_t = \frac{\partial}{\partial t} + u_j\frac{\partial}{\partial x_j}.$$

This is the Galilean covariant time derivative. We can use the Euler equation (7.28) once in the Euler equation (7.29) to simplify it. We can use (7.28) twice and (7.29) once in the Euler equation (7.30) to simplify it too. This gives us the five 'short Eulers':

$$D_t\rho + \rho\partial_j u_j = 0 \tag{7.41}$$

$$\rho D_t u_i + \partial_i P + \rho\partial_i\phi = 0, \qquad i = 1,2,3 \tag{7.42}$$

$$\frac{3k_B}{2m}D_t\Theta + \rho\frac{d\phi}{dt} + P\partial_j u_j = 0. \tag{7.43}$$

## 7.5 A Correct Navier-Stokes System

The law of conservation of entropy, which holds for the Euler equations, means that there is no dissipation in the solution, as time goes by. This fact is probably why the equations are very hard to solve; let us take initial values of the variables $\rho, u, \Theta$ to be smooth functions of $x$; can we find smooth solutions for all time? For various choices of the initial functions, numerical instabilities soon show up. This gives rise to one of the unsolved problems in the subject: can we prove that there are smooth solutions for all choices of smooth bounded initial conditions, or, conversely, can we exhibit a smooth bounded initial condition which leads to a singularity in a finite time? This, however, is an artificial problem. Fourier added a term to the energy equation which gives rise to the transfer of heat as time goes by; and Navier added viscosity terms to equations which express the law of conservation of momentum, which meant that a non-constant function $v(x, t)$ changed value as time goes by, due to the viscosity, while the total momentum remained conserved. These equations were extended by Stokes, who in 1856 [Stokes (1856)] also removed Navier's assumption that the fluid was composed of atoms; this unjustified assumption (at the time) was replaced by a derivation based on the assumptions of continuum physics. The equations were then dubbed the Navier-Stokes system.

The Navier-Stokes equations thus express the five laws of conservation, of particle-number, momentum and energy, as given by the time-derivatives of $\rho, p, \Theta$ in terms of their space-derivatives, up to second order. These equations are non-linear. One can show that a solution to the system has the property that the entropy is non-decreasing in time. However, it has not been possible, so far, to show that a strong smooth solution exists for all time. Indeed, this has been published as one of the seven millenium problems by the Clay Institute. Of course, the derivation of the Navier-Stokes system from an assumed solution of the Boltzmann equation is no proof that there are any solutions to the Navier-Stokes system: no global smooth solutions to the Boltzmann equation with general smooth initial conditions have been shown to exist either.

Yau [Yau (2000)] has derived the continuity equation from quantum mechanics in the form

$$\frac{\partial \rho}{\partial t} + \partial_j(\rho u_j) = K\nabla^2\rho, \tag{7.44}$$

where $K$ is positive. Unsatisfied with this, he took the scaling limit, by rescaling space and time, to remove the diffusive term $K\nabla^2\rho$, because he had set himself the task of deriving the usual system of equations. It has been said that the requirements of Galilean invariance and conservation of mass or particle number together lead to $K = 0$, but this is not so. Thus, consider the Galilean transformation

$$x' = x + vt \tag{7.45}$$

$$t' = t \tag{7.46}$$

where $v$ is a constant three-vector. A function $f$ of $x,t$ becomes a function $g$ of $x', t'$, as in the equation

$$g(x',t') = f(x(x',t'), t(x',t')) = f(x' - vt', t'),$$

so that, by the rules of calculus,

$$\frac{\partial g}{\partial x'_j} = \frac{\partial f}{\partial x_i}\frac{\partial x_i}{\partial x'_j} + \frac{\partial f}{\partial t}\frac{\partial t}{\partial x'_j} \tag{7.47}$$

$$\frac{\partial g}{\partial t'} = \frac{\partial f}{\partial x_i}\frac{\partial x_i}{\partial t'} + \frac{\partial f}{\partial t}\frac{\partial t}{\partial t'}. \tag{7.48}$$

The inverse map to the Galilean transformation (7.45) is

$$x = x' - vt' \tag{7.49}$$

$$t = t'. \tag{7.50}$$

This gives

$$\frac{\partial}{\partial x'_i} = \delta_{ij}\frac{\partial}{\partial x_j} = \frac{\partial}{\partial x_i} \tag{7.51}$$

$$\frac{\partial}{\partial t'} = \frac{\partial}{\partial x_i}(-v_i) + \frac{\partial}{\partial t}. \tag{7.52}$$

It follows from this that the Laplacian, $\frac{\partial}{\partial x_i}\frac{\partial}{\partial x_i}$ is invariant under Galilean transformations, and that the differential form of the equation,

$$\frac{\partial}{\partial t} + u_j\frac{\partial}{\partial x_j} = \frac{\partial}{\partial t'} + v_i\frac{\partial}{\partial x'_i} + u_j\frac{\partial}{\partial x'_j} = \frac{\partial}{\partial t'} + (u_i + v_i)\frac{\partial}{\partial x'_i},$$

takes the form of the continuity equation in the co-ordinates $x', t'$. Thus, the diffusive continuity equation is covariant under the Galilean group. Similarly, we may replace the diffusion terms chosen here by any term invariant under the Galilean group. Thus, we might consider the equation

$$\frac{\partial \rho}{\partial t} + \frac{\partial}{\partial x_i}(u_i\rho) = k\frac{\partial}{\partial x_i}\left[\rho^{-1}\frac{\partial}{\partial x_i}\left(\Theta^{\frac{1}{2}}\rho\right)\right]. \tag{7.53}$$

It should be stressed that there is no extra term at all if the system is in local equilibrium; then the distribution of the momentum is Gaussian. The dual variable to the momentum is then proportional to the velocity:

$$\zeta_i := \frac{\partial S}{\partial k_i}$$

and this gives $v = k/m$. This value is then the same as $\frac{\log \rho}{k_i}$, which was
the assumption made in [Streater (2003)]. However, the distribution of the
momentum is not quite Gaussian; it is the distribution of the mixture of the
particles arriving at the point $\mathbf{x}$ at time $t$ from the thermalising collisions,
and so is obviously Gaussian only if the distributions in the mixture are
independent. This is not true, even approximately, for particles that might
arrive at $\mathbf{x}$ from collisions at nearby times at nearby points on the lattice.
We expect, then, that the size of the correction terms will be small, since the
chance of two collisions occurring at nearby points of space-time is small.
For the momentum, we must modify the Euler equation (7.29) or the short
Euler (7.42) by adding diffusion terms, in the form of viscosity, which make
the system irreversible. To maintain Galilean invariance, the new terms
should be Euclidean tensors. In [Streater (2003)] we assumed that the lattice model of hard spheres, described in this chapter, was not quite in local
thermodynamic equilibrium, and found a correction to the Euler equation;
this led us to add the terms

$$\lambda \partial_i \left( \Theta^{\frac{1}{2}} \partial_j u_j \right) + \mu \partial_j \left( \Theta^{\frac{1}{2}} \left( \partial_j u_i + \partial_i u_j - \frac{2}{3} \partial_k u_k \delta_{ij} \right) \right) \tag{7.54}$$

to the right-hand side of (7.42). In fact, $\lambda = 0$ in this model; this term,
proportional to $\lambda$, is called the 'bulk viscosity', which is absent from the
approximation made there. In general, any non-negative values of $\lambda$ and $\mu$
are taken. Let us put $\lambda = 0$ and take the viscosity tensor to be

$$T_{ij} = \Theta^{\frac{1}{2}} \left( \partial_j u_i + \partial_i u_j - \frac{2}{3} \partial_k u_k \delta_{ij} \right).$$

For the energy, we can add the Fourier term $\kappa \partial_j \left( \Theta^{\frac{1}{2}} \partial_j \Theta \right)$ to the right of
Eq. (7.43), with $\kappa > 0$. Then the five variables $\rho, v_j, \Theta$ obey the system of
equations of the form

$$D_t \rho + \rho \partial_j v_j = K \partial_k \partial_k \rho \tag{7.55}$$

$$\rho D_t u_i + \partial_i P + \rho \partial_i \phi = \lambda \partial_i \left( \Theta^{\frac{1}{2}} \partial_j u_j \right) + \mu \partial_j T_{ij} \tag{7.56}$$

$$\frac{3k_B}{2m} D_t \Theta + \rho \frac{d\phi}{dt} + P \partial_j u_j = \kappa \partial_j \left( \Theta^{\frac{1}{2}} \partial_j \Theta \right). \tag{7.57}$$

This system has more chance than the usual ones, which are similar except
that there $K = 0$, of having smooth strong solutions for all time, for any
smooth initial conditions with compact support. For, in parabolic equations
such as these, the second derivatives in $\partial/\partial x_j$ dominate the nature of the

solution. In our system, thus the dominant term is given by a matrix of the form

$$
\begin{pmatrix}
K\partial_k\partial_k & \mathbf{0} & 0 \\
0 & \mu\hat{T} & 0 \\
0 & \mathbf{0} & \kappa\partial_k\left(\Theta^{\frac{1}{2}}\partial_k\right)
\end{pmatrix}. \tag{7.58}
$$

Here, $\hat{T}$ is the three-by-three matrix which gives Eq. (7.56) when inserted. Equation (7.58) is a negative definite matrix, when written as a quadratic form, which has a bounded inverse, called the propagator of the system. From this, we expect that our system has smooth, bounded solutions for all time. In the usual equations, $K = 0$ and the first line consists of five zeros, and so the second-order term has no inverse. This leads to the difficulty in bounding the approximatations. See Lions [Lions (1996)] for the treatment of the case when $K$ is taken positive at first, and then made to converge to zero, to get the usual theory, it was hoped, in the limit.

# PART 2
# Quantum Statistical Dynamics

# Chapter 8

# Introduction to Quantum Theory

The theory described in Part I is based on classical probability theory, and has applications to the theory of neural nets, and to artificial ecologies, as well as to physics and chemistry. However, molecules obey the statistical laws of quantum mechanics, and it is the aim of Part II to develop a parallel theory based on *quantum probability*, which is a true generalisation of probability theory. In Part I we were able to give a good account of thermodynamics by imitating the ideas of quantum mechanics, such as the field point of view, and taking the energy-levels to be discrete. This led to important improvements over the traditional particle theories. The classical theory suffers from two main deficiencies, even when improved in this way. The first is that there is no way to predict the size of atoms, even though there is no problem in putting the observed size into the lattice spacing of $\Lambda$. This contrasts with the quantum theory of the atom, whose size is determined by Schrödinger's theory. The second trouble is that the states can have infinite entropy, in the continuum limit. This problem can arise in the quantum case too, but will find its solution within quantum estimation theory.

In Chapter 9 we introduce quantum probability, through the concept of a probability algebra, taking the states to belong to the dual space. In the classical case we did it in the other order, and this is a possible way into the subject [Davies (1976)]. The state-space is a rather complicated object in quantum probability, and it is hard to justify its definition as a primary object. In this we are convinced by the argument of I. E. Segal [Segal (1963)]. The algebra, furnished with a conjugation, $A \mapsto A^*$ is postulated to be a $C^*$-algebra with identity, which is a natural generalisation of the function algebras occurring in Part I. Indeed, the algebra of bounded complex functions on a (countable) sample space is an abelian, that is, commutative,

$C^*$-algebra, with $^*$ given by complex conjugation, and whose real random variables are self-adjoint elements. To maintain the entirely elementary nature of this book, we shall concentrate on models with discrete space-time and usually choose models based on matrix algebras, though some functional analysis is needed for any infinite model. The idea of combining systems to form larger systems goes through very smoothly by using tensor products. This allows us to define algebras with local structure, and also to define the concepts of independence and marginal states. Traditional quantum mechanics is recovered by choosing $\mathcal{A}$ to be the set $B(\mathcal{H})$ of all bounded operators on a separable Hilbert space, $\mathcal{H}$, say. In any $C^*$-algebra there is the important subset, the positive cone, which is the convex set of elements of the form $A^*A$. In the finite-dimensional case the dual space $\mathcal{A}^d$ to the algebra is, as usual, the set of linear functionals on the algebra. Within $\mathcal{A}^d$ can be found the set of states $\Sigma$, which are the normalised elements of the cone dual to the positive cone in $\mathcal{A}$. It is a convex set, but is not a simplex unless $\mathcal{A}$ is abelian. This means that in quantum mechanics a mixed state can have many different expressions as a mixture over extreme points. We interpret the functional as the expectation of the observable when the system is in the given state. The extremal elements of $\Sigma$, the pure states, can be identified with expectations $\langle \psi, A\psi \rangle$ in some normalised vector as in wave mechanics. The non-extremal elements are the mixed states, corresponding to density operators. The desired thermal states will be among these.

In the commutative case the set of normal states $\Sigma$ can be identified with the set of probability measures on a set, and the functional as the expectation value in the classical sense. The quantum versions of independence and so on then coincide with those already met in Part I. In this way, we arrive at classical probability, but not classical mechanics, as a subtheory of quantum probability. Quantum statistical dynamics is then taken on a similar path to that taken in Part I. We must give an energy, which is a self-adjoint matrix in $\mathcal{A}$ in the finite-dimensional case, but could be an unbounded operator $H$ on the Hilbert space $\mathcal{H}$ if $\mathcal{A} = B(\mathcal{H})$ and $\dim \mathcal{H} = \infty$. In this case we make a similar assumption to that needed in the classical case, to ensure thermodynamic stability: $e^{-\beta H}$ has a finite trace. This allows us to define the canonical state. One idea that is essential in quantum theory is the relation between the states and the cyclic representations of the algebra. This goes under the names of Gelfand, Naimark and Segal, abbreviated to $GNS$, and is briefly reviewed. The Chapter ends with an account of von Neumann's entropy; we have analogues of several classical estimates for the entropy which will provide the technical results needed to do quantum statistical dynamics. In particular, the stoss map is defined, and is shown to be entropy-increasing.

In Chapter 10 we give a treatment of linear dynamics entirely parallel to that in Part I. It is called linear not because the equations of motion are linear in the dynamical variables, but because it is linear in the state, as in ordinary quantum mechanics. A reversible time-evolution is a homomorphism $\tau$ from the group $\mathbf{R}$, or $\mathbf{Z}$ if time is discrete, into the group of automorphisms of the algebra. Clearly, the dynamics given by a continuous one-parameter group commutes with its own generator (the Hamiltonian); so energy is conserved. Naturally, entropy is not changed by reversible dynamics; so we get both laws of thermodynamics, but in a trivial way. The simple act of forming random mixtures of reversible dynamics gives us an irreversible dynamics, in which entropy increases. Thus we get a realistic form of the second law. This is taken as the model for the concept of stochastic map and bistochastic map, and we follow the method of Part I to ensure that the first law holds. The problem is that a random unitary group, generated by $H_V = H_0 + V$, where $V$ is a random potential, will not commute with $H_0$. We get the first law by a different interpretation; the random unitary operators are scattering operators, corresponding to the evolution of the system, with Hamiltonian $H_V$, for a very large time (in microscopic units). It is known that for local potentials the scattering operator commutes with the asymptotic energy; it is this operator $H$ that is taken to be the energy that is conserved by the dynamics. The prototype of one step in linear dynamics is thus a mixture of energy-conserving automorphisms. Alternatively, the dynamics over a time-interval $(0, t)$ might be modelled as if $t$ were random; this also leads to an increase in entropy with time.

More generally, we consider one time-step to be given by a bistochastic map $T$ leaving each spectral projection of $H$ invariant. We stress the importance of adding one further property, which is automatic in the classical case, and that is complete positivity. We derive Kraus' form for a completely positive map in a simple way if the dimension is finite. This gives a way of constructing satisfactory quantum theories obeying both laws and converging to the microcanonical state at large time.

Chapter 11 starts with the Boltzmann map, which is the composition of the stoss map with a bistochastic map. The resulting dynamics is unusual, in that it is non-linear in the state. This brings in enough randomness to ensure that the system converges to the canonical state rather than the microcanonical state. Two versions are proved; the first uses Lyapunov's theorem, and the second, in which we assume a spectral gap in the dynamics, shows more, in that the rate of convergence is exponential. The next section 11.2 introduces the heat-particle, which are simply the excitations an oscillator, or a collection of such. We study the theory of bosonic oscillators on Fock space, using coherent states. We are able to construct the second quantising map, of which E. Nelson said, 'whereas first quantisation is a mystery, second quantisation is a functor'. We define the Weyl system, and the generating function of a state. This is the quantum analogue of the characteristic function of a probability measure. The $n$-point functions are then the analogues of the mixed moments of several random variables. There are also analogues of the cumulants. We define the quasifree states as states with zero cumulants beyond the second; they are therefore the analogues of the Gaussian distributions. We show that the canonical state of a quadratic Hamiltonian is quasifree. The quasifree state is the state of maximum entropy having the given two-point functions. There is also an explicit formula for the entropy of a quasifree state when the modes of the Hamiltonian are known. Section 11.3 does a similar analysis for fermions, using the real Clifford algebra with complex structure. Again, the quasifree state is the state of largest entropy having the given two-point functions. Thus, in both the bosonic and fermionic case the quasifree map, which replaces a state by the quasifree state with the same two-point functions, is entropy-increasing. This provides a natural way to add randomness to a state without changing its particle-density or its energy (if the latter is quadratic). It also leads to a reduced description, since one needs only to parametrise the quasifree states. One can incorporate the conservation of particles by looking at the subset of gauge-invariant states. However, in chemistry the number of molecules is not conserved, and we must introduce fields for the atoms rather than for the molecules. Then the quasifree states form a poor approximation to the canonical state, since even the free Hamiltonian is not quadratic in the atomic fields. This motivates a new algebra, isomorphic to $\mathbf{M}_{n+1}$, which we call the 'hard-core algebra', in which only one particle is allowed to occupy any site. This exhibits a more severe exclusion principle than fermions, since two or more different fermions are not excluded from being in the same state. Finally, it is shown that a bistochastic map followed by the quasi-free map can lead to the quantum Boltzmann equation for bosons and fermions.

Chapter 12 moves from the isolated dynamics to the isothermal dynamics by restoring the state of the heat-particle after each time-step to a canonical state of given $\beta$. The resulting dynamics is then linear in the state of the system. The main result is the quantum free-energy theorem. We show that the dynamics is a contraction in a norm, the quantum analogue of the Fisher metric. This enables us to prove a convergence theorem to equilibrium. Driven systems are then easy to formulate (but not to solve). We just restore a certain set of local intensive variables back to their original values after each time-step. We can also add source terms. This is illustrated by the Fröhlich pumped phonon model.

In Chapter 13 we are more ambitious; we study systems in continuous time and infinite volumes, but still with discrete space. We start with the algebra of the infinite system, defined by the inductive limit of the local algebras. Then we show how the Hamiltonian, defined only locally, can be used to give a reversible time-evolution as a continuous group of automorphisms of the algebra. We describe part of the theory of Haag, Hugenholtz and Winnink, on the thermal states of the infinite volume. I present an argument as to why a state obtained as a small perturbation of the thermal state always converges back to the same thermal state under the reversible dynamics. This is both a triumph and tragedy; a triumph because we have obtained irreversibility from a reversible dynamics (by the idealisation of the infinite-volume); a tragedy because we have lost the first law: added energy is lost to infinity, and does not lead to any change in beta. We then study dissipative systems in continuous time, and give a simple derivation of the form of the generator of a completely positive semigroup, due to Gorini, Kossakowski and Sudarshan, and Lindblad. There is a section on the Jaynes-Ingarden theory of information, and Ingarden's concept of temperatures of higher order. This enables us to invent entropy-increasing maps which are much more versatile than the quasi-free map, and which lead to a reduced description. These are used to define a dissipative dynamics for the Ising model, which is proved to return to equilibrium if disturbed by a small amount. The dynamics is non-linear, however, and a finite system in a canonical state will converge to a different beta if it is disturbed. This is an improvement over the theory in an infinite volume, which is physically ambiguous because of the lack of a boundary. We end with a dissipative dynamics for the Heisenberg ferromagnet in which the reversible dynamics is modified by lost information. The rate of convergence is then determined by the Hamiltonian and the level of description alone; there is no other dissipation. This leads to a differential equation determined by the initial state, whose solution might be called a non-linear quantum stochastic process.

In a new Chapter 14, not in the first edition, we prove that with probability tending to 1 as the system gets larger, the entropy increases along the reduced dynamics for a certain one-dimensional model; we follow the paper [De Roeck, Jacobs, Maes and Netočný (2003)]. This answers the question, which worried Maes and Lebowitz, namely, since the von Neumann entropy is invariant under the reversible time-translation, which entropy should increase? It is shown that the von Neumann entropy increases under the reduced dynamics of the model, and that the expectations of the slow variables change under the reduced dynamics in the same way as under the reversible dynamics, in the limit of large systems. Thus we can take the von Neumann entropy to be the thermodynamic entropy.

In the final Chapter 15, mostly new, we introduce information geometry, and show its relation to thermostatics, that is, to equilibrium thermodynamics. This was touched on in Chapter 13 of the first edition, under the name 'Jaynes-Ingarden theory'. Here we develop it into a general scheme for constructing dynamical models obeying the first and second laws of thermodynamics, inspired by the information dynamics of Ingarden, Kossakowski and others. They suggested in [Ingarden (1963); Kossakowski (1969)] that the true dynamics is a Hamiltonian flow, given by a dynamical group $U_t$ say; we write the action of $\{U_t\}$ on the right: $p \mapsto pU_t$. This flow does not change the statistical entropy of the state. The information manifold $\mathcal{M}$ is a subset of $\Sigma$ determined by a linear set $\mathcal{X}$ of distinguished random variables, called the 'slow variables'. These are those whose means can be measured in practice by a feasible experiment. Each state $p \in \Sigma$ for which the means are finite determines a unique point of $\mathcal{M}$, namely the point, denoted $pQ$, having the maximum entropy among all states with the same slow means as $p$. Obviously, $S(pQ) \geq S(p)$. Suppose that at time $t = 0$ the state $p$ lies in $\mathcal{M}$. At each time, the current state $pU_t$ can be projected onto $\mathcal{M}$; this projected state, $pT_tQ$, cannot be distinguished from the true state $pU_t$ by measurements of (the means of) the slow variables. According to Jaynes [Jaynes (1957)], $pU_tQ$ represents the observer's best guess as to what the state is, in view of his lack of information about any but the slow variables. It is obvious that $S(pU_tQ) \geq S(p)$ holds, for all $t$; the information in $p$ has moved from $\mathcal{M}$ to inaccessible degrees of freedom [Balian (2006b)]. The authors call $S(pU_tQ)$ the thermodynamic entropy, as opposed to the statistical entropy $S(pU_t) = S(p)$, which is invariant under time-evolution. Unfortunately, it is very difficult to find conditions on the dynamics such that 'thermodynamic' entropy is increasing along the orbit in $\mathcal{M}$, that is, such that $S(pU_tQ) \geq S(pU_sQ)$ for all $t > s$. For example, if the underlying

Hamiltonian dynamics is periodic, then the entropy of $pU_tQ$ must return to its initial value. The trouble is, the time-evolution, projected onto $\mathcal{M}$, is not a semigroup [Balescu (1997)]. In [Balian (2006a)] it is suggested that we follow the dynamics $U_t$ only for a small time $dt$, project onto $\mathcal{M}$ and take the newly projected state as the initial state for the next step. This gives us a discrete time evolution, a semigroup with increasing entropy. We might then let the time-step go to zero. This does not work: a theorem of Kossakowski [Kossakowski (1969)] says that in the limit of continuous time, the resulting dynamics converges to an isentropic dynamical law, as $dt$ converges to zero. That is, no entropy increase occurs in the limiting dynamics. We can understand this from our entropy estimates, which say that the difference in entropy $S(pQ) - S(p)$ is of second order in the 'distance' between $pQ$ and $p$ ; in the case here, this is also of second order in $dt$.

In this book, this trouble is avoided by choosing one step of the linear part of the dynamics to be a bistochastic map $T$ which increases the entropy at a non-zero rate. Thus we introduce dissipation by hand, rather than trying to deduce it from the fundamental reversible dynamics. The state $pT$ is projected onto the manifold, the operator $Q$ being interpreted as the physical process of thermalisation. We then get the dynamics as a path in the information manifold. We illustrate this in the dynamics of the Lenz-Ising model, and the Heisenberg model, following Chapter 13 of the first edition.

The Chapter and book end with a discussion of the theory of information manifolds. We give a proof of the Jaynes-Ingarden theory of maximum entropy, using a version of Amari's [Amari (1985)] account of estimation theory for classical probability, and then of the quantum version. Then comes a brief account of the infinite-dimensional classical theory, called non-parametric estimation, due to Pistone and Sempi [Pistone and Sempi (1995)]. We end with an account of our own version of non-parametric quantum estimation theory [Streater (2000, 2008b)], using a quantum Young function.

# Chapter 9

# Quantum Probability

## 9.1 Algebras of Observables

It is best to start with the idea of an observable, and the quantum analogue of the random variables, and to define the states as a derived concept, rather than the other way round. This is because the state space in quantum theory is very complicated. In place of the algebra of random variables on the chosen sample space, quantum theory uses a $C^*$-algebra. We shall denote the typical $C^*$-algebra by $\mathcal{A}$, and the typical element by $A$ or $B$. Thus $\mathcal{A}$ is a vector space over the complex field, obeying the following axioms:

**Definition 9.1.**

(1) $\mathcal{A}$ is furnished with a product, which is associative, and distributive relative to addition, and obeys the usual laws of algebra except possibly commutativity. The elements of $\mathcal{A}$ commute with the number field.
(2) $\mathcal{A}$ is furnished with a norm, which makes it into a Banach space; in particular, $\mathcal{A}$ is *complete*, which means that if $\{A_n\}$ is a sequence of elements such that $\|A_m - A_n\| \to 0$ as $m$ and $n$ go to $\infty$, then there exists a limit $A$ say, in $\mathcal{A}$ such that $A_n \to A$ as $n \to \infty$.
(3) $\mathcal{A}$ is furnished with a conjugation $*$; this is a map $A \mapsto A^*$ which is antilinear and obeys $(AB)^* = B^*A^*$ for all $A$ and $B$ in $\mathcal{A}$.
(4) The norm obeys the '$C^*$-property' $\|AA^*\| = \|A\|^2$.

The example that will be used in the next few chapters is $\mathcal{A} = \mathbf{M}_n$. This is the set of $n \times n$ complex matrices; the algebraic rules of addition and multiplication are those of matrix algebra, and the *-operation is that of taking the Hermitian conjugate. The norm is the largest singular value,

197

that is, the largest eigenvalue of $|A| = (A^*A)^{1/2}$. See Exercise 9.17. This algebra is also called the full matrix algebra of dimension $n$. More generally, if $\mathcal{H}$ is a Hilbert space, then the set of all bounded linear operators on $\mathcal{H}$ is a $C^*$-algebra, denoted $B(\mathcal{H})$. The $^*$ operation is the adjoint, also known as Hermitian conjugate, and the norm is the operator norm

$$\|A\| = \sup\{\|A\psi\| : \|\psi\| = 1\}.$$

If $\dim \mathcal{H} = n < \infty$ then $B(\mathcal{H})$ is isomorphic to $\mathbf{M}_n$; see Exercise 9.18. We always assume that our algebra contains the identity, denoted by 1.

An element $A$ of $\mathcal{A}$ is said to be Hermitian if $A = A^*$. The Hermitian elements will represent observables, and they take the place of the (real-valued) random variables in classical probability theory. If $A \in \mathbf{M}_n$, and is Hermitian, then $A$ has real eigenvalues, which form the spectrum of $A$. This is denoted $\operatorname{Spec} A$. The statistical interpretation of quantum mechanics forces us to the conclusion that any measurement of $A$ will lead to a value in $\operatorname{Spec} A$. In a more general $C^*$-algebra $\mathcal{A}$, the spectrum is defined as follows. The *resolvent set* of any element $A \in \mathcal{A}$ is the set of complex numbers $\lambda$ such that $(A - \lambda 1)$ has a bounded inverse. If $\lambda$ lies in the resolvent set, then $(A - \lambda 1)^{-1}$ exists and is an element of $\mathcal{A}$, called the *resolvent* of $A$. The complement in $\mathbf{C}$ of the resolvent set is called the spectrum of $A$. This agrees with the set of eigenvalues when $A$ is a matrix, since the eigenvalues are exactly the values of $\lambda$ for which $\det(A - \lambda 1) = 0$, so no inverse exists. In the general $C^*$-algebra, an operator may have a continuous part to the spectrum, as well as some isolated points. The interpretation of Hermitian elements as observables is supported by a theorem which says that the spectrum of a Hermitian element is a subset of $\mathbf{R}$. It is possible, even at this abstract level, to define certain analytic functions of an element. The definition of a function of several elements is however only possible if they mutually commute. Thus, if $A \in \mathcal{A}$ and $f = \sum f_n z^n$ is a function of a complex variable, analytic in a circle $\{z : |z| < r\}$, then we may define $f(A) = \sum f_n A^n$ if $\|A\| < r$. We use the fact that the series for $f(z)$ converges for $|z| < r$ to show that the partial sums in the series for $f(A)$ form a Cauchy sequence in the norm if $\|A\| < r$. Then the series converges to an element of $\mathcal{A}$, by the axiom of completeness. We note that $(z - \lambda)^{-1}$ is an analytic function of $z$ in the circle $\{z : |z| < |\lambda|\}$. In this way we can define the inverse of $A - \lambda 1$ if $|\lambda| > \|A\|$, proving that $\operatorname{Spec} A$ is a subset of $\{\lambda : |\lambda| \le \|A\|\}$. A $C^*$-algebra always contains Hermitian elements; indeed, any element $A$ can be written as $A = (A + A^*)/2 + i(A - A^*)/(2i)$, just as any complex number can be written as the sum of its real and imaginary parts. Another important class is the set of unitary elements; an operator $U \in \mathcal{A}$ is said to be unitary if it is invertible, and $U^{-1} = U^*$. The set of unitary elements in a $C^*$-algebra is a group (Exercise 9.19). If $A \in \mathcal{A}$ is Hermitian, then $e^{iA}$ is always defined, and is unitary, with inverse $e^{-iA}$. Thus any $C^*$-algebra always contains many unitary elements. The same cannot be said about (Hermitian) projections, that is elements $P \in \mathcal{A}$ obeying $P = P^*$ and $P^2 = P$. The spectrum of a projection consists of the two points, $\{0, 1\}$, unless $P = 0$ or $P = 1$. It therefore represents a 'question', that is, an observable with a yes-no answer.

The simplest $C^*$-algebra after $\{0\}$ and $\mathbf{C}$ is $\mathbf{M}_2$, the set of $2 \times 2$ complex matrices. This is four-dimensional, with a basis given by the matrix units

$$E_{11} = \begin{pmatrix} 1 & 0 \\ 0 & 0 \end{pmatrix} \quad E_{12} = \begin{pmatrix} 0 & 1 \\ 0 & 0 \end{pmatrix} \quad E_{21} = \begin{pmatrix} 0 & 0 \\ 1 & 0 \end{pmatrix} \quad E_{22} = \begin{pmatrix} 0 & 0 \\ 0 & 1 \end{pmatrix}. \quad (9.1)$$

Any $2 \times 2$ matrix is a sum of these with complex coefficients. An important alternative to this basis is that comprising the Pauli matrices and the unit matrix:

$$\sigma_0 = \begin{pmatrix} 1 & 0 \\ 0 & 1 \end{pmatrix} \quad \sigma_x = \begin{pmatrix} 0 & 1 \\ 1 & 0 \end{pmatrix} \quad \sigma_y = \begin{pmatrix} 0 & -i \\ i & 0 \end{pmatrix} \quad \sigma_z = \begin{pmatrix} 1 & 0 \\ 0 & -1 \end{pmatrix}. \quad (9.2)$$

The operators $s_x = \hbar\sigma_x/2$, $s_y = \hbar\sigma_y/2$ and $s_z = \hbar\sigma_z/2$ are used to represent the three components of the vector observable known as the spin, attributed to particles such as electrons, protons and neutrons, and some atoms. These matrices do not commute, but have the commutation relations of the Lie algebra $su(2)$: $[\sigma_x, \sigma_y] = i\hbar\sigma_z$. Let $\boldsymbol{\theta} \in \mathbf{R}^3$. Then the exponential $e^{i(\theta_x\sigma_x+\theta_y\sigma_y+\theta_z\sigma_z)}$ is unitary, forming the group $SU(2)$. The rest of the unitary group of $\mathbf{M}_2$ consists of multiples of the identity $e^{i\theta_0}1$. Since the eigenvalues of any of the components $s_x$, $s_y$, $s_z$ are $\pm\hbar/2$, we say that these particles 'have spin 1/2' (in units of $\hbar$).

There are many beautiful results within the theory of $C^*$-algebras; for example, it can be shown that any isomorphism is continuous, and that any derivation is bounded. However in two respects the theory of $C^*$-algebras is too general as a framework for quantum mechanics: first, a $C^*$-algebra may contain no projections except 0 (and 1 if it has a unit), whereas Hermitian projections play an important role in quantum probability; and secondly, the tensor product of two $C^*$-algebras can sometimes be ambiguous, in that there are several norms on the algebraic tensor product all with the $C^*$-property. Tensor products play an important role in constructing compound systems. Neither of these problems arises in the special class of $C^*$-algebras known as von Neumann algebras: such an algebra has so many projections that it is generated by them; and there is a natural choice for the norm of a tensor product. However spin systems on an infinite space $\Lambda$, as well as Bosons and fermions, are best described by certain $C^*$-algebras that are not von Neumann algebras. We do not meet these two problems, (lack of projections, and ambiguity of the norm in tensor products) in these special cases.

An element $A$ of a $C^*$-algebra of the form $A = B^*B$ is said to be positive. The spectrum of a positive element lies inside the set $\{x \geq 0\}$. The converse also holds (Kaplansky's theorem, [Bratteli and Robinson (1979)]). It can be proved that the set of positive elements is convex. It is obviously a cone, since if $A = B^*B$ is positive, then $\lambda A = (\lambda^{1/2}B)^*(\lambda^{1/2}B)$ is also positive, for any $\lambda > 0$. The positive cone in a $C^*$-algebra plays the role of positive random variables in the theory of probability.

There are two constructions with $C^*$-algebras which allow us to combine two algebras and get another; these are the direct sum and the tensor product. A $C^*$-algebra is a vector space (with further structure) and the *direct sum* of two, $\mathcal{A}_1$ and $\mathcal{A}_2$ is the set of pairs $(A_1, A_2)$, with multiplication by a number $\lambda$ given by $\lambda(A_1, A_2) = (\lambda A_1, \lambda A_2)$. We say that this is 'componentwise'. Addition and multiplication are also componentwise:

$$(A_1, A_2) + (A_1', A_2') = (A_1 + A_1', A_2 + A_2')$$
$$(A_1, A_2).(A_1', A_2') = (A_1 A_1', A_2 A_2').$$

We denote the element $(A_1, A_2)$ of the direct sum by $A_1 \oplus A_2$. The norm of an element $A_1 \oplus A_2$ is defined to be $\max\{\|A_1\|, \|A_2\|\}$. The direct sum of two vector spaces $\mathcal{A}_1$ and $\mathcal{A}_2$, which in our discussion happen also to be algebras, is denoted $\mathcal{A}_1 \oplus \mathcal{A}_2$. It can be shown that any finite-dimensional $C^*$-algebra with an identity is the direct sum of full matrix algebras. If the algebra is an algebra of matrices, this result means that any element of the algebra can be written in block diagonal form.

The vector space underlying the *algebraic* tensor product of two $C^*$-algebras is just the tensor product of the underlying vector spaces. *The tensor product is the completion of the algebraic tensor product in a suitable norm.* If at least one of the algebras is of finite dimension, the completion is not needed, as the algebraic tensor product is already complete. So let us now give the definition of the tensor product (over the complex field) of two vector spaces of any dimension. We start with vector spaces $V_1$ and $V_2$, and form the set $V = V_1 \times V_2$ of ordered pairs of elements. Next we form the space Span $V$, of formal finite sums of elements of $V$, with complex coefficients. We have already met this construction in Sec. 2.1. This space is a vector space of enormous dimension, one for each pair of vectors, taken from $V_1$ and $V_2$, and not just from a chosen basis in each. We get the space we want, the tensor product, by forming the quotient vector space of Span $V$ by a rather large subspace, formed by combinations in Span $V$ that should be zero by the rules of the tensor calculus. Under these rules, the following elements of Span $V$ are equivalent to zero, for any vectors $v_i$ and numbers $\lambda$:

$$(\lambda v_1, \, v_2) - \lambda(v_1, \, v_2) \, ,$$
$$(v_1, \, \lambda v_2) - \lambda(v_1, \, v_2) \, ,$$
$$(v_1 + v_1', \, v_2) - (v_1, \, v_2) - (v_1', \, v_2) \, ,$$
$$(v_1, \, v_2 + v_2') - (v_1, \, v_2) - (v_1, \, v_2') \, . \tag{9.3}$$

Let $V_0$ be the vector subspace of Span $V$ spanned by all elements of the form of one of the four lines in Eq. (9.3). Then we define the *algebraic tensor product* to be the quotient space:

$$V_1 \otimes V_2 = \text{Span} \, (V_1 \times V_2)/V_0 \, . \tag{9.4}$$

(I did not believe it either, when I first saw it written, even though it was in a book.) The equivalence class containing $(v_1, \, v_2)$, and thus all elements equivalent to it, is denoted $v_1 \otimes v_2$. As usual, we add equivalence classes by adding representative elements, one from each, the addition here being in $V$, and then we find the equivalence class of the sum. This procedure is independent of the choice of representative. Similarly, we can define the multiplication of an equivalence class by a complex number. That this all works out is part of the theory of the quotient space $V/V_0$. Because of the definition of $V_0$ by Eq. (9.3), the tensor product satisfies the corresponding relations

$$(\lambda v_1) \otimes v_2 - \lambda(v_1 \otimes v_2) = 0 \, ,$$
$$v_1 \otimes (\lambda v_2) - \lambda(v_1 \otimes v_2) = 0 \, ,$$
$$(v_1 + v_1') \otimes v_2 - v_1 \otimes v_2 - v_1' \otimes v_2 = 0 \, ,$$
$$v_1 \otimes (v_2 + v_2') - v_1 \otimes v_2 - v_1 \otimes v_2' = 0 \, .$$

If now we have two $C^*$-algebras, $\mathcal{A}_1$ and $\mathcal{A}_2$, then the structure of $\mathcal{A}_1 \bigotimes \mathcal{A}_2$ as a vector space is just that of the tensor product of the vector spaces $\mathcal{A}_1$ and $\mathcal{A}_2$. In particular, every element of $\mathcal{A}_1 \bigotimes \mathcal{A}_2$ is a finite sum of dyadic tensors, that is, elements of the form $A \otimes B$. To furnish the tensor product with a *-operation and a product, we define them first on dyadic elements:

$$(A \otimes B)^* = A^* \otimes B^*$$

$$(A \otimes B)(A' \otimes B') = AA' \otimes BB'$$

by choosing representatives. We extend these by anti-linearity (for *) and bilinearity (for the product) to all elements of the tensor product. The easy but rather long calculations needed to verify that the product is the same, whichever representatives are chosen, will be done (by you) in Exercise 9.21.

To put a norm on $\mathcal{A}_1 \bigotimes \mathcal{A}_2$, we encounter an ambiguity unless one of the algebras is of finite dimension. We shall only discuss here the case when $\mathcal{A}_1 = \mathbf{M}_n$ and $\mathcal{A}_2 = \mathbf{M}_m$. Then it can be shown that $\mathbf{M}_n \bigotimes \mathbf{M}_m$ is isomorphic to $\mathbf{M}_{nm}$, and also to the algebra of $n \times n$ matrices whose entries are $m \times m$ matrices, which is denoted $\mathbf{M}_n(\mathbf{M}_m)$. By symmetry, all these are isomorphic to $\mathbf{M}_m(\mathbf{M}_n)$. These algebras have a natural $C^*$-norm, $\|A\| =$ the maximum eigenvalue of $(A^*A)^{1/2}$. Now any $C^*$-algebra has a unique norm, so this must be it. The reason why this does not work when both $\mathcal{A}_1$ and $\mathcal{A}_2$ are infinite-dimensional is that the algebraic tensor product is then not complete in any $C^*$ norm, and the uniqueness theorem mentioned only holds for $C^*$-algebras. Remember, Axiom (2) requires the algebra to be a Banach space, which includes the completeness axiom. Finite-dimensional vector spaces are always complete.

The direct sum and tensor product can be done for more than two algebras, and is associative. This means that $(\mathcal{A}_1 \oplus \mathcal{A}_2) \oplus \mathcal{A}_3$ is isomorphic to $\mathcal{A}_1 \oplus (\mathcal{A}_2 \oplus \mathcal{A}_3)$, and that $(\mathcal{A}_1 \otimes \mathcal{A}_2) \otimes \mathcal{A}_3$ is isomorphic to $\mathcal{A}_1 \otimes (\mathcal{A}_2 \otimes \mathcal{A}_3)$. Distributivity also holds: $\mathcal{A}_1 \otimes (\mathcal{A}_2 \oplus \mathcal{A}_3)$ is isomorphic to $\mathcal{A}_1 \otimes \mathcal{A}_2 \oplus \mathcal{A}_1 \otimes \mathcal{A}_3$.

The physical meaning of the tensor product of algebras is the combination of the two systems they represent into one system. The two systems might be mixed up, or might still be well separated in distance; it is just that we wish to consider them both as part of the same theory. This construction is possible, whether or not the energy of the combined system is the simple sum of the energies of its parts (non-interacting systems) or whether a further energy of interaction must be added. The tensor construction does not imply that the two systems are statistically independent; this depends on whether the state (to be defined in the next section) factorises. But what we do need is that the observer, in contemplating making a measurement on the combined system, must have available all the measurements that are possible on say the first system. This is certainly true of the tensor product; the operators of the form $A \otimes 1$ as $A$ runs over $\mathcal{A}_1$ make up a sub-algebra of $\mathcal{A}_1 \bigotimes \mathcal{A}_2$ that is isomorphic to $\mathcal{A}_1$. Similarly, the set of elements of the form $1 \otimes A$ with $A \in \mathcal{A}_2$ form a subalgebra of the tensor product isomorphic to $\mathcal{A}_2$. The mapping, $A \mapsto A \otimes 1$ is an ampliation, similar to that of regarding a given function of one variable, say $x$, as a function of two variables, say $x, y$, which happens to be independent of the second variable. This construction is common in classical probability theory. It is this tensor construction that does not seem to be possible in certain attempts to construct even more general theories than quantum mechanics; for example, there is no tensor product in the general theory of Jordan algebras, which have been advocated as a possible generalisation of $C^*$-algebras. Without the ampliation, a theory fails to be *local* as used in the philosophical debate over Bell's inequalities; a more recent and better word is *non-contextual*. If the mathematical object used to describe a certain measurement on a system, $S_1$, while not considering another system $S_2$, which happens to be far away, is $A_1$, but the mathematical object used to describe the *same* measurement on $S_1$, while contemplating both systems $S_1$ and $S_2$, is essentially different, say $A_1'$, then the assignment of $A_1$ to that particular measurement is called 'contextual'. In a contextual assignment, the choice between $A_1$ and $A_1'$ might depend on what observable is being measured on $S_2$, even though it is miles away. It is intended in these theories that $A_1$ and $A_1'$ have the same statistical predictions for the first observable, that is, the same probability distribution. Even so, contextual theories are very odd: it is universal in mathematical modelling that the physical acts of measuring something are replaced at an early stage by a well-defined mathematical object. The theory is a theory of these objects, and is expressed in terms of these objects. If the assignment of mathematical objects to physical processes is allowed to slide around, and change with the context, then the theory is capricious; no definite conclusions can be drawn in such theories. Contextual variables were introduced into the debate on quantum mechanics as a result of Bell's inequalities, which say that the predictions of quantum mechanics (in a theory on a Hilbert space of dimension at least four) cannot be obtained from *any* classical probability theory. Landau's proof of this [Landau (1987)] does not use the assumption of locality that was needed in Bell's original proof. We shall present Landau's proof in the next Section. Instead of admitting that a more general probability theory (quantum mechanics) is needed, Bohm argued that we should use contextual random variables in an otherwise classical theory; this is an even more radical departure from probability theory than quantum mechanics itself. Indeed, contextual assignments of random variables have a drastic effect on the interpretation of the sample space: an outcome may be represented by different sample points (even in different sample spaces [Streater (2000)]) depending on what measurements are being made at other parts of space. This hardly meets the criterion of *reality* laid down by Einstein, Podolsky and Rosen [Einstein, Podolsky and Rosen (1935)], which was one of the original motives for this type of work. This is referred to as the EPR-experiment. In this sense, quantum mechanics is non-contextual, and realistic: the same mathematical object, the matrix $A_1$, or its ampliations, is always assigned to a given measuring process, and is called the 'observable' in question. For example (and this is what is used in Bell's inequalities), if we have two far-separated particles of spin $1/2$, the $C^*$-algebra used is $\mathbf{M}^2 \otimes \mathbf{M}^2$; the $z$-component of the spin of the first particle is described by the element $\hbar\sigma_z/2 \otimes 1$ irrespective of whether an observer is measuring the spin of the second particle in the $z$-direction, the $x$-direction, or not measuring it at all.

Quantum probability is a true generalisation of classical probability, and is not merely a different theory. We shall show this for any countable sample space $\Omega$. Indeed, the set of complex bounded functions on $\Omega$, the complex random variables, form the elements of a (commutative) $C^*$-algebra, and so form a special case of the set-up given here. In the next section we introduce quantum states, and show that any probability measure on $\Omega$ is a special type of quantum state, known as 'normal', acting on a special type of algebra (an abelian one).

In quantum statistical dynamics, we have a finite set $\Lambda$ of points in space, and to each $x \in \Lambda$ we are to assign a $C^*$-algebra $\mathcal{A}_x$, representing the observables that pertain to $x$. To the region $\Lambda$ we associate the algebra

$$\mathcal{A}(\Lambda) = \bigotimes_{x \in \Lambda} \mathcal{A}_x .$$

If each algebra is the same full matrix algebra, this is the $C^*$-algebra of the lattice spin system. We shall see later that the existence of the ampliations between partial tensor products allows us to define the $C^*$-algebra for spins on an infinite lattice.

## 9.2 States

A given system can be in many different states, just as in classical probability the same sample space admits many probability measures. The state, then, is the description of the statistical properties of the system when prepared repeatedly in the same way. Following I. E. Segal [Segal (1963)] a *state* on a $C^*$-algebra is an assignment of the expectation value for every element of the algebra. This should obey the natural laws of linearity and positivity. Thus, let $\mathcal{A}$ be a $C^*$-algebra with identity; then a state on $\mathcal{A}$ is a linear map $\rho : \quad \mathcal{A} \to \mathbf{C}$; we shall write $\rho \cdot A$ for $\rho(A)$, to conform with our convention in the classical case. Then, to be a state, $\rho$ must obey

- $\rho \cdot 1 = 1$.
- $\rho \cdot (\lambda A) = \lambda \rho \cdot A$ for all $\lambda \in \mathbf{C}$ and all $A \in \mathcal{A}$.
- $\rho \cdot (A + B) = \rho \cdot A + \rho \cdot B$ for all $A$ and $B$ in $\mathcal{A}$.
- $\rho \cdot A \geq 0$ for all $A$ in the positive cone of $\mathcal{A}$.

As an example of a $C^*$-algebra and a state, the expectation or mean in a classical theory is a state: let $\Omega$ be a countable sample space and let $\mathcal{A}$ be the $C^*$-algebra of all bounded complex-valued functions $F$ on $\Omega$. Then any probability $p \in \Sigma(\Omega)$ defines a state by $\rho \cdot F = p \cdot F$, of which the definition is $\sum_\omega p(\omega)F(\omega)$. The concept of state is more general than that of probability measure, since a state does not need to be countably additive. As another example, let $\mathcal{H}$ be a Hilbert space with scalar product $\langle \bullet, \bullet \rangle$, and put $\mathcal{A} = B(\mathcal{H})$; let $\psi \in \mathcal{H}$ be a normalised vector. Then

$$\rho_\psi \cdot A = \langle \psi, A\psi \rangle \qquad (9.5)$$

is a state on $\mathcal{A}$. Such a state is called a *vector state*. If $\mathcal{A}_0 \subseteq B(\mathcal{H})$ is a $C^*$-subalgebra, then $\rho_\psi$, restricted to $\mathcal{A}_0$, is a state of $\mathcal{A}_0$; it is also a vector state of $\mathcal{A}_0$; rather, it is a vector state relative to the representation of $\mathcal{A}_0$ by operators on $\mathcal{H}$.

A third example is that of density matrix; let $\mathcal{A} = \mathbf{M}_n$ and let $\delta$ be an $n \times n$ positive semidefinite matrix with unit trace. It is easy to see that the set of density matrices is a convex subset of $\mathbf{M}_n$. We get a state, called $\rho_\delta$, by the map $A \mapsto \rho_\delta \cdot A = Tr \, \delta A$; see Exercise 9.23. By the cyclic property of the trace, we can also write $\rho_\delta \cdot A = Tr \, A\delta$. This example can be generalised to infinite dimensions. Let $\mathcal{H}$ be a separable Hilbert space, and $\mathcal{A} = B(\mathcal{H})$. Let us say that a bounded operator $\delta$ is of trace-class if the operator $(\delta^* \delta)^{1/2}$ has a complete set of normalised eigenfunctions $\psi_i$ with eigenvalues $\{\lambda_i\}$ of finite multiplicity (except that 0, if an eigenvalue, might have infinite multiplicity), and if $\sum_i \lambda_i < \infty$, where we have counted each eigenvalue with its multiplicity. Then we define the trace of an operator $\delta$ of trace-class to be

$$Tr \, \delta = \sum_i \delta_{ii} \qquad \text{where } \delta_{ii} = \langle \psi_i, \delta\psi_i \rangle.$$

It can be proved that this always converges absolutely, and that the answer is independent of the basis in which the calculation is done. A *density operator* is a positive semi-definite operator of trace-class, whose trace is equal to 1. The density operators form a convex subset of $B(\mathcal{H})$. One can show that if $A$ is bounded and $\delta$ is of trace-class, then $A\delta$ and $\delta A$ are of trace class, and their traces are equal. We can now use a density operator $\delta$ to define a state: $\rho_\delta \cdot A = Tr(\delta A)$. Not all states are of this form unless $\dim \mathcal{H} < \infty$, in which case they are. See Exercise 9.24. We are then able to define the inverse of the map $\delta \mapsto \rho_\delta$, and associate a density matrix, denoted by $\Delta\rho$, to the state $\rho$. In infinite dimensions, a state given by a density operator is said to be *normal*. So if $\rho$ is normal, we denote by $\Delta\rho$ the (unique) density operator such that $\rho_{\Delta\rho} = \rho$.

Let $\mathcal{A}$ be a $C^*$-algebra with identity and let $\Sigma(\mathcal{A})$ denote the set of states on $\mathcal{A}$. We can regard $\Sigma(\mathcal{A})$ as a subset of the space $\mathcal{A}^d$ of all linear functionals on $\mathcal{A}$ which has a natural linear structure. Then $\Sigma(\mathcal{A})$ is a convex subset of $\mathcal{A}^d$. Thus, if $0 < \lambda < 1$, and $\rho_1$ and $\rho_2$ are states, then $\lambda\rho_1 + (1 - \lambda)\rho_2$ is a state; see Exercise 9.25. We say that the state space has an *affine* structure. We interpret this combination as the incoherent statistical mixing of the two states, in the proportion $\lambda$ to $1 - \lambda$. It is easy to do this in practice: we choose a sample in the state $\rho_1$ with probability $\lambda$ and a sample with state $\rho_2$ with probability $1 - \lambda$. In the same way, we can set up the mixture $\rho = \sum_i^n \lambda_i \rho_i$, where $0 < \lambda_i < 1$ and $\sum_i \lambda_i = 1$. Clearly, the set of normal states has the same affine structure as the set of density operators, when they are related as above.

We say that a state $\rho$ of $\Sigma$ is *mixed* if $\rho$ can be written as $\lambda\rho_1 + (1-\lambda)\rho_2$ where $0 < \lambda < 1$ and $\rho_1$ is different from $\rho_2$. If a state is not mixed, it is said to be pure. The pure states are thus the extreme points of the convex set $\Sigma$. In [Bratteli and Robinson (1979)] a different definition of pure is given: $\rho$ is pure if and only if it does not dominate any other state; it is shown to be equivalent to ours (Theorem 2.3.15). According to Minkowski's theorem, in finite dimensions any point in a convex set is a mixture of the extreme points of the set. In general, the pure states generate the whole of state space, in that the smallest closed convex set containing all pure states is $\Sigma(\mathcal{A})$. This is the Krein-Milman theorem. Closed here means closed in the weak* topology.

We meet a significant difference between quantum probability in general and the special case of probability measures: the classical state space (of probability measures on a sample space) is a simplex, while $\Sigma(\mathcal{A})$ is not, provided that $\mathcal{A}$ is not abelian. Recall that a convex subset $\Sigma$ of $\mathbf{R}^m$ is said to be a simplex if the expression of any element $\rho \in \Sigma$ as a convex sum of extreme points $\rho_i \in \Sigma$ is unique. If $\mathcal{A}$ is abelian, then it can be identified with the set of continuous complex-valued functions on a compact space, and $\Sigma(\mathcal{A})$ is the set of finitely additive measures on it. This is the Gelfand isomorphism theorem. See Theorem 2.1.11A of [Bratteli and Robinson (1979)], where the more general case of $C^*$-algebra without identity is included. The case of a finite-dimensional abelian algebra is not hard; try Exercise 9.26.

Let us concentrate on the case of $\mathbf{M}_n$. In this case an extremal state is given by an extremal density matrix, and this must be a one-dimensional projection; see Exercise 9.27. It follows from this that the pure states are exactly the vector states in this case, the one-dimensional projection giving the same state as any vector in the subspace it defines; see Exercise 9.28. We can get the projection onto the subspace spanned by a single column vector $\psi$ by forming the matrix product $\psi\psi^*$ of the matrix with its transposed complex conjugate. This can also be regarded as a tensor product $\psi \otimes \psi^*$. A projection of dimension $r \leq n$ also defines a density matrix, when normalised by dividing by $r$, so that its trace is 1. In particular, the matrix $n^{-1}1$, where 1 is the $n \times n$ identity matrix, is a density matrix. This is the quantum analogue of the uniform distribution, also known as the microcanonical state when the energy is a constant. The density operator of this state can be written as

$$n^{-1}1 = \sum_i^n (n^{-1})E_{ii}$$

in terms of the matrix units $E_{ii}$. We can write the matrix 1 as a sum of one-dimensional projections in many different ways: we just need to choose any orthogonal basis, and use the projections onto its elements. So the decomposition is not unique, and $\Sigma(\mathbf{M}_n)$ is not a simplex. In the case of $n = 2$ the uniform state describes the spin of a completely unpolarised beam of electrons, say. The density matrix $E_{11}$ describes a beam of particles all completely polarised in the 'up' direction, in that they are all eigenstates of $s_z$ with eigenvalue $+\hbar/2$. The density matrix $E_{22}$ similarly describes a beam completely polarised in the 'down' direction. We now mix these together, taking a particle from one beam or the other with probability 1/2. The resulting beam is described by the density matrix diag $(1/2, 1/2)$, the uniform state. Exactly the same result would have been obtained if we had similarly mixed two other vector states which are orthogonal, such as the eigenstates of $s_x$ with eigenvalues $\pm\hbar/2$. There is no way that anyone can discover from the mixture what were the constituents by any experiment on the mixed beam: the classical mixing, with probability 1/2, gets confused with the quantum superposition, which allows us to write the eigenstates of $s_z$ as coherent mixtures of those of say $s_x$. The rules of matrix or tensor products do the rest.

It is sometimes necessary to admit that two different observers should assign two different states to the same exemplar of the system. This leads to the 'paradox' of Wigner's friend. In this paradox, an unpolarised beam of electrons is subjected to a Stern-Gerlach measurement by Wigner's friend, but he does not tell Wigner the result. What state should Wigner assign to the exemplar that resulted from the experiment? A betting man in Wigner's position would have to say that it is still unpolarised, even though his friend would rightly assign it an eigenstate. This has been thought to be a paradox, since the state appears to be a subjective entity. However, the same 'problem' arises in classical probability, when some information due to conditioning is available to one person and not another. The formal way to handle this is through *information sets*, which keeps track of who knows what in the theory of games. We can generalise this to the *quantum game*; this is a game with several players, each with a strategy, and some chance elements; the difference with the usual game is that the probabilities with which pure strategies are chosen, and the outcomes of the chance elements, are governed by quantum, not classical, probability. Several elaborations of the *EPR* experiment are of this type; paradoxes arise if we do not treat the information sets correctly. For an introduction, see [Meyer (1999)].

Apart from the uniform or microcanonical state, we shall meet the canonical and grand canonical states, which are defined as follows.

**Definition 9.2 (Canonical State).** *Let $H$ be a self-adjoint operator on the Hilbert space $\mathcal{H}$ such that $e^{-\beta H}$ is of trace class; then the density operator on $B(\mathcal{H})$:*

$$\Delta \rho_\beta = Z_\beta^{-1} \exp\{-\beta H\} \qquad (9.6)$$

*where $Z_\beta = Tr(\exp\{-\beta H\})$ defines the canonical state $\rho_\beta$ with beta, or inverse temperature, equal to $\beta$ and Hamiltonian $H$. $Z_\beta$ is called the partition function.*

As in the case of classical statistical dynamics, we want to show why many systems tend to converge to the canonical state for large times. Sometimes there is one or more conserved quantities, whose presence in the theory preclude states from converging to the canonical state. In such models we need a generalisation:

**Definition 9.3 (Grand Canonical State).** *Let $H$ and $N$ be self-adjoint operators whose spectral projections commute, and are such that $e^{-\beta(H-\mu N)}$ is of trace class; then the density operator*

$$\Delta \rho_{\beta,\mu} = Z_{\beta,\mu}^{-1} \exp\{-\beta(H - \mu N)\} \qquad (9.7)$$

with $Z_{\beta,\mu} = Tr(\exp\{-\beta(H - \mu N)\})$ *defines the grand canonical state* $\rho_{\beta,\mu}$ *with beta* $\beta$ *and chemical potential* $\mu$, *Hamiltonian* $H$ *and number operator* $N$. $Z_{\beta,\mu}$ *is called the grand partition function.*

We shall see that as in the classical case, these states maximise the entropy subject to certain constraints. Representation theory is a large industry in $C^*$-algebras. A representation of $\mathcal{A}$ is a *-homomorphism $\pi$ from $\mathcal{A}$ into $B(\mathcal{H})$ for some Hilbert space $\mathcal{H}$. The representation is said to be faithful if $\pi(A) = 0$ only if $A = 0$. A state $\rho$ is said to be faithful if $\rho(A^*A) > 0$ unless $A = 0$. Gelfand showed that every $C^*$-algebra has at least one faithful representation, and so might be regarded as an operator algebra without loss of generality. Two representations, $\pi_1$ on $\mathcal{H}_1$ and $\pi_2$ on $\mathcal{H}_2$ are said to be equivalent if there exists a unitary operator $U : \mathcal{H}_1 \to \mathcal{H}_2$ such that $U\pi_1(A) = \pi_2(A)U$ for every $A \in \mathcal{A}$. We shall need the concepts of irreducible representation, and cyclic representation. Let $\mathcal{X}$ be any set of bounded operators on a Hilbert space $\mathcal{H}$. We say that a subspace $\mathcal{H}_1$ is invariant under $\mathcal{X}$ if $X\psi \in \mathcal{H}_1$ for all $\psi \in \mathcal{H}_1$ and all $X \in \mathcal{X}$. We say that $\mathcal{X}$ is irreducible if the only subspaces invariant under $\mathcal{X}$ are the zero subspace and $\mathcal{H}$ itself. There is then the useful lemma of Schur. Suppose that whenever $\mathcal{X}$ contains $X$ it also contains $X^*$; then we say that $\mathcal{X}$ is self-adjoint. We define the *commutant* of $\mathcal{X}$, denoted by $\mathcal{X}'$, as the set of operators in $B(\mathcal{H})$ that commute with all elements of $\mathcal{X}$. Schur's lemma says that a self-adjoint set of operators is irreducible if and only if its commutant consists of multiples of the identity. This is Proposition 2.3.8 of [Bratteli and Robinson (1979)]. The theorem also says that these are equivalent to the fact that every vector is cyclic (unless $\mathcal{A} = 0$ and $\mathcal{H} = \mathbf{C}$). We say that a vector $\psi \in \mathcal{H}$ is cyclic for $\mathcal{X}$ if the set $\{X\psi : X \in \mathcal{X}\}$ is dense in $\mathcal{H}$. We say that a representation $\pi$ of a $C^*$-algebra $\mathcal{A}$ is *cyclic* with cyclic vector $\psi$ if $\psi$ is a cyclic vector for $\mathcal{X} = \{\pi(A) : A \in \mathcal{A}\}$. Cyclic representations are very common in quantum mechanics: we build up all states by acting on $\psi$, often the vacuum or equilibrium state, with the operators at our disposal. Two cyclic representations $(\mathcal{H}_1, \pi_1, \psi_1)$ and $(\mathcal{H}_2, \pi_2, \psi_2)$ are said to be *cyclic equivalent* if there exist an isometric map $U : \mathcal{H}_1 \to \mathcal{H}_2$ from $\mathcal{H}_1$ into $\mathcal{H}_2$ such that $U\pi_1 = \pi_2 U$ and $U\psi_1 = \psi_2$. We say that two cyclic representations, $\{\mathcal{H}, \pi, \psi\}$ and $\{\mathcal{K}, \sigma, \phi\}$ are *cyclic unitarily equivalent* if there exists a unitary $W : \mathcal{H} \to \mathcal{K}$ such that $W\pi W^{-1} = \sigma$ and $W\psi = \phi$.

We now come to the most important idea, the Gelfand-Naimark-Segal construction.

**Theorem 9.4 (GNS).** *Let $\mathcal{A}$ be a $C^*$-algebra and $\rho$ a state on $\mathcal{A}$. Then there exists a Hilbert space $\mathcal{H}_\rho$, a representation $\pi_\rho$ and a cyclic vector $\psi_\rho$ for $\pi_\rho$, such that*

$$\rho \cdot A = \langle \psi_\rho, \pi(A)\psi_\rho \rangle \quad \text{for all } A \in \mathcal{A}.$$

*The representation is unique up to cyclic unitary equivalence. Moreover, $\pi$ is irreducible if and only if $\rho$ is pure.*

For the proof, see [Bratteli and Robinson (1979)], Theorems 2.1.16 and 2.1.19. The great thing about the Gelfand-Naimark-Segal theorem is that it is constructive. We shall show this in the case when $\rho$ is a faithful state; this will be so for the canonical states of interest to us. Thus let $\mathcal{A}$ be a $C^*$-algebra with identity and let $\rho$ be a faithful state. We now define a scalar product on the vector space $\mathcal{A}$ by

$$\langle A, B \rangle_\rho = \rho \cdot (A^*B).$$

This is linear in the second variable and antilinear in the first; it is also definite, in that $\|A\|^2 = \langle A, A \rangle_\rho > 0$ unless $A = 0$. This holds since it expresses that $\rho$ is faithful. The Hilbert space $\mathcal{H}_\rho$ mentioned in the theorem is then the completion of $\mathcal{A}$ in the metric defined by the norm. The representation $\pi_\rho$ mentioned in the theorem is the action on $\mathcal{A}$ by itself by left multiplication. This can be shown to give rise to bounded operators, which have unique extensions to $\mathcal{H}_\rho$ by continuity. The cyclic vector $\psi_\rho$ is the identity element of $\mathcal{A}$. Naturally, if $\dim \mathcal{A} < \infty$ then the completion and extensions are not necessary.

If $\mathcal{A}$ is a subalgebra of $B(\mathcal{H})$ for some Hilbert space $\mathcal{H}$, and $\rho$ is a normal faithful state (given by density matrix $\Delta(\rho)$), then we may define a norm on certain elements in the vector span of states by

$$\|\sigma\|_\rho = [Tr\,(\Delta(\rho)^{-1}\sigma^*\sigma)]^{1/2}. \tag{9.8}$$

This is the dual to the norm on $\mathcal{A}$ given by the state $\rho$. It will turn out to be a useful norm which may be used to define a metric on the space of states.

There is a natural definition of independence for quantum probability. The analogue of Boolean subring is subalgebra $\mathcal{A}_1 \subseteq \mathcal{A}$. Let $\mathcal{A}$ be finite dimensional and let $\mathcal{A}_1$ and $\mathcal{A}_2$ be two $C^*$-algebras. Suppose that $\mathcal{A} = \mathcal{A}_1 \otimes \mathcal{A}_2$ and suppose we are given $\rho_1 \in \Sigma(\mathcal{A}_1)$, $\rho_2 \in \Sigma(\mathcal{A}_2)$. We write $\rho = \rho_1 \otimes \rho_2$ for the unique bilinear extension to $\mathcal{A}$ of the definition on dyads:

$$\rho(A_1 \otimes A_2) = \rho_1(A_1)\rho_2(A_2). \tag{9.9}$$

Then we say that the subalgebras, the ampliations of $\mathcal{A}_1$ and $\mathcal{A}_2$ are independent in the state $\rho$. The $\infty$-dimensional case needs a more careful treatment of the tensor product. We shall deal with that problem if it arises.

Of great importance in classical statistical dynamics was the concept of marginal distribution, and we need an analogue of this. So let $\mathcal{A}$ be a tensor product $\mathcal{A}_1 \otimes \mathcal{A}_2$, and let $\rho \in \Sigma(\mathcal{A})$. The marginal state of $\rho$, $\rho\mathcal{M}_1$ on the first factor, often described as the 'trace of $\rho$ over the second factor' is defined, for each element $A_1$ of the first algebra, as

$$(\rho\mathcal{M}_1) \cdot A_1 = \rho \cdot (A_1 \otimes 1). \tag{9.10}$$

The marginal map obviously obeys

$$(\rho_1 \otimes \rho_2)\mathcal{M}_1 = \rho_1.$$

Similarly the marginal state on the second factor is defined. So the marginal map $\mathcal{M}_1 : \Sigma(\mathcal{A}_1 \otimes \mathcal{A}_2) \to \Sigma(\mathcal{A}_1)$ is the dual of the ampliation, which maps the algebras in the reverse direction.

Quantum probability is a more general model than probability for modelling systems subject to chance; it cannot be usefully described by hidden variables in a classical theory, unless the algebra of observables is abelian; for example, for a single non-relativistic particle, the algebra generated by the three position operators at a given time $t$ is abelian. In such a case, the simultaneous diagonalisation of the position variables can be achieved, and a theory such as that of Bohm makes sense at time $t$; however, position operators at different times do not commute, and a classical description involving time evolution is not possible; to try would be like trying to model non-Euclidean geometry by using figures of a strange shape in Euclidean space. It can't be done; this is what Bell really proved in his famous theorem.

The original argument of Bell posited the existence of some hidden variables related to the spin measurements of *EPR* by some local equations, in that the spin on the right was not a function of any variables localised on the left, and *vice versa*. Omnès [Omnès (1999)] gives an account of Bell's argument; see also [Omnès (2005)]. However, Landau has given a proof without any assumption about the existence and properties of hidden variables. Landau's proof uses only that the four spin variables are random variables on a common sample space. In this version, then, we can say that the quantum result cannot be explained by *any* classical probability theory (with or without hidden variables having whatever properties). In this form, Bohm's theory does not escape: it does not give the same answers as quantum mechanics; it gives the wrong correlation functions for position variables at unequal times. Omnès allows that it might agree with quantum mechanics, because Bell's derivation used locality, which Bohm's theory does not obey; in this, Omnès is too polite.

Let us follow [Landau (1987)]. Let $P, Q$ be non-commuting projectors, and also let $P', Q'$ be non-commuting projectors, while $P$ is compatible with $P'$ and with $Q'$, and $Q$ is compatible with $P'$ and $Q'$. For example, in the *EPR* experiment, $P$ and $Q$ could be projectors onto the value $+1$ of the $x$ and $z$-components of the spin of the particle on the left, and $P'$ and $Q'$ similar projectors on the right. Define $A = 2P - I$, $B = 2Q - I$, and similarly for $A'$ and $B'$. For any state $\rho$ define $R$ by

$$R := \rho(AA' + AB' + BB' - BA') = \rho(C)$$

where $C = A(A' + B') + B(B' - A')$. Then $A^2 = B^2 = A'^2 = B'^2 = 1$, so

$$C^2 = 4 + [A, B][A', B'] = 4 + 16[P, Q][P', Q']. \tag{9.11}$$

Since $\|A\| = \|B\| = \|A'\| = \|B'\| = 1$, it follows that

$$\|[A, B][A', B']\| \leq 4,$$

so $C^2 \leq 8$ and $|R|^2 = |\rho(C)|^2 \leq \rho(C^2) \leq 8$. So in quantum theory, $|R| \leq 2\sqrt{2}$. This result was also found by Tsirelson [Tsirelson (1985)]. If there is a joint probability space on which we can describe $A, \ldots, B'$ by the random variables $f, \ldots g'$ taking the values $\pm 1$, and a measure $p$ on it, then $R = E_p[h]$ where

$$h = f(f' + g') + g(g' - f').$$

Then these random variables commute, so Eq. (9.11) becomes $h^2 = 4$, and

$$|R^2| = E_p[h]^2 \leq E_p[h^2] = 4.$$

So $|R| \leq 2$, (Bell's inequality). Bell showed that the entangled states of the Bohm-*EPR* set-up give a $\rho$ such that $R = 2\sqrt{2}$, violating this. Thus no description by classical probability is possible.

The famous experiment by Aspect et al. tested Bell's inequalities [Aspect (Grangier and Roger)]. This involves observing a system (in a pure entangled state) in a long run of measurements; the correlations singled out by Bell, between several compatible pairs of spin observables, were measured. The experiments showed that $R$ was just less than $2\sqrt{2}$, in agreement with the quantum predictions.

We see that this proof does not involve Bell's locality assumptions; only the non-commutativity of the algebra of observables is used.

## 9.3  Quantum Entropy

Let $\rho$ be a normal state on $B(\mathcal{H})$ and $\Delta\rho$ the corresponding density operator; in this section, we shall use the symbol $\rho$ to denote $\Delta\rho$. In his book [von Neumann (1932)], von Neumann defined the entropy of $\rho$ to be

$$S(\rho) = -Tr(\rho \log \rho). \tag{9.12}$$

In this definition, $\log \rho$ is defined by the functional calculus for self-adjoint operators. If $\dim \mathcal{H} = \infty$ then there are some operators $\rho$, even of trace class, for which this does not converge. Some authors write $S(\rho) = \infty$ in this case. This entropy, like its classical version, Def. 2.6, is extensive. Thus, suppose that the algebra of observables $\mathcal{A}$ has a local structure over a space $\Lambda$, and is a tensor product: $\mathcal{A} = \bigotimes_x \mathcal{A}_x$. Suppose too that the state $\rho$ is independent over $\Lambda$, by which we mean that

$$\rho = \bigotimes_{x \in \Lambda} \rho_x.$$

Then

$$S(\rho) = \sum_{x \in \Lambda} S(\rho_x).$$

The proof is very similar to that for classical probability, and you may do it yourself in Exercise 9.29. Thirring [Thirring (1980)] p. 58, gives a list of properties of $S$ which determine $S$ uniquely. We see that the entropy of a pure state is zero. This is because we use the convention in the definition of $S$ that $0 \log 0 = 0$ and $\log 1 = 0$; this means that $P \log P = 0$ as an operator if $P$ is a one-dimensional projection. This is in spite of the fact that in quantum mechanics any pure state gives rise to some uncertainty (Heisenberg's uncertainty principle). Thus, $S(\rho)$ measures the lack of information that is in addition to the intrinsic quantum uncertainty. Note that the classical entropy of the probability distribution $|\psi(x)|^2$ of a normalisable wave-function in one dimension is infinite, if it is defined as the limit of the Shannon entropy of the coarse-grained approximations to it. Since it represents a pure state, its quantum entropy is zero. This is one of many ways in which quantum probability improves on classical probability.

In the finite-dimensional case, the microcanonical state is the state that maximises the entropy. To see this, use a Lagrange multiplier to express that the trace of a state is 1. Thus we seek to maximise

$$-Tr(\rho \log \rho) + \lambda Tr \rho$$

as we roam over the set of positive matrices $\rho$. The trace is invariant under unitary transformations; let us diagonalise $\rho$ by a unitary matrix. So it suffices to consider the same problem where $\rho$, and therefore also $\log \rho$, is diagonal. But then the product $\rho \log \rho$ involves only the product of these diagonal elements, and the problem reduces to the classical theorem. The result is that all components along the diagonal are equal, so that $\rho$ is a multiple of the identity matrix. All the unitary transforms are then equal to each other, so the answer is unique.

The crux is the following

**Theorem 9.5.** *Let $\mathcal{H}$ be a Hilbert space and $H \geq 0$ an unbounded self-adjoint operator with eigenvalue $0$ such that $e^{-\beta H}$ is of trace class for all $\beta > 0$. Let $\rho_\beta$ be the canonical state defined by the density operator $Z_\beta^{-1} e^{-\beta H}$. Let $E > 0$ be given. Then $\rho_\beta$ is the unique state that maximises $S(\rho)$ among all normal states $\rho$ on $B(\mathcal{H})$ satisfying the conditions that $\rho H$ is of trace class and $Tr(\rho H) = E$. This equation then determines $\beta$ uniquely.*

This theorem will easily follow from a quantum version [Streater (1985)] of Kullback's lemma:

**Lemma 9.6.** *Let $\rho$ and $\sigma$ be density operators. Then*

$$S(\rho|\sigma) := Tr\left(\rho(\log \rho - \log \sigma)\right) \geq \frac{1}{2} Tr(\rho - \sigma)^2.$$

Note. The left-hand side is called the relative entropy of $\sigma$, given $\rho$.

**Proof.**   [proof of lemma] Let $(\phi_i, a_i)$ and $(\psi_i, b_i)$ be the orthonormal eigen-vectors and eigenvalues of $\rho$ and $\sigma$ respectively. Then $0 \leq a_i, b_j \leq 1$. Let $f(x) = x \log x$ and $c_{ij} = \langle \phi_i, \psi_j \rangle$. Then $\sum_j |c_{ij}|^2 = \langle \phi_i, \phi_i \rangle = 1$ and

$$\langle \phi_i, \{f(A) - f(B) - (A - B)f'(B) - (1/2)(A - B)^2\}\phi_i \rangle$$
$$= \langle \phi_i, f(a_i)\phi_i \rangle - \sum_j \langle \phi_i, \psi_j \rangle \langle \psi_j, f(b_j)\phi_i \rangle$$
$$- \sum_j \langle \phi_i, (a_i - b_j)f'(b_j)\psi_j \rangle \langle \psi_j, \phi_i \rangle$$
$$- \frac{1}{2}\left\{ \langle \phi_i, a_i^2 \phi_i \rangle - 2\sum_j a_i b_j \langle \phi_i, \psi_j \rangle \langle \psi_j, \phi_i \rangle + \sum_j \langle \phi_i, b_j^2 \psi_j \rangle \langle \psi_j, \phi_i \rangle \right\}$$
$$= \sum_j |c_{ij}|^2 \left\{ f(a_i) - f(b_j) - (a_i - b_j)f'(b_j) - \frac{1}{2}(a_i - b_j)^2 \right\}$$

and this is non-negative by Taylor's theorem with remainder. Summing over $i$ gives

$$Tr\left\{A\log A - B\log B - (A-B)(\log B + 1) - \frac{1}{2}(A-B)^2\right\} \geq 0 \quad (9.13)$$

that is

$$Tr\,(A(\log A - \log B)) \geq \frac{1}{2}Tr\,(A-B)^2. \qquad \square$$

Note. This is a sharpened form of a result in [Ruelle (1969)], proved by the same method. There is a more general form with the same proof, in which $A$ and $B$ are positive operators with eigenvalues in $[0, 1]$. Then the result is adjusted by $Tr\,A - Tr\,B$, as can be seen from Eq. (9.13).

**Proof.** [proof of theorem] Let $\sigma = \rho_\beta$ in the lemma. Then

$$Tr\,(\rho(\log\rho - \log\rho_\beta)) = Tr\,(\rho\log\rho) + Tr\,(\rho\beta H) + \log Z_\beta$$
$$\geq \|\rho - \rho_\beta\|_2^2/2$$

by the lemma. Rearranging gives

$$-Tr\,(\rho\log\rho) + \|\rho - \rho_\beta\|_2^2/2 \leq \beta Tr\,(\rho H) + \log Z_\beta$$
$$= \beta Tr\,(\rho_\beta H) + \log Z_\beta$$
$$= S(\rho_\beta).$$

To show that $\beta$ is uniquely determined, note that the mean energy is infinite if $\beta = 0$, so we can choose $\beta$ so that the mean energy is larger than the given $E$. But then differentiation shows that

$$\frac{\partial}{\partial\beta}Tr\,(\rho_\beta H) = -Tr(\rho_\beta(H - \overline{H})^2) < 0$$

where $\overline{H}$ is the mean energy in the state $\rho_\beta$. So it is strictly decreasing. Since the mean energy at infinite $\beta$ is zero, the energy of the ground state, we can find a unique $\beta$ with $\overline{H} = E$. $\qquad\square$

We see from the proof that under the conditions of the theorem, $S(\rho)$ is finite. It also follows that the operator $H\rho$ is of trace-class; we shall write $\rho.H$ for its trace; this replaces the notation $\rho(A)$ when $A$ is unbounded.

A small elaboration on this proof shows that the grand canonical state is the unique state maximising the entropy subject to a given mean energy and a given mean particle number. In fact we can include any finite number of constraints, in terms of the given means of Hermitian operators $X_1, \ldots, X_m$, not necessarily commuting. We have

**Theorem 9.7.** *Suppose that $X_1, \ldots, X_m$ are positive operators with a common domain, such that for all positive $\beta_i$, $\exp\{-\sum_i \beta_i X_i\}$ is of trace class. Then the great grand canonical state*

$$\Delta \rho = Z^{-1} \exp\{-\sum_i \beta_i X_i\} \tag{9.14}$$

*is the unique state that maximises $S(\rho)$ among all normal states such that $\rho X_i$ is of trace class and*

$$\rho.X_1 = E_1, \qquad \ldots \qquad \rho.X_m = E_m.$$

**Proof.** First, solve a problem we know how to do: let $\lambda_1, \ldots, \lambda_m$ be any positive numbers, and put $H = \sum \lambda_i X_i$. Now maximise $S(\rho)$ subject to one condition, $\rho.H = E$. The result is unique: $\rho = \rho_\beta$, where $\Delta \rho_\beta = Z^{-1} e^{-\beta H}$, and $\beta$ is determined by $\rho_\beta.H = E$. Now the observables $X_i$ will have certain expectations in this state, say $\rho_\beta.X_i = E_i$; these must satisfy the condition $\sum \lambda_i E_i = E$. Conversely, put $\beta_i = \beta \lambda_i$; then for a given possible set of values $E_i$, the $\beta_i$ are unique (by the fact that entropy is a strictly decreasing function of each variable). Moreover, $\rho_\beta$ is the state of maximum entropy with these values of $E_i$; for if not, we would get a state with the same value of $E$ but a bigger entropy, contrary to the definition of $\rho_\beta$.     $\square$

In Jaynes's theory, it is argued that if we measure the means of quantum variables $X_1, \ldots, X_n$, but nothing else, then the best estimate for the state is that of maximum entropy, given the means. He takes a Bayesian view of the question. In Chap. 15 we justify this answer, using the non-Bayesian method of estimation theory.

We see that quantum entropy is not decreased by the stoss map:

**Theorem 9.8.** *Let $\mathcal{H}$ be a Hilbert space and $\mathcal{A} = \mathcal{A}_1 \bigotimes \mathcal{A}_2 \subseteq B(\mathcal{H})$, with marginal maps $\mathcal{M}_1$ and $\mathcal{M}_2$. Then for any normal state $\rho \in \Sigma(\mathcal{A})$ with finite entropy we have*

$$S(\mathcal{M}_1 \rho \otimes \mathcal{M}_2 \rho) \geq S(\rho) + \frac{1}{2} \| \rho - \mathcal{M}_1 \rho \otimes \mathcal{M}_2 \rho \|_2^2. \tag{9.15}$$

**Proof.** Just put $\sigma = \mathcal{M}_1 \rho \otimes \mathcal{M}_2 \rho$ in Kullback's lemma, (9.6) and use the idempotent property of the partial trace:

$$Tr\,(\rho \log \mathcal{M}_1 \rho) = Tr_1 \mathcal{M}_1 \rho \log \mathcal{M}_1 \rho$$

for example.     $\square$

We can use Kullback's lemma (quantum version), (9.6) to estimate the concavity of the entropy; we get the same result as Theorem 2.2:

**Theorem 9.9 (Concavity of the Quantum Entropy).** *Let $\rho$ and $\sigma$ be density operators of finite entropy on a Hilbert space. Then*

$$S(\lambda\rho + (1-\lambda)\sigma) - \lambda S(\rho) - (1-\lambda)S(\sigma) \geq \frac{1}{2}\lambda(1-\lambda)\|\rho - \sigma\|_2^2 \quad (9.16)$$

*for all $\lambda \in [0, 1]$.*

**Proof.** Kullback's lemma says that

$$-Tr\,\rho\log\sigma \geq -Tr\,\rho\log\rho + \frac{1}{2}\|\rho - \sigma\|_2^2;$$

try this with $\sigma$ replaced by $\lambda\rho + (1-\lambda)\sigma$. It gives

$$-Tr\,\rho\log(\lambda\rho + (1-\lambda)\sigma)$$
$$\geq -Tr\,\rho\log\rho + \frac{1}{2}\|\rho - (\lambda\rho + (1-\lambda)\sigma)\|_2^2$$
$$= S(\rho) + \frac{1}{2}(1-\lambda)^2\|\rho - \sigma\|_2^2.$$

Now use it with $\rho$ replaced by $\sigma$ and $\sigma$ replaced by $\lambda\rho + (1-\lambda)\sigma$:

$$-Tr\,\sigma\log(\lambda\rho + (1-\lambda)\sigma)$$
$$\geq -Tr\,\sigma\log\sigma + \frac{1}{2}\|\sigma - (\lambda\rho - (1-\lambda)\sigma)\|_2^2$$
$$= S(\sigma) + \frac{1}{2}\lambda^2\|\rho - \sigma\|_2^2.$$

Take $\lambda$ times the first estimate and $(1-\lambda)$ times the second, and add, to get

$$S(\lambda\rho + (1-\lambda\sigma)) - \lambda S(\rho) - (1-\lambda)S(\sigma)$$
$$\geq \frac{1}{2}(\lambda(1-\lambda)^2 + (1-\lambda)\lambda^2)\|\rho - \sigma\|_2^2$$
$$= \frac{1}{2}\lambda(1-\lambda)\|\rho - \sigma\|_2^2.$$

$\square$

**Corollary.** Any mixture $\tau = \sum_i \lambda_i\tau_i$ of automorphisms $\tau_i$ gives rise to a dual action $\tau^*$ which is entropy non-decreasing.

Many further properties of quantum entropy can be found in [Thirring (1980)] and [Petz (2002)].

## 9.4   Exercises

**Exercise 9.17.** Show that $\mathcal{A} = \mathbf{M}_n$ obeys the axioms of a $C^*$-algebra.

Exercise 9.18. Let $\mathcal{A}_1$ and $\mathcal{A}_2$ be $C^*$-algebras. A linear map, $T$, from $\mathcal{A}_1$ to $\mathcal{A}_2$ is called a *homomorphism* if it preserves the structure of $C^*$-algebra. This means that

- $T(AB) = T(A)\,T(B)$ for all $A$, $B$ in $\mathcal{A}$;
- $T(A^*) = (TA)^*$ for all $A \in \mathcal{A}$.

We say that $T$ is an isomorphism if it is bijective; in this case we say that $\mathcal{A}_1$ and $\mathcal{A}_2$ are *isomorphic*.

Show that if $\dim \mathcal{H} = n < \infty$, then $B(\mathcal{H})$ and $\mathbf{M}_n$ are isomorphic.

Exercise 9.19. Show that the set of unitary elements in a $C^*$-algebra with identity forms a group.

Exercise 9.20. Show that the star operation defined on dyads and extended by anti-linearity gives the same answer whichever representatives are chosen.

Exercise 9.21. Show that the definition of product on $\mathcal{A}_1 \otimes \mathcal{A}_2$ is independent of the representatives chosen.

Exercise 9.22. Let $G$ be a finite group, with $n$ elements. Define a multiplication on $\mathrm{Span}\,G$ by

$$\left(\sum_i \lambda_i g_i\right) \times \left(\sum_j \mu_j g_j\right) = \sum_{ij}(\lambda_i \mu_j)(g_i g_j).$$

In this, the symbol $g_i g_j$ denotes the product in the given group. Define a conjugation by $g^* = g^{-1}$, extended by anti-linearity to $\mathrm{Span}\,G$. Show that this makes $\mathrm{Span}\,G$ into a *-algebra, $\mathcal{A}(G)$ say. Now let $\mathcal{A}(G)$ act on $\mathrm{Span}\,G$ by left multiplication, (the *regular* representation). Show that in this way, each element of $\mathcal{A}$ defines a linear operator on $\mathrm{Span}\,G$. Take the elements of $G$ as a basis in $\mathrm{Span}\,G$, and for each $A \in \mathcal{A}$ let $M(A)$ be the matrix of this linear map in this basis. Show that $A \mapsto M(A)$ is an injective homomorphism from $\mathcal{A}$ into $\mathbf{M}_n$. Use this fact to furnish $\mathcal{A}(G)$ with a $C^*$-norm.

Exercise 9.23. Let $\mathcal{A} = \mathbf{M}_n$ and let $\rho$ be a positive semi-definite $n \times n$ matrix with unit trace. Show that the map $A \mapsto Tr\,(\rho A)$ defines a state on $\mathcal{A}$.

Exercise 9.24. Show that any state $\rho$ on $\mathbf{M}_n$ is of the form $Tr\,[\Delta(\rho)A]$ for some density matrix $\Delta(\rho)$. [Hint: Try defining $\Delta(\rho)$ as the matrix whose $i, j$ matrix-elements are the expectations of the matrix units $E_{ji}$ in the state

$\rho$. We shall often use the symbol $\rho$ for the corresponding density matrix $\Delta(\rho)$ in the following text.]

Exercise 9.25. Show that the state space of any $C^*$-algebra is a convex set.

Exercise 9.26. Let $\mathcal{A}$ be an abelian $C^*$-algebra of finite dimension, containing the identity. Show that $\mathcal{A}$ is isomorphic to the set of functions on a finite set. [Hint: let $\Omega$ be the space of homomorphisms from $\mathcal{A}$ into $\mathbf{C}$.]

Exercise 9.27. Show that a density matrix is extremal in $\Sigma(\mathbf{M}_n)$ if and only if it is a projection of dimension 1. [Hint: show that the spectral resolution of a density matrix expresses it as a mixture of projections.]

Exercise 9.28. Show that a vector state on $B(\mathcal{H})$ given by a vector $\psi$ is the same as that defined by the density operator $P_\psi$, the projection onto the subspace spanned by $\psi$.

Exercise 9.29. Let $\rho = \otimes_{x \in \Lambda} \rho_x$ be a state on $\mathcal{A} = \otimes_{x \in \Lambda} \mathcal{A}_x$. Show that $S(\rho) = \sum_x S(\rho_x)$. [Hint: the logarithm of a density matrix, defined by the spectral theorem, has all the usual properties.]

# Chapter 10

# Linear Quantum Dynamics

## 10.1 Reversible Dynamics

The dynamics of a quantum system is either expressed in the Heisenberg picture or in the Schrödinger picture. The first concerns the $C^*$-algebra and the second the state space. The first seems to be the more basic. One time-step in the algebra, like any symmetry transformation, is expressed as a *-algebraic automorphism of the algebra of observables; this is simply an isomorphism $\tau$ say, from $\mathcal{A}$ onto $\mathcal{A}$. We write the action on the left, to concur with what we did in the classical case: $A \mapsto \tau A$. Because an isomorphism preserves the * operation, $\tau$ must take Hermitian elements to Hermitian elements. Any automorphism has an inverse, which is also an automorphism. Obviously, the identity map is an automorphism. Also, the composition of two automorphisms is an automorphism. We denoted the composition by $\circ$: $\tau_1 \tau_2 A = (\tau_1 \circ \tau_2)A$ for any $A \in \mathcal{A}$. Thus the set of automorphisms of a $C^*$-algebra is a group denoted $AUT\mathcal{A}$. We may throw the action of the automorphism group onto the states, by duality. Again, this is written as a right-action, thus: given $\tau$ and a state $\rho$, we define the transformed state $\rho\tau$ by

$$\rho\tau \cdot A = \rho \cdot (\tau A) \text{ for all } A \in \mathcal{A}.$$

Given the transformation $\tau$ for one time step, we define the transformation for $n$ time steps as $\tau^n = \tau \circ \tau \ldots \tau$ ($n$ factors). A time-evolution (with discrete time) is then a homomorphism (here $n \mapsto \tau^n$) from the group $\mathbf{Z}$ into $AUT\mathcal{A}$. In the same way, we can argue that a time-evolution with continuous time is a homomorphism $t \mapsto \tau_t$ from $\mathbf{R}$ into $AUT\mathcal{A}$. To be a homomorphism means that the group law $\tau_s \circ \tau_t = \tau_{s+t}$ holds for all real $s$ and $t$. In this case it is natural to postulate some sort of continuity for the homomorphism.

Because we are led to a dynamical group, and not just a semigroup, automorphisms lead to reversible dynamics: there is no dissipation.

In a $C^*$-algebra $\mathcal{A}$, a unitary element $U \in \mathcal{A}$ defines an automorphism by the map $A \mapsto \tau_U A = U A U^{-1}$ for all $A \in \mathcal{A}$. Such an automorphism is called *inner*; the automorphism is often denoted by $\mathrm{Ad}\, U$. One can prove that $\tau_U \circ \tau_V = \tau_{UV}$, $\tau_1 = Id$ and $(\tau_U)^{-1} = \tau_{U^{-1}}$. See Exercise 10.15. Indeed the set of inner automorphisms form a group, denoted $INN\mathcal{A}$. One can prove that every automorphism of $\mathbf{M}_n$ is inner. If $\theta$ is a real number, and $V = e^{i\theta} U$, then $\tau_U = \tau_V$. More generally, let $\mathcal{Z}(\mathcal{A})$ denote the *centre* of $\mathcal{A}$, that is, the set of elements of $\mathcal{A}$ that commutes with all elements of $\mathcal{A}$, and let $Z \in \mathcal{Z}$ be unitary. Then $\tau_{ZU} = \tau_{UZ} = \tau_U$. In fact this is the extent of the ambiguity in $U$; see Exercise 10.16. Obviously for a commutative algebra only the identity automorphism is inner. Thus there are $C^*$-algebras for which not all automorphisms are inner; in infinite dimensions, this is usually the case, even for non-abelian algebras. In that case the following question becomes interesting: given a representation $\pi$ of $\mathcal{A}$ on a Hilbert space $\mathcal{H}$, and an automorphism $\tau$ of $\mathcal{A}$, can we find a unitary operator $U \in B(\mathcal{H})$ such that

$$\pi(\tau A) = U \pi(A) U^{-1} \text{ for all } A \in \mathcal{A}?$$

If so, we say that $\tau$ is *spatial in* $\pi$. If $\tau$ is inner, say $\tau = \tau_V$, then $\tau$ is spatial in all representations, since in the representation $\pi$ we may take $U = \pi(V)$. There is an important case when we can show that an automorphism $\tau$ is spatial in a representation; this is when the representation is of the form $\pi_\rho$, where $\rho \in \Sigma(\mathcal{A})$ is invariant under $\tau$. For then the operator on $\mathcal{H}_\rho$ given by $\pi_\rho(A)\psi_\rho \mapsto \pi_\rho(\tau A)\psi_\rho$ is unitary; see and do Exercise 10.17. The same remarks apply to any automorphism of $\mathcal{A}$, whether or not it represents time-evolution. In particular, in a non-relativistic theory, a *symmetry* of the system is an automorphism $\alpha$ commuting with the time-evolution; this might or might not be spatial in a representation $\pi$, even if $\tau_t$ is spatial in $\pi$. If $\alpha$ is a symmetry which is not spatial in $\pi$, we say that it is spontaneously broken in $\pi$. Note that the concept of spontaneously broken symmetry depends on the representation chosen, and so is not a purely algebraic property. Our definition of spontaneously broken symmetry [Streater (1965)] differs from that used by some authors, who use the term to describe the situation where the state $\rho$ used in the $GNS$ construction is invariant under the time-evolution but not under the symmetry automorphism $\alpha$. This leaves open the question as to whether $\alpha$ is spatial in $\pi_\rho$ or not. The Heisenberg ferromagnet exhibits a spontaneous breakdown of symmetry in both senses. The ground state is not invariant under rotations; and also the rotation-automorphisms commute with time-evolution, but are not spatial in the vacuum representations. This model is studied in Sec. 15.4. In [Streater (1990)] a strange example is constructed; time is continuous and space is discrete, and the space-time automorphisms commute. A state $\rho$ is presented which is invariant under time-evolution but not space translations, but where nevertheless space-time translations are spatial in $\pi_\rho$, being given by $U(t)$ and $V(n)$ say. The remarkable fact is that in this example there is no choice of implementing operators $U(t)$ and $V(n)$ that commute with each other. This situation is called an *anomaly*. In this example, the symmetry group of space-time translations is the group $G := \mathbf{Z} \times \mathbf{R}$, and an element $g \in G$ is the pair $g = (n, t)$. We give an example of a representation $\pi$ of the observable algebra, in which $\tau_g$ is spatial, given by operators $U_g$, say. So,

$$U_g \pi(A) U_g^{-1} = \pi\left(\tau_g(A)\right) \text{ for all } g \in G \text{ and all } A \in \mathcal{A}.$$

However, each $U_g$ can be replaced by its multiple by any unitary in the commutant, $\pi(\mathcal{A})'$; thus, the $U_g$ do not obey the group law of $G$, but instead obey

$$U_g U_h = \omega(g, h) U_{gh}, \tag{10.1}$$

where $\omega$ is a unitary operator in the commutant of $\pi(\mathcal{A})$. It forms a cocycle in a more general cohomology theory than that used by Wigner [Streater (1990, 2007)]. Indeed, in [Streater (2007)] it is proposed that by allowing a reducible representation of the observables, obtained by a non-invertible endomorphism of the free representation, we might get non-abelian multipliers and gauge fields, which could describe an interacting system of relativistic particles.

In a relativistic theory, with relativity group $G$, we are given a homomorphism $\tau$ from $G$ into $AUT\mathcal{A}$, so that $\tau_g$ is an automorphism for each $g \in G$, and $\tau_g \tau_h = \tau_{gh}$. Let $(\mathcal{H}, \pi)$ be a representation of $\mathcal{A}$ in which each $\{\tau_g, g \in G\}$ is spatial, being implemented by unitaries $\{U_g\}$; then we say that $G$ is a symmetry group of the model. Note that in general $g \mapsto U_g$ is not a group homomorphism, but obeys Eq. (10.1), where $\omega$ is a unitary cocycle. If $\rho$ is invariant under the dual action of the group on the states, then each $\tau_g$ is spatial in $\pi_\rho$, and the implementing operators $U_g : g \in G$ say, also obey the group law: $U_g U_h = U_{gh}$. In this case there are no multipliers or anomalies. In particular, the time-evolution forms a one-parameter group, $U_t$, which may under reasonable assumptions be taken to be continuous. Thus, $U_t$ satisfies

**Definition 10.1 (Continuous One-parameter Group).** *(1) $U_0 = 1$*
*(2) $U_t \psi \to \psi$ in $\mathcal{H}$ as $t \to 0$.*
*(3) $U_s U_t = U_{s+t}$ for all $s$, $t \in \mathbf{R}$.*

Stone's theorem asserts that $U_t \psi$ is differentiable for a dense set $\mathcal{D}$ of $\psi$, with self-adjoint generator, called the energy, $H$. Thus, in the sense of norm convergence in Hilbert space,

$$-iH\psi = \lim_{t \to 0} t^{-1}(U_t - 1)\psi \text{ if } \psi \in \mathcal{D}. \tag{10.2}$$

It is clear that the energy operator depends on the representation in which we are working (unless $\tau_t$ is inner, which is rare if $\dim \mathcal{A} = \infty$). Even the property of being bounded below is not algebraic; usually, the generator of the time-evolution, the energy, is positive in the representation generated by a certain state, called the ground state. It will turn out that for infinite systems, the energy is never bounded below in the thermal states (a general term for one of the canonical states). Even when the algebra is of finite dimension, it will be convenient to choose the zero-point of energy to be in the centre of the range of eigenvalues, when we are in a thermal state. We see the beginnings of this in the dilation theorem for Markov chains, Theorem 5.14.

Just as in the classical case, it is easy to see that the entropy is unchanged by any spatial automorphism: the trace is invariant under any unitary transformation, since it is independent of the base used in its calculation. Thus let $\rho$ be a normal state and $\Delta\rho$ the density operator corresponding to it; the functional calculus for self-adjoint operators implies that $\log(U \Delta\rho U^{-1}) = U \log(\Delta\rho) U^{-1}$. Thus if $\rho'$ is such that $\Delta\rho' = U \Delta\rho U^{-1}$, we have

$$\begin{aligned}
S(\rho') &= -Tr\left[U \Delta\rho U^{-1} \log(U \Delta\rho U^{-1})\right] \\
&= -Tr\left[U \Delta\rho U^{-1} U \log(\Delta\rho) U^{-1}\right] \\
&= -Tr\left[U(\Delta\rho \log \Delta\rho)U^{-1}\right] = S(\rho).
\end{aligned}$$

Thus we have the same problem as in classical statistical dynamics, that the microscopic theory does not lead to dissipation. We therefore follow the same steps as we used there, and next consider random unitary dynamics.

## 10.2 Random Quantum Dynamics

In this section, we will take a representation of the algebra of observables $\mathcal{A}$ in which time-evolution is spatial, with a continuous unitary representation

$U_t$ of the group $\mathbf{R}$, with generator $H$. We will assume that the spectral projections of $H$ lie in $\mathcal{A}$. The idea of quantum ergodic theory is that a measurement of an observable takes a time very much longer than the time-scale for microscopic dynamics. Instead of measuring an observable represented by an operator $A \in \mathcal{A}$, we measure the mean of very many time-translates of $A$; thus we measure $A_N = N^{-1} \sum_{n=0}^{N} \tau^n A$. (This is still an operator in $\mathcal{A}$, not its quantum mean in some state.) Here, $\tau$ is the time translation over a small time-step, typical of the microscopic system. We then let $N$ tend to infinity. On the other hand, $N$ is to be a small time on the scale of the slow variables, whose dynamics we wish to study. We note that for each term in the sum, $U(n)$ commutes with the energy operator $H$, and so $\tau$ and $\tau^{-1}$ leave fixed the spectral projections $\{P_E\}$ of $H$ (which lie in the algebra). The map $A \mapsto A_N$ is linear, and can be thrown by duality onto $\Sigma(\mathcal{A})$. It is easy to see that the resulting motion through the states conserves the mean energy. See Exercise 10.18. By the corollary to Theorem 9.9, this dynamics increases entropy; it therefore obeys both the first and second laws of thermodynamics.

We are not obliged to adopt the ergodic theorist's interpretation of measurement. These ideas can be incorporated into a more general scheme; we let $H$ be a self-adjoint operator, taken to be the energy, and consider $N$ unitary operators $U_1, \ldots, U_N$, each of which commutes with every spectral projection of $H$. Let $\tau_n A := U_n A U_n^{-1}$, and consider the mixture $\tau = \sum_i \lambda_i \tau_i$, where $0 < \lambda_i < 1$ and $\sum_i \lambda_i = 1$. Then the dynamics defined by $\tau$ obeys the first and second laws of thermodynamics. We call $\tau$ a *random reversible dynamics*, (and also limits of such, as the number of terms goes to infinity). In the classical case, conservation of mean energy was a result of the property that the transformation leaves the energy-shell invariant. Equivalently, its action on the algebra of random variables leaves invariant the indicator functions of the energy-shells. The quantum version of this is that each $\tau_i$ from which $\tau$ is composed, and the inverses, should leave the spectral projections $P_E$ of $H$ fixed. Equivalently, they should map the subalgebras $P_E A P_E$ to themselves, for all eigenvalues $E$ of $H$. It will follow from the assumption that $e^{-H}$ is of trace class that the spectrum of $H$ is discrete and that each eigenvalue has finite multiplicity. There is also a restriction on the number of eigenvalues $E_n$ as $n$ becomes large. Similar restrictions were found to be necessary in the classical case, where for example, we needed to assume that each energy-shell was finite.

The interpretation of $\tau_i$ as the dynamics due to a random potential causes problems, since there seems no physical reason to expect all these random Hamiltonians to commute with the energy represented by $H$. We therefore prefer to regard the $\tau_i$ as the scattering automorphisms caused by the random interactions; on the time scale of the fast variables, one time-step of our dynamics is indeed very large, and the observables at the end of this time-span are related to the same at the beginning of the time-span by the scattering operator for the fast variables: $A_+ = SA_-S^{-1}$. It is a fact from scattering theory that the scattering operator conserves the energy of the asymptotic particles, for a wide range of potentials; so we are not out of line in asssuming that $H$ represents the asymptotic energy, and that each $\tau_i$ leaves its spectral projections invariant. Indeed, the collision term in the Boltzmann equation described scattering and it conserves the asymptotic energy.

We shall be interested in finding examples of dynamics for which $\rho\tau^n$ converges to a limit as $n \to \infty$ for every state $\rho$. For this, we need that $\tau$ mix up each algebra $\mathcal{A}_E = P_E \mathcal{A} P_E$ thoroughly. Since $\tau$ is a mixture of automorphisms $\tau_i$, we can always write it as

$$\tau = \tau_1 \left[ \lambda_1 1 + \sum_{i=2} \lambda_i \tau_1^{-1} \tau_i \right];$$

here, the choice of $\tau_1$ was arbitrary. The prefactor $\tau_1$ does not contribute at all to the mixing, since entropy is invariant under automorphisms; this part of the dynamics is reversible. It is the remaining part that must cause the convergence to equilibrium. This idea leads us to the following formulation:

**Definition 10.2.** Suppose that $\mathcal{A}$ is a represented $C^*$-algebra with an implemented time-evolution, with generator $H$, whose spectral projections lie in $\mathcal{A}$. Suppose that a random dynamics $\tau$ is a mixture of automorphisms

$$\tau = \lambda_0 1 + \sum_i \lambda_i \tau_i, \text{ where } 0 < \lambda_i < 1, \qquad \sum_0^n \lambda_i = 1$$

each $\tau_i$ leaving the spectral projections of $H$ invariant. We say that $\tau$ is *ergodic relative to $H$* if the only elements of $\mathcal{A}$ invariant under $\tau$ are functions of $H$.

Being a 'function of $H$' is in the sense of the functional calculus for operators, and is equivalent to lying in the $C^*$-algebra generated by the spectral projections $P_E$. The ergodic condition then expresses that there is no non-trivial subalgebra of $\mathcal{A}_E$ that is mapped to itself by the dynamics, so each 'energy-shell' really does get fully mixed up.

**Theorem 10.3.** *Let* $\dim \mathcal{A}_E = N < \infty$ *and suppose that* $\tau$ *is a mixture as in the definition, and is ergodic relative to* $H$. *Then for any initial state* $\rho \in \Sigma(\mathcal{A}_E)$, *we have*

$$\lim_{r \to \infty} \rho^r \to N^{-1}1.$$

**Proof.** In finite dimensions the set of states is compact. The entropy is a strict Lyapunov function for the dynamics, since by Theorem 9.9,

$$S(\rho \tau^n) = S\left(\rho(\lambda_0 1 + \sum_{i=1}^{n} \lambda_i \rho \tau_i)\right)$$

$$= S\left(\lambda_0 \rho + (1 - \lambda_0) \sum_{i=1}^{n} \frac{\lambda_i}{1 - \lambda_0} \rho \tau_i\right)$$

$$\geq \lambda_0 S(\rho) + (1 - \lambda_0) S\left(\sum_{i=1}^{n} \frac{\lambda_i}{1 - \lambda_0} \rho \tau_i\right) + \varepsilon \qquad (10.3)$$

where

$$\varepsilon = \frac{\lambda_0 (1 - \lambda_0)}{2} \left\| \rho - \sum_{i=1}^{n} \frac{\lambda_i}{1 - \lambda_0} \rho \tau \right\|_2^2. \qquad (10.4)$$

We note that $\varepsilon = 0$ only if $(1 - \lambda_0)\rho = \sum \lambda_i \rho \tau_i$, which reduces to $\rho \tau = \rho$. By the ergodic condition, this is possible only if $\rho$ is a multiple of the identity, that is, $\rho$ must be the microcanonical state. We can therefore afford to leave out this sharp form of the inequality in the further entropy estimates. We note that

$$\sum_{i=1}^{n} \frac{\lambda_i}{1 - \lambda_0} \rho \tau_i = \frac{\lambda_1}{1 - \lambda_0} \rho \tau_1 + \frac{1 - \lambda_0 - \lambda_1}{1 - \lambda_0} \sum_{i=2}^{n} \frac{\lambda_i}{1 - \lambda_0 - \lambda_1} \rho \tau_i. \qquad (10.5)$$

It follows from this and Theorem 9.9 that

$$S\left(\sum_{i=1}^{n} \frac{\lambda_i}{1 - \lambda_0} \rho \tau_i\right) \geq \frac{\lambda_1}{1 - \lambda_0} S(\rho \tau_1)$$

$$+ \frac{1 - \lambda_0 - \lambda_1}{1 - \lambda_0} S\left(\sum_{i=2}^{n} \frac{\lambda_i}{1 - \lambda_0 - \lambda_1} \rho \tau_i\right).$$

Put this in (10.3) to get

$$S(\rho \tau) \geq \lambda_0 S(\rho) + \lambda_1 S(\rho \tau_1) + (1 - \lambda_0 - \lambda_1) S\left(\sum_{i=2}^{n} \frac{\lambda_i}{1 - \lambda_0 - \lambda_1} \rho \tau_i\right). \qquad (10.6)$$

Proceeding in the same way, and using the fact that $\tau_i$ are automorphisms, we arrive at

$$S(\rho\tau) \geq (\lambda_0 + \lambda_1 + \ldots + \lambda_n)S(\rho) + \varepsilon = S(\rho) + \varepsilon.$$

This shows that $S$ is a strict Lyapunov function, since $\varepsilon = 0$ only at a fixed point. Since the fixed point is unique, and the space is compact, the theorem follows from Lyapunov's direct method. $\qquad\square$

## 10.3 Quantum Dynamical Maps

A random dynamics $\tau$ generated by a mixture of automorphisms $\tau_i$ of a $C^*$-algebra $\mathcal{A}$ is a *stochastic map*, that is, it obeys

(1) $\tau$ is linear;
(2) $\tau$ is positive (takes positive elements to positive elements);
(3) $\tau 1 = 1$ ($\tau$ preserves the identity).

Note that it is (3), (in which $\tau$ acts on the left, remember) that ensures that the dual action on the states preserves the normalising condition $\rho \cdot 1 = 1$. This dual action can be taken as the starting point of a theory of quantum stochastic processes [Davies (1976)]. We shall adopt the notation $T$ for a general stochastic map, and write it on the left, thus: $A \mapsto TA$. We shall reserve the use of $A \mapsto \tau A$ for the special case of random dynamics. The set of stochastic maps on an algebra forms a convex set. A nice result of Størmer [Størmer (1963)] shows that a stochastic map $T$ is a contraction in the norm, that is,

**Theorem 10.4.** $\|TA\| \leq \|A\|$.

This is the analogue of the classical result, Theorem 3.7.

*Proof.* Note that if $T$ is stochastic and $A$ commutes with its adjoint, then $\|TA\| \leq \|A\|$; for we can simultaneously diagonalise both $A$ and $A^*$, and the problem reduces to an abelian algebra, to which we can apply (3.7). In particular, $T$ is contractive on unitary operators, and therefore on the convex span of the unitaries. But any operator on the unit sphere in $\mathcal{A}$ is the norm limit of a convex sum of unitaries; see [Bratteli and Robinson (1979)], Vol. I, Cor. 3.2.6. $\qquad\square$

The dynamics generated by a stochastic map is a very general concept, similar to the general Markov chain in classical theory, in which we lose both laws of thermodynamics. When $\mathcal{A} = \mathcal{B}(\mathcal{H})$, a random dynamics has a further property, that of preserving the trace: $Tr\,\tau A = Tr\,A$. This makes sense directly if $\dim \mathcal{A} < \infty$; in general, we can look at this condition only if a trace can be defined on the algebra. A stochastic map on $\mathcal{B}(\mathcal{H})$ that also preserves the trace is called *bistochastic*. There is no problem in formulating the trace condition if we are interested only in maps that leave the spectral projections $P_E$ of the energy invariant, when these are of finite dimension. We can then impose the condition for bistochasticity directly on the action of the map on the algebras $\mathcal{A}_E$, which you recall, was defined as $P_E \mathcal{A} P_E$. More generally, let $\mathcal{A}$ be any algebra of finite dimension; the trace can be used to provide it with a scalar product

$$\langle A,\, B \rangle = Tr\,(A^* B). \tag{10.7}$$

The corresponding norm is called the Hilbert-Schmidt norm. Indeed, let $A \in \mathcal{A}$; then the map $A \mapsto (\dim A)^{-1} Tr\,A$ is a state $\rho$ on $\mathcal{A}$, the micro-canonical state, and the scalar product (10.7) on $\mathcal{A}$ is that given in the GNS theory, up to normalisation. The state $\rho$ is faithful, since if $A$ is a positive matrix and $Tr\,A = 0$, then $A = 0$. The stochastic map $T$ then defines a map $\pi_\rho(A) \mapsto \pi_\rho(TA)$ on the GNS Hilbert space $\mathcal{H}_\rho$. The Hermitian conjugate of this map then coincides with the action $T^\dagger$ on the states, identified as density matrices. It is therefore positive as well as trace preserving. Thus we may say that a map $T$ is bistochastic if and only if both $T$ and $T^\dagger$ are stochastic. We may regard $T^\dagger$ as acting on the states in general, rather than on the density operators; we may then write it as right action $\rho \mapsto \rho T$, so that $\rho T \cdot A = \rho \cdot TA$ holds. Thus, $T$ is bistochastic if and only if both the left and right actions are stochastic. In finite dimensions this is equivalent to the condition that $T$ is stochastic and $T^\dagger$ has the microcanonical state as a fixed point. If $\dim \mathcal{A} = \infty$ we shall limit discussion to the case when $T$ has the spectral projections of the energy as fixed points, and they are of finite rank; we denote the subspace of Hilbert-Schmidt operators in $\mathcal{A}$ by $\mathcal{A}_2$; this is the set of operators with finite Hilbert-Schmidt norm. We say that an operator on $\mathcal{A}_2$ is bistochastic if its restriction to each *-subalgebra of finite rank is bistochastic. The set of bistochastic maps on $\mathcal{A}_2$ is clearly convex (and obviously closed if the dimension is finite). Suppose now that $\dim \mathcal{A} = N < \infty$. Any bistochastic map is therefore a mixture of the extreme points (Minkowski's theorem in finite dimensions). The reversible dynamics, given by unitary conjugation, are extreme points of this set. One might conjecture that an analogue of Birkhoff's theorem holds, but we are out of luck: not every bistochastic map on an algebra of dimension bigger than two is a mixture of unitary conjugations [Tregub (1986); Landau and Streater (1993)]. The general form of the extreme points is not known. Nevertheless, the main property of random dynamics, that it increases the entropy, does hold for the more general case of bistochastic maps [Streater (1985)]. We shall show that the increase in entropy is strict, using (9.9) and the following theorem, to be found in [Alberti and Uhlmann (1981)], Theorem 2-2:

**Theorem 10.5.** *Let $A \in \mathcal{A}$ be given, with $\dim \mathcal{A} < \infty$, let $T$ be a bistochastic map acting on $\mathcal{A}$, and let $B = TA$. Then there exists a random dynamics $\tau = \sum_i \lambda_i \tau_i$ such that $B = \tau A$.*

Remark: This does not mean that any bistochastic map is a mixture of unitary conjugations. In the theorem, the random dynamics $\tau$ only does the job of $T$ for the single matrix $A$, and not all matrices.

**Theorem 10.6.** *Let $T$ be a bistochastic map on $\mathcal{A}$ and let $\rho$ be a density matrix such that there is no unitary operator $U$ with $\rho T = U \rho U^{-1}$. Then $S(\rho T) > S(\rho)$.*

***Proof.*** Write
$$\rho T = \sum \lambda_i \rho \tau_i = \rho \tau$$
with at least two non-zero terms, as $T$ is not unitary. Choose one of the $\tau_i$ to play the role of $\tau_0$, and write $\tau = \tau_0 \tau'$. We can drop the factor $\tau_0$ since it does not change the entropy. The dynamics $\tau'$ has the form $\lambda_0 1 + \sum \lambda_i \tau_i'$, and then Theorem 9.9 gives the result.                                      □

The hypothesis that the dynamics is ergodic relative to $H$ means that there is a unique invariant state within each $\Sigma(\mathcal{A}_E)$, and since the dynamics maps each such space, the energy-shell, to itself, we can in the ergodic case apply Lyapunov's direct method to show that the dynamics generated by $T$ converges to equilibrium for any initial state. The limit state is a mixture of microcanonical states.

Before we give you yet another estimate [Streater (1985)] for the gain in entropy under a bistochastic map, we note that a bistochastic map $T$, and its adjoint $T^\dagger$ relative to the Hilbert-Schmidt scalar product, are contractions in the operator norm. This is because they are stochastic maps, so we can apply the result of Størmer [Størmer (1963)], Theorem 10.4. We now show that they are contractions in the Hilbert-Schmidt norm too.

**Theorem 10.7.** *For any bistochastic map $T$ on an algebra $\mathcal{A}$, with $\dim \mathcal{A} = N < \infty$, and $A \in \mathcal{A}$, we have $\|TA\|_2 \leq \|A\|_2$.*

**Proof.** The Hilbert-Schmidt norm is related to that given by the micro-canonical state $\omega_0$ by a factor $N^{1/2}$, thus:

$$\|A\|_2 = (Tr\,(A^*A))^{1/2} = N^{1/2}\,(\omega_0(A^*A))^{1/2} = N^{1/2}\|A\|_0.$$

It is therefore enough to show the contractive property for $\|\bullet\|_0$. We find

$$\begin{aligned}
\langle TA,\,TA\rangle_0 &= \langle A,\,T^\dagger(TA)\rangle_0 \\
&\leq \|A\|_0\,\|T^\dagger(TA)\|_0 \\
&= \|A\|_0\left(\langle T^\dagger(T(A)),\,T^\dagger(T(A))\rangle_0\right)^{1/2} \\
&= \|A\|_0\langle A,\,T^\dagger\left(T(T^\dagger(T(A)))\right)\rangle_0^{1/2} \\
&\leq \|A\|_0\,\|A\|_0^{1/2}\|T^\dagger\left(T(T^\dagger T(A))\right)\|_0^{1/2} \leq \cdots \\
&\leq \|A\|_0^{2-2^{-n}}\|(T^\dagger T)^{2^n}A)\|_0^{2^{-n}} \text{ for any } n \in \mathbf{N}.
\end{aligned}$$

Now, both $T^\dagger$ and $T$ are norm contractions, and so is their product any number of times, and putting $B = (T^\dagger T)^{2^n}A$, we see that $\|B\|_0 \leq \|B\| \leq \|A\|$. Letting $n \to \infty$, we get

$$\|TA\|_0^2 \leq \|A\|_0^2$$

since $\lim \|A\|^{2^{-n}} = 1$.

In an infinite dimensional Hilbert space $\mathcal{H}$, the operators of finite rank are dense in the space of Hilbert-Schmidt operators; the contraction property on those of finite rank can therefore be extended to all Hilbert-Schmidt operators. $\qquad\square$

In this context, we now prove our gap estimate.

**Theorem 10.8.** *[Streater (1985)] Let* $\dim \mathcal{A} < \infty$, *and let* $T$ *be bistochastic on* $\mathcal{A}_2$; *suppose that the operator* $TT^\dagger$ *has a spectral gap* $\gamma > 0$ *(so that the spectrum of* $TT^\dagger$ *lies in* $[0,\,1-\gamma]\cup\{1\}$), *then for all density operators* $\rho$,

$$S(\rho T) - S(\rho) \geq \frac{\gamma}{2}\|P^\perp\Delta_\rho\|_2^2. \tag{10.8}$$

Here, $P^\perp = 1 - P_1$, where $P_1$ is the projection, in the Hilbert space of Hilbert-Schmidt operators, onto the eigenvalue 1 of $TT^\dagger$.

**Proof.** First, note that $\Delta_\rho$ has finite rank. Then by Theorem 10.5, we may write $B := \Delta_\rho T = \sum \lambda_i A_i$, with $A_i = U_i\Delta_\rho U_i^{-1}$. Then for each $i$, by the quantum Kullback inequality,

$$Tr\,A_i\,(\log A_i - \log B) \geq \frac{1}{2}Tr\,(A_i - B)^2.$$

Multiply by $\lambda_i$ and sum, noting that $Tr\,(A_i \log A_i) = Tr\,\Delta_\rho \log \Delta_\rho$, we get

$$S(B) - S(\Delta_\rho) = Tr\,(\Delta_\rho \log \Delta_\rho - B \log B)$$

$$\geq \frac{1}{2} \sum_i \lambda_i Tr\,\left(A_i^2 - A_i B - B A_i + B^2\right)$$

$$= \frac{1}{2} \left( \sum_i \lambda_i Tr\,\Delta_\rho^2 - \sum_i Tr\,(\lambda_i A_i B) \right.$$

$$\left. - Tr\,B \sum_i \lambda_i A_i + Tr\,B^2 \right)$$

$$= \frac{1}{2} \left( \langle \Delta_\rho,\, \Delta_\rho \rangle - \langle B,\, B \rangle \right)$$

$$= \frac{1}{2} \left( \langle \Delta_\rho,\, \Delta_\rho \rangle - \langle \Delta_\rho T,\, \Delta_\rho T \rangle \right)$$

$$= \frac{1}{2} \langle \Delta_\rho,\, \left(1 - TT^\dagger\right) \Delta_\rho \rangle.$$

Now writing $P_x$ for the spectral resolutions of $T^\dagger T$, we have

$$1 - TT^\dagger = 1 - P_1 - \int_0^{1-\gamma} x\,dP_x \geq 1 - P_1 - (1-\gamma) \int_0^{1-\gamma} dP_x$$

$$= 1 - P_1 - (1-\gamma)P^\perp = \gamma P^\perp.$$

Thus

$$S(\Delta_\rho T) - S(\Delta_\rho) \geq \frac{\gamma}{2} \langle \Delta_\rho,\, P^\perp \Delta_\rho \rangle \tag{10.9}$$

which gives the result. □

We shall be interested in maps that map each finite-dimensional algebra of the form $\mathcal{A}_E := P_E \mathcal{A} P_E$ to itself, where $P_E$ is the spectral projection of the energy onto states of energy $E$. Then Theorem 10.8 means that if the dynamics is ergodic on each energy-shell $\mathcal{A}_E$, with a gap $\gamma_E > 0$, for all $E$, the entropy increases to that of the microcanonical state and the system converges in Hilbert-Schmidt norm to the microcanonical state. It is sometimes more useful than (10.3) in that the rate of convergence is determined by $\gamma$, which is independent of $A$.

Another property true of random dynamics but not true for the general bistochastic map is complete positivity. This concept was introduced into analysis by Stinespring [Stinespring (1955)]; its importance in quantum mechanics was discovered by Kraus [Kraus (1971)]. We have mentioned that an essential property of quantum probability is the existence of the ampliation, which is a map which takes an observable $A \in \mathcal{A}$ of a system into the observable $A \otimes I \in \mathcal{A} \otimes \mathcal{B}$ where $I$ is the unit operator of a second system, described by the algebra $\mathcal{B}$, but which plays no role. Now a dynamical map $\tau$ must be positivity preserving, or else its dual action on the states will not map states to states. The ampliation expresses that we can regard our system as a part of the compound system, and for this to be true of its dynamics, we must require that $\tau \otimes I$ should also be positive, for any choice of the second algebra. This leads to

**Definition 10.9 (Complete Positivity).** *We say that a linear map $\tau$ on an algebra $\mathcal{A}$ is n-positive if $\tau \otimes I$ is a positive map on $\mathcal{A} \bigotimes \mathbf{M}_n$. We say that $\tau$ is completely positive if it is n-positive for all $n$.*

In classical probability, every positive map is completely positive, so every stochastic process can be expanded by ampliation. One can see this by writing the time-step as the multiplication (in the natural basis in Span $\Omega$) of the probability by a matrix with non-negative entries. The matrix of the map $\tau \otimes I$ in the product basis also has non-negative entries, and so is positive. Hence the original matrix was completely positive. We can also see that, in quantum probability, any implemented reversible dynamics is completely positive. For, if $\tau = \operatorname{Ad} U$, then put $V = U \otimes I$, and $\operatorname{Ad} V$ implements $\tau \otimes I$ and is positive. Moreover, any random reversible dynamics is also completely positive. To see this, simply note that the set of completely positive maps is convex; see Exercise 10.20. It is annoying that there are completely positive bistochastic maps that are not expressible as random dynamics, if the dimension is greater than 2. This was shown in dimension greater than 3 by Maassen and Kümmerer, [Kümmerer and Maassen (1987)], and by L. J. Landau and the author in dimension 3 [Landau and Streater (1993)]. This puzzling result may be due to the fact that we define a mixture of two maps using classical probability, whereas it might be more natural to use quantum probability in the definition.

If $X \in \mathbf{M}_n$, then the map $T_X : A \mapsto X^*AX$ is completely positive; for, $T_X \otimes 1_m$, acting on $\mathbf{M}_n \bigotimes \mathbf{M}_m$, is the map

$$A \mapsto (X \otimes 1)^* A (X \otimes 1), \text{ with } A \in \mathbf{M}_n \bigotimes \mathbf{M}_m$$

which is positive. Thus any convex sum of such maps is completely positive. The harder converse is due to Kraus [Kraus (1971)]. We [Landau and Streater (1993)] shall now show more, that any $n$-positive map has this form, thus proving Choi's result [Choi (1972)] on the way (that any $n$-positive map on $\mathbf{M}_n$ is completely positive). We first introduce some notation.

Let $m$ be any positive integer. To any $A \in \mathbf{M}_m$ we associate the linear functional $f_A : \mathbf{M}_m \to \mathbf{C}$ given by

$$f_A(B) = Tr(AB) = Tr(BA) = (A, B) \text{ say}, \qquad (10.10)$$

and any linear functional is uniquely represented in this manner. We shall write $\Delta$ for the inverse of the map $A \mapsto f_A$, so that if $f$ is a linear functional, $\Delta f$ is the matrix such that $f(B) = Tr(B\Delta f)$. We have already met this for states in Exercise 9.24. We shall use this for $m = n^2$. If $T$ is a linear map on $\mathbf{M}_n$ we denote by $T^t$ the transpose, defined as the unique map obeying

$$(T^t A, B) = Tr(T^t AB) = Tr(ATB) = (A, TB) \text{ if } A, B \in \mathbf{M}_n. \quad (10.11)$$

We note that $T^{tt} = T$ and that $T^t$ is Hermitian if $T$ is Hermitian, and that $T^t$ is positive if $T$ is positive (see Exercise 10.21). Moreover, $T^t$ coincides with the adjoint $T^\dagger$ relative to the scalar product $\langle A, B \rangle = n^{-1} Tr A^* B = \rho_0(A^*B)$ where $\rho_0$ is the tracial state on $\mathbf{M}_n$. See Exercise 10.22. We also need the correspondence between linear maps $T$ from $\mathbf{M}_n$ to itself and linear functionals $L$ on $\mathbf{M}_n \otimes \mathbf{M}_n$, based on the definition

$$L_T(A \otimes B) = n^{-1} Tr\left(B^t T A\right) \qquad (10.12)$$

where $B^t$ denotes the transpose of the matrix $B$. Equation (10.12) defines $L_T$ by linear extension to the whole of $\mathbf{M}_n \otimes \mathbf{M}_n$.

We have seen in the Gelfand-Naimark-Segal construction that an algebra acts on itself by left multiplication. This idea can be extended to both left and right actions, which always commute. Define an action of $\mathbf{M}_n \otimes \mathbf{M}_n$ on $\mathbf{M}_n$ by $(A \otimes B)C = ACB^t$, extended by linearity to sums of elements of the form $\sum A \otimes B$. Define the linear functional $L_{C,D}$ on $\mathbf{M}_n \otimes \mathbf{M}_n$ by the formula

$$L_{C,D}(M) = \langle C, MD \rangle$$

for $M \in \mathbf{M}_n \otimes \mathbf{M}_n$ regarded as an operator on $D \in \mathbf{M}_n$. Any linear functional on $\mathbf{M}_n \otimes \mathbf{M}_n$ can be expressed as a linear combination of such functionals, and defines the matrix $\Delta L \in \mathbf{M}_n \otimes \mathbf{M}_n$. We note that $L_{C,D} = L_T$ if and only if $T(A) = C^*AD$. It is enough to check this for $M = B \otimes B$. Then $\langle C, MD \rangle = \rho_0(C^*ADB^t) = n^{-1} Tr(B^t(C^*AD)) = L_T(A \otimes B)$. Clearly, the functional

$$M \mapsto L_{1,1}(M) = \langle 1, M1 \rangle$$

is a state on $\mathbf{M}_n \otimes \mathbf{M}_n$, being the expectation in the unit vector $1 \in \mathbf{M}_n$. It is therefore a positive functional. The corresponding density matrix,

$\Delta L_{1,1}$ is, as always for pure states, the projection in $\mathcal{B}(\mathbf{M}_n)$ onto the unit vector 1.

**Definition 10.10.** Let us say that a linear functional $L$ on $\mathbf{M}_n \otimes \mathbf{M}_n$ is

(1) *weakly positive* if $L(A \otimes B) \geq 0$ whenever $A \geq 0$ and $B \geq 0$.
(2) *positive* if $L(M) \geq 0$ for all positive $M \in \mathbf{M}_n \otimes \mathbf{M}_n$.

We note that if $T$ is a positive map, then $L_T$ is weakly positive, as follows from the fact that the trace of the product of two positive operators is non-negative. The main part of the proof of Kraus's formula is contained in the following

**Theorem 10.11.** $L_T$ *is positive if* $T$ *is* $n$-*positive.*

**Proof.** Let $\hat{T} = T \otimes \text{Id}$. Then $\hat{T}$ is a positive map. Also

$$L_T(A \otimes B) = \rho_0(B^t T(A)) = L_{1,1}(T(A) \otimes B) = L_{1,1}\left(\hat{T}(A \otimes B)\right). \quad (10.13)$$

Thus

$$L_T = L_{1,1} \circ \hat{T}. \quad (10.14)$$

Hence $L_T$ is positive, being the composition of two positive maps.

If $T$ is an $n$-positive stochastic map then $L_T$ is a state. It can therefore be written as a convex combination of pure states, each of which is of the form $L_{C,C}$. Otherwise, if $T$ does not preserve the identity, we can normalise $L_T$ and then apply this decomposition. Thus any $n$-positive map $T$ is a convex combination of maps $A \mapsto C^*AC$, with at most $n^2$ terms in the sum. It follows that $T$ is completely positive (Choi's theorem [Choi (1972)]), and has the form given by Kraus. We can also regard the decomposition, which is not unique, into the decomposition of the density matrix $\Delta L_T$ into a spectral resolution of one-dimensional projections. We can therefore arrange to do this into elements $C_i$ that are mutually orthogonal (in the sense of Hilbert-Schmidt).

Given algebras $\mathcal{A}_i$ with completely positive dynamics $\tau_i$, then the combined system $\bigotimes \mathcal{A}_i$ can be given the dynamics $\otimes \tau_i$. This is positive. For it is the product of the positive maps $\tau_i \otimes I$, where $I$ denotes the identity on the rest of the factors, other than $\mathcal{A}_i$. We shall take the theory further in connection with continuous semigroups of completely positive maps in Chapter 12.

## 10.4    Exercises

Exercise 10.15. Show that the map $U \mapsto \tau_U$ is a homomorphism from $UNI$ $\mathcal{A}$ onto $INN\mathcal{A}$. Here, $UNI\mathcal{A}$ denotes the group of unitary elements of $\mathcal{A}$.

Exercise 10.16. Let $\tau_U$ be an inner automorphism of $\mathcal{A}$, and let $V \in \mathcal{A}$ be such that $\tau_U = \tau_V$. Prove that there exists a unitary element $Z$ of the centre $\mathcal{Z}(\mathcal{A})$ such that $U = ZV$.

Exercise 10.17. Let $\mathcal{A}$ be a $C^*$-algebra and $\tau$ an automorphism of $\mathcal{A}$. Let $\rho \in \Sigma(\mathcal{A})$ be invariant under the right-action of $\tau$, and define a sesquilinear form on $\mathcal{A}$ by $\langle A, B \rangle = \rho(A^*B)$. Show that $\langle \bullet, \bullet \rangle$ is invariant under $\tau$.

Exercise 10.18. Let $U$ be a unitary operator commuting with the spectral projections of the self-adjoint operator $H$. Let $\rho \mapsto U^{-n}\rho U^n$ be the time-evolution in the Schrödinger picture. Show that any mixture of this dynamics for various $n$ conserves the mean value of $H$.

Exercise 10.19. Suppose that $\dim \mathcal{A} < \infty$. Show that the set of random dynamics is a convex set. Show that its extreme points are the automorphisms. Hint: use the strict concavity of the entropy.

Exercise 10.20. Show that the set of completely positive maps on a $C^*$-algebra is convex.

Exercise 10.21. Let $T : \mathbf{M}_n \to \mathbf{M}_n$ be a linear map. Show that $T^t$, defined in (10.11) obeys $T^t(A^*) = (T^tA)^*$, and that $T^t$ is positive if $T$ is positive.

Exercise 10.22. Show that $\langle T^tA, B \rangle = \langle A, TB \rangle$, the scalar product being that given by the tracial state.

Exercise 10.23. Show that on $\mathbf{M}_n$ the map $T \mapsto T^d$, its matrix transpose, is not completely positive if $n > 1$.

# Chapter 11

# Isolated Quantum Dynamics

## 11.1 The Quantum Boltzmann Map

Suppose that our system consists of two parts, described by algebras $\mathcal{A}_1$ and $\mathcal{A}_2$, and we wish to consider the combined system. As in classical theory, we must distinguish three levels of separateness: if there is no functional relationship between the elements of the two algebras, it is traditional to say that they are 'independent' degrees of freedom, and we can use the algebra $\mathcal{A} = \mathcal{A}_1 \bigotimes \mathcal{A}_2$ as the algebra of the combined system. If this holds, we can discuss the second level of separateness, which is whether the systems are 'non-interacting' or interacting. This involves the form of the energy. If they are non-interacting, then we can write the total energy as a sum of operators, one for each subsystem:

$$H = H_1 \otimes 1 + 1 \otimes H_2. \tag{11.1}$$

If the algebra is a tensor product, and whether or not the systems are non-interacting, we can discuss whether they are statistically independent. This concept depends on the state, and not on the energy. The systems are statistically independent in the combined system if the state is a tensor product $\rho = \rho_1 \otimes \rho_2$. In this chapter, we will assume that the algebra $\mathcal{A}$ is a tensor product, and the energy has the form given in (11.1); one step in the time-evolution will be given by a bistochastic map on $\mathcal{A}$; it can mix up states with the same total energy, and a state which is a tensor product at time $t = 0$ will not in general be such after one time-step. This happens even when the total energy is a sum, as in (11.1), because the dynamical map is only required to conserve the total energy, and not the values of the energy of the two parts separately. The map that replaces a state $\rho \in \Sigma(\mathcal{A})$ by the tensor product of its marginals, $\mathcal{M}_1 \rho \otimes \mathcal{M}_2 \rho$ will be called the quantum stoss map, or just the stoss map. By Theorem 9.8, the map is entropy-increasing, and the increase is strict. This means that the entropy increases unless $\rho$ is already a product state. For non-interacting systems, the stoss map does not change the mean energy:

**Theorem 11.1.** *If $H$ has the form (11.1), and $\rho$ has finite mean energy, then*

$$\rho \cdot H = \rho \mathcal{M}_1 \cdot H_1 + \rho \mathcal{M}_2 \cdot H_2. \tag{11.2}$$

**Proof.**  We have $\rho \mathcal{M}_1 \cdot H_1 = Tr_1 \, Tr_2 \, (\Delta_\rho \cdot (H_1 \otimes 1)) = \rho \cdot (H_1 \otimes 1)$, and similarly for the second term.                                                                □

We have seen that a dynamics generated by an ergodic bistochastic map $\tau$ acting on $\mathcal{A}$, in a way that conserves energy, will converge to a mixture of microcanonical states. It is not able to mix the different energy-shells; for this, some more stochasticity is needed. Just as in the classical case, in quantum statistical dynamics this is provided by the stoss map. Thus, let $\tau$ be an energy-conserving bistochastic map and define for $\rho \in \Sigma(\mathcal{A})$ the map

$$\rho \mapsto (\rho \tau \mathcal{M}_1) \otimes (\rho \tau \mathcal{M}_2). \tag{11.3}$$

This we call the Boltzmann map defined by $\tau$. A similar idea was introduced and used by [Alicki and Messer (1983)], who are also responsible for some of the basic properties. The Boltzmann map is non-linear; this might be why it was not considered by [Thirring (1980)], who discusses (p. 148) only the linear map $\rho_1 \mapsto (\rho_1 \otimes \rho_2) \tau \mathcal{M}_1$. We use such maps in the next Chapter; there, $\rho_2$ is kept constant in time. This is ruled out in [Thirring (1980)], on the grounds that in real systems the first system has an influence on the second system; for example, some energy might be transferred to it. We shall see that it is nevertheless a good model for isothermal systems, in which the second system is rapidly restored to its former state after each time-step. In our ideal isothermal systems, this restoration is done instantaneously.

The dynamics given by iterating a Boltzmann map obeys the first and second laws of thermodynamics: each of the two steps conserves the mean energy and does not decrease the entropy. Any canonical state $\rho_\beta$ is given by a density operator $\Delta_{\rho_\beta}$:

$$\Delta \rho_\beta = Z_\beta^{-1} e^{-\beta H} = Z_{1\beta}^{-1} e^{-\beta H_1} Z_{2\beta}^{-1} e^{-\beta H_2}.$$

It is a fixed point of a Boltzmann map, since it is a function of $H$ as well as being a tensor product of states on each of the two algebras. We have already got some of the ingredients allowing us to prove that the iterated dynamics takes any initial state to equilibrium. The remaining properties that we need are that the fixed point be unique, that the increase in entropy be strict, and that the orbit be compact.

We are interested in theories with local structure. Thus, we have a finite set $\Lambda$, with $x \in \Lambda$ denoting the typical point. It may be that correlations between different points $x$ are destroyed by random effects, which we can model by the local stoss map:

$$\rho \mapsto \otimes_x (\rho \mathcal{M}_x) \,, \tag{11.4}$$

which increases the entropy unless $\rho$ is already a product state. We then define the *Boltzmann map with local independence* to be the map

$$\rho \mapsto \otimes_{x \in \Lambda} (\rho \tau \mathcal{M}_x) \,, \tag{11.5}$$

which obeys both laws of thermodynamics. A simpler description arises if we use the Boltzmann map with $LTE$: after the map (11.5) we replace each factor with the canonical state on $\mathcal{A}_x$ with the same mean energy as $\rho \tau \mathcal{M}_x$. This increases entropy even more, and obviously conserves mean energy. The state of the system is described by a beta field, which obeys self-contained equations of motion determined by $\tau$. This would be a natural model if $\tau$ is ergodic relative to $H$. If there is a conserved quantity, such as the number of atoms, then we would use the grand canonical state with the same energy and particle number instead of the canonical state. Then the state is, at any time, described by a beta field $\beta(x)$ and a potential field $\mu(x)$, or equivalently, by the extensive fields $E(x)$ and $N(x)$. In either case the states follow an orbit inside a compact set (we assume that $\Lambda$ is finite).

We can now formulate conditions under which the iterated map, applied to any initial state, converges to equilibrium. Suppose that on each algebra $\mathcal{A}_E = P_E A P_E$ the map $\tau$ is ergodic, and is a mixture of the identity map and another automorphism. Then the entropy is a strict Lyapunov function and any left fixed point of $\tau_E$ is the uniform state on $\mathcal{A}_E$. Because of strictness, to be a fixed point of the Boltzmann map with $LTE$, the state must be invariant under both stages, and so must be a product of equilibrium states, as well as being the uniform state on each $\mathcal{A}_E$. We can now show that there is a unique $LTE$ product state with these properties under a further assumption, which is very natural, namely, that the spectrum of each $H_x$ is the same; this is a consequence of invariance under translation, which we have not yet imposed. We may suppose that each spectrum begins at zero, and is $\{0 < E_1 < \ldots\}$. The $H_x$ all commute, since they are operators on different factors in a tensor product. They may therefore be simultaneously diagonalised. If the density matrix of the state is a constant on the energy shell $\mathcal{A}_{E_1}$, and at the same time is a product of canonical states at possibly different betas at different points $x$ then all the betas must be equal. This follows immediately from the comparison of the coefficients of the projection onto the simultaneous eigenvectors of total energy $E_1$. It is even true that the only product state that is constant on each energy shell is a canonical state; see Exercise 11.58. Thus there is a unique $LTE$-state with a given mean energy. We can therefore prove that in models whose dynamics is given by an ergodic $T$, followed by the $LTE$-map, the system converges to equilibrium from any initial state, by Lyapunov's theorem. Although the state-space is not compact, the set of product $LTE$-states with fixed mean energy is.

In Krylov's thesis [Krylov (1979)] it is pointed out that more should be demanded of the theory than the convergence for large times; we should also get a prediction of the relaxation time. This is the time needed for the system, whatever its initial state, to get very close to equilibrium. Otherwise, we may simply be talking about the astronomical times of ergodic theory. It is an experimental fact that for many systems the convergence to equilibrium is exponentially fast, and the rate is independent of the initial state. The rate is controlled by the gap in the spectrum of $TT^{\dagger}$; if we have a gap $\gamma_E$ on each energy-shell, then the convergence is in the sense of the Hilbert-Schmidt norm. In the classical isothermal case, in Theorem 5.12 we introduced a stronger norm (called the KMS norm) and showed that convergence could be expressed in terms of this. In the next section we obtain a related result in the quantum case.

## 11.2   The Quantum Heat-Particle

The quantum heat-particle is a quantum oscillator, and the effect of introducing it into a model is very similar to what happens in the classical case. In particular, it is fair to describe noise as thermalised heat-particles. The heat-particles themselves can represent the excitations of a mode of the electromagnetic field, or of the phonon field. The Hilbert space of the oscillator is $\ell^2$, the space of sequences $c$ of complex numbers $\{c_j\}$, $j = 0, 1, 2 \ldots$ such that $\sum |c_j|^2 < \infty$. The energy is the multiplication operator $H_\gamma$:

$$(H_\gamma c)_j = j\varepsilon c_j \text{ where } \varepsilon \text{ is the energy of one quantum} \qquad (11.6)$$

so that $H_\gamma$ is diagonal in the natural basis in $\ell^2$. The suffix $\gamma$ indicates that it is the energy of the heat-particle. The algebra of observables for the heat-particle is $\mathcal{A}_\gamma = B(\ell^2)$, the $C^*$-algebra of all bounded operators on $\ell^2$. We could introduce any number of heat-particles, each with its own energy per quantum; $\mathcal{A}_\gamma$ is then the tensor product of the algebras, one of each type. We avoid the ambiguity in the tensor product of general $C^*$-algebras by choosing the spatial tensor product. That is, given two Hilbert spaces $\mathcal{H}_1$ and $\mathcal{H}_2$, we define $B(\mathcal{H}_1) \bigotimes B(\mathcal{H}_2)$ to be $B(\mathcal{H}_1 \otimes \mathcal{H}_2)$.

Consider the case with one heat-particle. The thermal state with beta $\beta$ is easily computed, in the basis where $H_\gamma$ is diagonal, to be the diagonal matrix $Z^{-1}e^{-\beta E_i}$. By Theorem 9.3 this is the state of maximum entropy, given the mean value of the energy. There is no grand canonical state for the heat-particle, as the number of quanta is simply proportional to the energy. In interactions, the number of heat-particles is not conserved; the total energy of the system is conserved, but the energy of the heat-particles is not.

The heat-particles are bosons, designed to carry energy from one atom of material to another. The Hilbert space of a single oscillator is separable, and so is isomorphic to $L^2(\mathbf{R})$. While there are many isomorphisms between separable Hilbert spaces, in this case there are natural ones, which identify the basis vectors $\psi_0 = (1,0,0,\ldots)$, $\psi_1 = (0,1,0,\ldots)\ldots$ (up to a phase) with the eigenvectors of the number operator of the harmonic oscillator,

$$\mathcal{N} = \frac{1}{2}(p^2 + q^2 - 1)$$

where $p$ and $q$ are the Schrödinger operators for momentum and position, and we have put $\hbar = 1$. These $p$, $q$ are the generators of the Weyl operators, $U(a)$ and $V(b)$, defined for $a \in \mathbf{R}$ and $b \in \mathbf{R}$ as one-parameter unitary groups on $L^2(\mathbf{R})$ by

$$(U(a)\psi)(x) = \psi(x+a); \quad (V(b)\psi)(x) = \exp\{ibx\}\psi(x). \tag{11.7}$$

The unitary operators $U(a) = \exp\{ipa\}$ and $V(b) = \exp\{iqb\}$ obey the Weyl relations

$$U(a)V(b) = \exp\{iab\}V(b)U(a) \tag{11.8}$$

which yields the Heisenberg commutation relations $[q, p] = i1$ on differentiation. Weyl suggested that his was the more rigorous way to formulate the commutation relations, since (11.8) does not involve unbounded operators. We can express the number operator as $\mathcal{N} = a^*a$, where $a$ and its adjoint $a^*$ are the annihilation and creation operators, related to $p$, $q$ by

$$q = 2^{-1/2}(a^* + a), \qquad p = i2^{-1/2}(a^* - a).$$

The Heisenberg relations are then equivalent to $[a, a^*] = 1$. The annihilation operator $a$ kills the ground state, that is, it obeys the 'Fock condition' $a\psi_0 = 0$. The eigenvalues of $\mathcal{N}$ are $\{0, 1, 2, \ldots\}$, and are simple. The map which takes the normalised eigenfunction of $\mathcal{N}$ with eigenvalue $n$ to the vector $\psi_n \in \ell^2$ is well defined up to a phase, and is the natural isomorphism we were talking about above. Under this isomorphism $\mathcal{A}_\gamma$ becomes identified with $B\left(L^2(\mathbf{R})\right)$.

In a system with local structure, we may attach a heat-particle to each site $x \in \Lambda$, all with the same energy per quantum. The algebra is then taken to be

$$\mathcal{A} = \bigotimes_x \left(\mathcal{A}_{c,x} \otimes \mathcal{A}_{\gamma,x}\right).$$

The physical interpretation of $H_{\gamma,x}$ could then be the kinetic energy of all the material located at $x$, added to the energy of the electromagnetic field around $x$. The same arguments used in the classical case hold here too: the kinetic energy in liquids thermalises much faster than the rate of chemical reactions, and the dynamics can be modelled by the Boltzmann map which decouples the heat-particles from the chemicals. Thus the dynamics is determined by an energy-conserving bistochastic map $\tau$, and one time-step is the map

$$\rho \mapsto \mathcal{M}_c\left(\tau^*\rho\right) \otimes \mathcal{M}_\gamma\left(\tau^*\rho\right).$$

The map $\tau$ can cause a molecular reaction at $x$, by changing the number of heat-particles at $x$, producing them in an exothermic reaction and absorbing them in an endothermic reaction. It can also contain a hopping term, either for the molecules or the heat-particles.

If we want a model with a simpler description than this and with even more randomness, we can follow the Boltzmann map by the stoss map

$$\rho \mapsto \otimes_x \mathcal{M}_x \rho.$$

In these models the microstate of the heat energy at $x$ at any time is described by the state $\mathcal{M}_{x\gamma}\rho$. This contains much more detail than is needed for chemistry, and we now discuss ways to simplify the treatment without spoiling the two laws of thermodynamics. If the state at any time, restricted to the algebra $\mathcal{A}_{\gamma,x}$, is a thermal state, we can define a local, non-equilibrium value for beta at $x$. The hypothesis of local thermodynamic equilibrium can be expressed by saying that the local kinetic energy at $x$ thermalises so fast that in one time step, on the scale of the chemical reactions, the local state of the heat-particle evolves into the thermal state with the same energy as it had after the map $\tau$. We can implement this mathematically by following the Boltzmann map and the stoss map by the *LTE* map

$$\rho_{\gamma,x} \mapsto \rho_{\beta(x)},$$

where $\rho_{\beta(x)}$ is the thermal state with the same mean energy as $\rho_{\gamma,x}$. Models of this sort describe the heat content by a field $\beta(x)$, and do not keep tabs on the details of which molecules possess the heat. It is usually adequate for chemical reactions not involving the specific excitation of certain individual atoms, *e.g.* by lasers.

It is also possible to localise the heat-particle on the bond, or link, between neighbouring sites, as we did in the classical case. This is natural in modern quantum field theory, where gauge fields are thought to reside on the links between the particles.

We now develop the theory of bosons further. We can identify $\ell^2$ with the Fock space over the one-dimensional Hilbert space $\mathbf{C}$. To include the case with several heat-particles, let us give the construction of Fock space for any Hilbert space $\mathcal{H}$, which is called the *one-particle space*. In our case, a vector in $\mathcal{H}$ describes a single heat-particle with a wave-function spread over the points of $\Lambda$. If several particles are present, we must remember that bosons, spoken of as particles, are indistinguishable, and their configurations require the field description. This is achieved in quantum mechanics by the quantised field. A pure state of several particles is described by a vector in the symmetric tensor product

$$\mathcal{H}_n = (\mathcal{H} \otimes \mathcal{H} \otimes \ldots \otimes \mathcal{H})_s \qquad n \text{ factors.}$$

The suffix $s$ indicates that only vectors in the tensor product that are invariant under the permutation of the factors are to be included. In doing Exercise 11.59 you will prove that the space of symmetric tensors is a vector subspace of the full tensor product, and is spanned by symmetric product vectors, that is, vectors of the form $\otimes^n \psi = \psi \otimes \psi \ldots \otimes \psi$, with $\psi \in \mathcal{H}$. Then the *Fock space over* $\mathcal{H}$ is the direct sum

$$F(\mathcal{H}) = \mathbf{C} \oplus \sum_1^\infty {}^\oplus \mathcal{H}_n. \tag{11.9}$$

In particular, if $\mathcal{H} = \mathbf{C}$, then $F(\mathcal{H}) = F(\mathbf{C}) = \ell^2$, since all tensor powers of $\mathbf{C}$ give back just $\mathbf{C}$ itself. Thus the algebra of a single heat-particle acts on the Fock space over $\mathbf{C}$. The first term in the direct sum in (11.9) is $\mathbf{C} = \mathcal{H}^0$, and the vector $1 \in \mathbf{C}$ is called the Fock vacuum. We shall denote this by $\Psi_F$.

Fock spaces have a 'functorial' property,

$$F(\mathcal{H}_1 \oplus \mathcal{H}_2) = F(\mathcal{H}_1) \otimes F(\mathcal{H}_2). \tag{11.10}$$

This means that the Hilbert space of two boson degrees of freedom, which is the *unsymmetrised* tensor product of two oscillator spaces, is the Fock space over $\mathbf{C}^2$; similar remarks hold for any number of degrees of freedom. This answers a question that has puzzled some students: why is the tensor product (not symmetrised) of two bose systems correctly obtained by introducing an internal label to the particle, and then using the symmetric Fock construction? In the fermion case, this was done instinctively by Heisenberg when he introduced isotopic spin, after which he treated the proton and the neutron as two states of the same particle. We shall see that fermion space has a similar functorial property. The functorial property has induced some people to use the notation $e^{\mathcal{H}}$ for $F(\mathcal{H})$; Eq. (11.10) is easily proved using the very useful concept of *coherent state*, which in Fock space has a simple description, emphasised by Klauder [Klauder (1959)]:

**Definition 11.2.** For $\psi \in \mathcal{H}$, the coherent state with wave-function $\psi \in \mathcal{H}$ is the vector denoted $e^{\psi}$, in $F(\mathcal{H})$, given by

$$e^{\psi} = 1 \oplus \psi \oplus (2!)^{-1/2} \psi \otimes \psi \oplus (3!)^{-1/2} \otimes^3 \psi \oplus \dots. \tag{11.11}$$

**Theorem 11.3.** *The coherent states form a total set in $F(\mathcal{H})$.*

**Proof.** [Araki (1976)]. It suffices to show that the Hilbert space spanned by the coherent vectors contains each $\otimes^n \psi$, for any $\psi \in \mathcal{H}$, since such vectors span the symmetric tensor product (Exercise 11.59). Proceed by induction. For $n = 0$, it is clear that $1 = e^0$ spans $\mathbf{C} = \mathcal{H}^0$. Now assume that $(\otimes^j \mathcal{H})_s$ is spanned by the vectors $e^{\psi}$ for $j = 0, \dots, n-1$. Consider the family of vectors

$$\Psi_n(\lambda) = e^{\lambda \psi} - \sum_{j=0}^{n-1} (j!)^{-1/2} (\otimes \lambda \psi)^j.$$

We have

$$\lim_{\lambda \to 0} \Psi_n / (\lambda^n) = (n!)^{-1/2} (\otimes \psi)^n$$

in the sense of strong convergence. Since the limit lies in the closure of the span of the coherent vectors, we have the induction hypothesis for $n$. $\qquad \square$

The notation $e^{\psi}$ is suggested by the easy relation

$$\langle e^{\phi}, e^{\psi} \rangle = \exp \langle \phi, \psi \rangle. \tag{11.12}$$

This is proved by summing the exponential series, noting that direct sums are orthogonal. Now the functorial property is easy:

**Theorem 11.4.** *There is an isomorphism $U$ between $F(\mathcal{H}_1 \oplus \mathcal{H}_2)$ and $F(\mathcal{H}_1) \otimes F(\mathcal{H}_2)$ such that*

$$Ue^{\psi_1 \oplus \psi_2} = e^{\psi_1} \otimes e^{\psi_2}.$$

**Proof.**

$$\|e^{\psi_1 \oplus \psi_2}\|^2 = e^{\|\psi_1 \oplus \psi_2\|^2} = e^{\|\psi_1\|^2 + \|\psi_2\|^2}$$
$$= e^{\|\psi_1\|^2} . e^{\|\psi_2\|^2} = \|e^{\psi_1} \otimes e^{\psi_2}\|^2.$$

Hence $U$ is an isometric and hence bijective map between total sets, and so can be extended to a unitary map. □

We note that the Fock vacuum is the coherent state corresponding to zero wave-function: $\Psi_F = e^0$. To any unitary operator $U$ on $\mathcal{H}$, we define its 'second quantisation' $\Gamma U$ on $F(\mathcal{H})$ by its action on the coherent vectors

$$\Gamma U e^{\psi} = e^{U\psi}.$$

We can check that this is unitary on each pair of coherent states, and thus on the linear span. It can therefore be extended by continuity to the whole Fock space, since the coherent states are total. This can be done for each element of a one-parameter group, say $U(t)$, acting on $\mathcal{H}$. This might be the time-evolution of a single particle, for example, with infinitesimal generator given by the one-particle Hamiltonian, $h$ say. Then $\Gamma U(t)$ is a one-parameter unitary group acting on $F(\mathcal{H})$, whose generator is called the second quantisation $d\Gamma h$ of $h$. This notation and most of the theory of this is due to Segal [Segal (1961, 1963)]. By differentiating $\Gamma U(t)$ in time we find that $d\Gamma h$ is the correct free Hamiltonian for a many-particle system:

$$d\Gamma h(\psi \otimes \psi \otimes \ldots \otimes \psi) = \sum_j \psi \otimes \psi \ldots \otimes h\psi \otimes \psi \ldots \otimes \psi \qquad (11.13)$$

where the one-particle operator $h$ acts on the $j$th factor. As an example, consider the simplest one-parameter group on $\mathcal{H}$, namely $u(t) = e^{it}1$, where $1$ is the unit operator on $\mathcal{H}$. Then the generator of $\Gamma u(t)$ is the number operator on $F(\mathcal{H})$. The coherent states $e^{\psi}$ we have defined are not normalised. This is easily remedied by dividing by $e^{-\|\psi\|^2/2}$. The Weyl operators $U(a)$, $V(b)$ have a neat version in terms of their action on coherent states. To each $\psi \in \mathcal{H}$, define the 'Segal-Weyl' operator $W(\psi)$ by

$$W(\psi)e^{\phi} = \exp\left\{\frac{1}{2}(\|\phi\|^2 - \|\phi + \psi\|^2) + i\,Im\langle\phi, \psi\rangle\right\}e^{\phi + \psi}. \qquad (11.14)$$

In Exercise 11.60, you will check that this is a unitary map when acting on the coherent states. Since these are total, we can extend $W(\psi)$ uniquely to a unitary operator on the whole of Fock space. The operators $W(\psi)$ obey the Segal-Weyl relations

$$W(\psi)W(\phi) = \exp\{iIm\langle\phi, \psi\rangle\}W(\psi + \phi). \tag{11.15}$$

To prove this, just verify it on a general coherent state $e^\chi$ (Exercise 11.61). The secret behind this is that the Segal-Weyl operators $W(\psi)$ are unitary, obey Eq. (11.15), and that $W(\psi)$ creates the coherent state $e^\psi$ when acting on the vacuum, except for normalisation. If we choose the phase so that

$$W(\psi)e^0 = \exp\{-\|\psi\|^2/2\}e^\psi,$$

this is enough to determine the operator $W(\psi)$. We see that the product of two Segal-Weyl operators is a complex multiple of another. Since obviously $W(\psi)^* = W(-\psi)$, we conclude that the linear span of Segal-Weyl operators, as $\psi$ runs over the one-particle space, is a $*$-algebra. We can obtain the Weyl operators $U(a)$ and $V(b)$, if $\dim\mathcal{H} = 1$, by choosing a particular complex conjugation on $\mathcal{H}$. For example, if we identify $\mathcal{H}$ with $\mathbf{C}$, we could take complex conjugation on $\mathbf{C}$. Then we can speak about real vectors in $\mathcal{H}$. This is the quantum analogue of choosing a Lagrangian subspace in classical Hamiltonian theory. The point is that Weyl's (or Schrödinger's) form of the commutation relations depends on a selection of one canonical variable as the co-ordinate $q$, and its conjugate, $p$, as momentum. This is needed in Lagrange's equations of classical mechanics, and in the Lagrangian approach to quantum mechanics. Hamilton's form is invariant under canonical transformations; in quantum mechanics, linear canonical transformations are exactly those preserving the bilinear form $Im\langle\phi, \psi\rangle$. Thus if $\mathcal{H} = \mathbf{C}$, we can identify $F(\mathcal{H})$ with $\ell^2$ and for $a + ib \in \mathbf{C}$ put $U(a) = W(-a/\sqrt{2})$, $V(b) = W(ib/\sqrt{2})$. Then these obey the Weyl relations. This provides a link between Segal's canonically covariant formalism and the Schrödinger theory. By the isomorphism Theorem 11.4, we can identify the Hilbert space $L^2(\mathbf{R}^n)$ of $n$ oscillators ($n < \infty$) with $F(\mathbf{C}^n)$.

It is known that the Schrödinger operators $x$ and $-id/dx$ form an irreducible representation of the Heisenberg commutation relations on $L^2(\mathbf{R})$. It follows that the Segal-Weyl operators form an irreducible family on Fock space (by the isomorphism between the spaces). Thus every bounded operator on Fock space is in the strong closure of the algebra generated by the Segal-Weyl operators. In this Chapter we limit discussion to a finite number of degrees of freedom. Then we can take the $C^*$-algebra $\mathcal{A}_\gamma$ of the bose field to be $B(F(\mathcal{H}))$, the set of all bounded operators on Fock space, just as we did above. The Stone-von Neumann theorem tells us that if $\dim \mathcal{H} < \infty$, every continuous irreducible representation of the Segal-Weyl relations is unitarily equivalent to the Fock representation. We need to clarify this; all Fock spaces are isomorphic, and so we need to distinguish between models with different numbers of degrees of freedom and oscillator frequencies, but in an algebraic way valid for all representations of $\mathcal{A}_\gamma$. Segal does this by defining a Weyl system not simply as $\mathcal{A}_\gamma$, but to be a regular *map* $W$ from the one-particle space $\mathcal{H}$ into $\mathcal{A}_\gamma$ obeying the Segal-Weyl relations Eq. (11.15). Two representations $\pi_1$ and $\pi_2$ of the Weyl system, on Hilbert spaces $\mathcal{H}_1$ and $\mathcal{H}_2$, are unitarily equivalent if there exists a unitary map, $U$, from $\mathcal{H}_1$ to $\mathcal{H}_2$, such that

$$\pi_1(W(\psi)) = U\pi_2(W(\psi))U^{-1} \qquad \text{for all } \psi \in \mathcal{H}.$$

A state $\rho$ on $\mathcal{A}_\gamma$ is said to be *regular* if in the corresponding cyclic representation $\pi_\rho$ the map $\psi \to W(\psi)$ is continuous. In this case by Stone's theorem its generators, related to the operators $p$ and $q$, are self-adjoint, and could be taken to represent observables. Because of the Segal-Weyl relations, any operator in $\mathcal{A}_\gamma$ is the limit of *sums* of the form $\sum c_i W(\psi_i)$ with complex constants $c_i$. Thus any regular state on $\mathcal{A}_\gamma = B(F(\mathcal{H}))$ is determined by the expectation $\mathcal{F}$ it gives for the Weyl operators:

$$\mathcal{F}(\psi) = \rho \cdot W(\psi) := \rho(W(\psi)) \qquad \text{(the generating function given by } \rho\text{)}.$$
$$(11.16)$$

This is the analogue of the 'characteristic function' of a probability measure $P(x)$ on the line, as in Eq. (2.21):

$$\mathcal{F}(t) = \int \exp\{itx\}dP(x)$$

and it has similar properties. Thus $\mathcal{F}(\psi)$ is a sort of non-abelian Fourier transform. The following is a version of the Araki-Segal twisted Bochner theorem [Araki (1960); Segal (1961)]:

**Theorem 11.5.** *Let $M$ be a finite-dimensional real vector space with a non-degenerate symplectic form $B(\bullet, \bullet)$, and let $\psi \mapsto \pi_\rho(W(\psi))$ be a representation of the Weyl-Segal relations, given by the regular state $\rho$. Then $\mathcal{F}(\psi) = \rho \cdot W(\psi) = \langle \psi_\rho, \pi_\rho(W(\psi))\psi_\rho \rangle$ obeys*

*(1) F(0)=1;*
*(2) for each finite set $\{c_j\} \subseteq \mathbf{C}$, $\{\psi_j\} \subseteq M$,*

$$\sum_{j,k} \mathcal{F}(\psi_j - \psi_k) \exp\{iB(\psi_j, \psi_k)\}\overline{c}_k c_j \geq 0;$$

*(3) the map $\psi \mapsto \mathcal{F}(\psi)$ is continuous.*

*Conversely, a functional $\mathcal{F}$ obeying these properties comes from a unique regular state $\rho$.*

The necessity of the conditions is not hard; (2) follows from the positivity of the norm of $\sum c_j W(\psi_j)\psi_\rho$; for the sufficiency, we note that $\mathcal{F}$ supplies us with a positive linear functional on the linear span of the symbols $W(\psi)$, which we have noted form an algebra when they obey the Segal-Weyl relations. This defines a putative norm which gives $\|1\| = 1$, and so the linear functional is bounded, and therefore continuous. The extension by continuity to the completion then gives us the claimed regular state on the Weyl system. If the state $\rho$ is a bit more than regular (being an analytic vector for the number operator is enough, for example) then by differentiating the generating function along one-parameter subgroups we can generate the 'Wightman' functions; for one degree of freedom these are the moments of the operators $p$ and $q$ in the cyclic vector $\psi_\rho$:

$$\langle \psi_\rho, \, p^n q^m \psi_\rho \rangle = (-i)^{m+n} \frac{\partial^m}{\partial a^m} \frac{\partial^n}{\partial b^n} \rho(V(b)U(a)) \tag{11.17}$$

and $\rho \cdot (V(b)U(a)) = \rho \cdot (W(ib/(\sqrt{2})W(-a/\sqrt{2})))$ can be expressed in terms of $\mathcal{F}$. Moments with the operators in any other order can be obtained similarly. It is known that the logarithm of $\mathcal{F}$ generates the correlation functions (also known as the cumulants, or the truncated functions, $\rho_T$). For any state $\rho$, these are defined recursively in terms of the moments, or Wightman functions, of any collection of operators $A_1, \ldots, A_m$ by

$$\rho(A_1 \ldots A_n) = \rho_T(A_1 \ldots A_n) + \sum_I \rho_T(I_1)\rho_T(I_2)\ldots\rho_T(I_r). \tag{11.18}$$

Here, $I$ runs over all partitions $I = I_1 \cup I_2 \cup \ldots \cup I_r$ of the set $\{A_1, \ldots, A_n\}$ into $r$ parts, and then $r$ runs from 2 to $n$. In each partition, $I_k$ stands for the product of $A_j \in I_k$ in the order of increasing $j$. We start the induction with the one-point function $\rho_T(A_k) = \rho(A_k)$. The correlation functions do not satisfy the positivity conditions of a state; they correspond in perturbation theory to the sum of the connected Feynman diagrams.

In classical probability, a Gaussian measure leads to a characteristic function which is the exponential of a quadratic form. Its logarithm is therefore a quadratic polynomial, and all correlations beyond the second order vanish. In a Segal-Weyl system, a similar class of states is the *quasi-free* class [Robinson (1965b)], with generating function

$$\mathcal{F}(\psi) = \exp\{-\frac{1}{2}\|\psi\|^2 - A(\psi,\psi) + iIm\langle f, \psi \rangle\} \tag{11.19}$$

where $A$ is a positive quadratic form on $M$, and $f \in M$. Since for a quasifree state, $\log \mathcal{F}$ is a quadratic polynomial, we see that in a quasifree state, all correlation functions of the fields of order greater than 2 vanish. It was shown in [Robinson (1965a)] that if we have a state of the *CCR*-algebra, and all correlation functions vanish beyond some order, then the state is necessarily quasifree. This is the quantum analogue of the theorem of [Marcinkiewicz (1939)].

Let $\rho$ be a quasifree state of the *CCR* with $f = 0$, and let $(\pi_\rho, \mathcal{H}_\rho, \psi_\rho)$ be the Gelfand-Naimark-Segal representation determined by $\rho$. There is the following explicit realisation: let $F(\mathcal{H})$ be the Bosonic Fock space over $\mathcal{H}$, $\psi_0$ the Fock vacuum and $W_0(\psi)$ the Fock representation of the Segal-Weyl algebra on $F(\mathcal{H})$. Let $J$ be a conjugation on $\mathcal{H}$ commuting with $A$. Then we may put

$$\mathcal{H}_\rho = F(\mathcal{H}) \otimes F(\mathcal{H});$$
$$\psi_\rho = \psi_0 \otimes \psi_0;$$
$$\pi_\rho(W(\phi)) = W_0\left(-A^{1/2}J\phi\right) \otimes W_0\left((1+A)^{1/2}\phi\right).$$

This tensor structure is typical of the *KMS* theory of quasi-free systems, and was first made clear in [Araki and Woods (1963)]. A coherent state $e^\phi$ is quasifree, with $A = 0$ and $f = 2\phi$, as can be seen by computing its generating function

$$\mathcal{F}(\psi) = \exp\left\{-\|\phi\|^2\right\} \langle e^\phi, W(\psi)e^\phi\rangle \qquad (11.20)$$

$$= \exp\left\{-\frac{1}{2}\|\psi\|^2 - 2iIm\langle\psi, \phi\rangle\right\}. \qquad (11.21)$$

It is a pure state, and in fact, all pure quasifree states are of this form [Manuceau and Verbeure (1968)]. The equilibrium state of a single harmonic oscillator is also quasifree; its generating function can be found explicitly. The main part of the calculation is to find the expectation of the anti-Wick ordered Weyl operator (in a Wick-ordered operator, all creators lie to the left, and all annihilators to the right; we want just the opposite):

**Theorem 11.6.** *Let $H = \varepsilon a^* a$ acting on $L^2(\mathbf{R})$. Then $\Delta\rho = Z^{-1}e^{-\beta H}$ is of trace class, and defines a quasifree state $\rho$ such that*

$$\rho \cdot (\exp\{az\} \exp\{a^*\overline{z}\}) = \exp\{z\overline{z}/(1-\zeta)\} \qquad (11.22)$$

$z, \overline{z} \in \mathbf{C}$ *where* $\zeta = \exp\{-\beta\varepsilon\}$.

**Proof.**　Let $\psi_n \in F(\mathbf{C})$ be the normalised $n$-particle state

$$\psi_n = 1 \otimes 1 \ldots \otimes 1 = (n!)^{-1/2} a^{*n} \psi_0 = (n!)^{-1/2} \frac{\partial^n}{\partial \bar{z}^n} \exp\{a^* \bar{z}\} \psi_0 |_{\bar{z}=0}.$$

Then

$$\rho \cdot \left( e^{az} e^{a^* \bar{z}} \right) = Z^{-1} Tr \exp\{-\beta \varepsilon a^* a\} \exp\{az\} \exp\{a^* \bar{z}\}$$

$$= Z^{-1} \sum_n \exp\{-n\beta\varepsilon\} \langle \psi_n, \exp\{az\} \exp\{a^* \bar{z}\} \psi_n \rangle$$

$$= Z^{-1} \sum_n \zeta^n (n!)^{-1/2} (n!)^{-1/2} \frac{\partial^n}{\partial z^n} \frac{\partial^n}{\partial \bar{z}^n} \langle \psi_0, \exp\{az\} \exp\{a^* \bar{z}\} \psi_0 \rangle$$

$$= Z^{-1} \sum_n \frac{\zeta^n}{n!} \frac{\partial^n}{\partial z^n} \frac{\partial^n}{\partial \bar{z}^n} \exp\{z\bar{z}\}$$

$$= Z^{-1} \sum_n \frac{\zeta^n}{n!} \frac{\partial^n}{\partial z^n} (z^n \exp\{z\bar{z}\})$$

$$= \frac{\exp\{z\bar{z}\}}{Z} \sum_n \frac{\zeta^n}{n!} \left( n! + n.\frac{n!}{1!} z\bar{z} + \frac{n(n-1)}{2!} \frac{n!}{2!} (z\bar{z})^2 + \cdots \right)$$

$$= \frac{e^{z\bar{z}}}{Z} \left( \sum_n \zeta^n + z\bar{z}\zeta \sum_n n\zeta^{n-1} + \frac{(z\bar{z}\zeta)^2}{(2!)^2} \sum_n n(n-1)\zeta^{n-2} + \cdots \right)$$

$$= \frac{e^{z\bar{z}}}{Z} \left( \frac{1}{1-\zeta} + z\bar{z}\zeta \frac{\partial}{\partial \zeta} \frac{1}{1-\zeta} + \frac{1}{(2!)^2} (z\bar{z}\zeta)^2 \frac{\partial^2}{\partial \zeta^2} \frac{1}{(1-\zeta)^2} + \cdots \right)$$

$$= \frac{\exp\{z\bar{z}\}}{Z(1-\zeta)} \left( 1 + \frac{z\bar{z}\zeta}{1-\zeta} + \frac{(z\bar{z}\zeta)^2}{2!(1-\zeta)^2} + \cdots \right)$$

$$= \exp\{z\bar{z}\} \exp\{z\bar{z}\zeta/(1-\zeta)\}$$

since $Z(1 - \zeta) = 1$. Collecting the exponents together gives the result. □

This is the hard part. Note that we have treated $z$ and $\bar{z}$ as independent variables. Putting $iz$ for $z$, $i\bar{z}$ for $\bar{z}$ and using the Segal-Weyl relations (11.15) gives us the generating function. Its logarithm is quadratic in $z$, so the state is quasifree: all truncated functions beyond the second vanish. We can read off the two-point function, $\rho \cdot (aa^*) = (1 - \zeta)^{-1}$, which leads to Planck's distribution

$$\rho \cdot (a^* a) = \rho \cdot (aa^*) - 1 = \zeta/(1 - \zeta).$$

More generally, if we have $N$ non-interacting harmonic oscillators then the Hamiltonian can be written as a sum of modes, and the trace in the definition of the generating function of the canonical state is a product of traces.

That is, in the equilibrium state, the degrees of freedom are statistically independent. The generating function is thus a product of exponentials, each quadratic, and so the canonical state is quasifree. We now consider the general quadratic Hamiltonian in $N$ degrees of freedom, with bilinear couplings between the modes. This can be reduced to diagonal form by a canonical transformation, in which we replace the $a_i^*, a_i$, $1 \leq i \leq N$, with the creators and annihilators of the normal modes, called quasiparticles. If the Hamiltonian is not positive, this is not an elementary result [Moshinski and Winternitz (1980)]. We note that the one-particle space $\mathcal{H}$ is a real vector space of dimension $2N$, and is furnished with the non-degenerate symplectic form $B(\psi, \phi) = 2Im\langle\psi, \phi\rangle$. This is non-degenerate in the sense that $B(\psi, \phi) = 0$ for all $\psi$ implies that $\phi = 0$. It is *symplectic* because it is **R**-bilinear and antisymmetric: $B(\psi, \phi) = -B(\phi, \psi)$. A canonical transformation is then a real linear map $S : \mathcal{H} \to \mathcal{H}$ that preserves this form:

$$Im\langle S\psi, S\phi\rangle = Im\langle\psi, \phi\rangle.$$

Such a map $S$ is also called a symplectic transformation. A special case is a unitary map, which preserves the real as well as the imaginary part of the scalar product. The property of being unitary needs the choice of a complex structure $J$ on the symplectic space. A *complex structure* is an **R**-linear map $J$ on a real symplectic space $(\mathcal{H}, B)$ obeying

(1) $J^2 = -I$;
(2) $B(J\phi, \psi) = -B(\phi, J\psi)$;
(3) $(\phi, \psi)_J = B(\phi, J\psi)$ is a positive definite bilinear form.

Then we can form a complex Hilbert space by defining multiplication by $i$ to be $\psi \mapsto J\psi$, and a complex scalar product to be

$$\langle\psi, \phi\rangle = (\psi, \phi)_J + iB(\psi, \phi).$$

Each choice of $J$ leads to the construction of a Fock space, which is based on a complex Hilbert space, not a symplectic space. Thus the definitions of $a$ and $a^*$ depend on $J$. A symplectic transformation is complex-linear if it commutes with $J$. That $S$ is not obliged to be complex-linear is what permits canonical transformations to mix up $p$ and $q$, or $a^*$ and $a$. The set of all symplectic maps is a real Lie group, denoted $Sp(2N, \mathbf{R})$, and called the real symplectic group. The maximal compact subgroup of $Sp(2N, \mathbf{R})$ is isomorphic to $SU(N)$. How this subgroup sits inside the symplectic group depends on which choice of complex structure we have made. For a choice of $J$ allows us to define a real symmetric form which can act as the real part of a Hilbert scalar product.

If $N = 1$ a symplectic map is a real linear map of $\mathbf{R}^2$ that preserves the area. We see from this that the dimension of the group $Sp(2, \mathbf{R})$ is three, as it is given by the four parameters of a matrix in $GL(2, \mathbf{R})$ subject to one condition, that the area is preserved. We can think of $Sp(2N, \mathbf{R})$ as the set of $2N \times 2N$ real matrices $S$ preserving the 'symplectic metric' $G$:

$$\begin{pmatrix} S_{11}^t & S_{21}^t \\ S_{12}^t & S_{22}^t \end{pmatrix} \begin{pmatrix} 0 & I \\ -I & 0 \end{pmatrix} \begin{pmatrix} S_{11} & S_{12} \\ S_{21} & S_{22} \end{pmatrix} = \begin{pmatrix} 0 & I \\ -I & 0 \end{pmatrix} = G.$$

Here, $S_{ij}$ are $N \times N$ real matrices, and $I$ is the unit $N \times N$ matrix. The metric $G$ arises from the Heisenberg commutation relations; write $z = \{q_1, q_2, \ldots, q_N\}$. The $CCR$ can be written

$$[z_i, z_j] = i\hbar G_{ij};$$

then the new coordinates $z' = Sz$ obey the commutation relations if and only if $SGS^t = G$, as in Exercise 11.64, and if we write $z$ as a complex coordinate $\zeta = q + ip$, $p \in \mathbf{R}^N$, $q \in \mathbf{R}^N$, then $Im\langle \zeta', \zeta \rangle = z'Gz$. So the symplectic maps are canonical, and preserve the Segal-Weyl relations. We can think of a symplectic map as a change in the complex structure from one $J$ to another. We immediately see that $S$ induces an automorphism of the $C^*$-algebra $\mathcal{A} = B(F(\mathcal{H}))$, where $\mathcal{H} = \mathbf{C}^N$. Thus if $\psi \mapsto W(\psi)$ obeys the Segal-Weyl relations, then so does $\psi \mapsto W(S\psi)$, and therefore gives a representation of $\mathcal{A}$. By the Stone-von Neumann theorem, this representation is unitarily equivalent to the original one, so there exists a unitary operator $U(S)$ in $\mathcal{A}$ such that $W(S\psi) = UW(\psi)U^{-1}$ for all $\psi \in \mathcal{H}$. This is true of any representation, as the automorphism is inner. We can throw the automorphism on the states, by duality; it is clear that the states transform by the inverse unitary map

$$\Delta\rho \mapsto U^{-1}\Delta\rho U.$$

This converts density operators to density operators, so we do not leave the class of normal states. A unitary conjugation does not change the spectrum of any operator; nor does it change the trace, and so a canonical transformation does not change the entropy of any state. The implementing operator $U$ coincides with $\Gamma S$, the second-quantised operator determined by $S$, when $S$ is unitary. A quadratic Hamiltonian will mean an operator of the form

$$H = z^t h z, \qquad \text{where } z^t = (q_1 \ldots, q_N, p_1, \ldots, q_N)$$

and $h$ is a matrix which in the (Lagrangian) splitting above is of the form

$$h = \begin{pmatrix} C & A^t \\ A & B \end{pmatrix} \qquad B = B^t, \qquad C = C^t.$$

If we can find $S$ which diagonalises the matrix $h$ then the corresponding $U$ will diagonalise the operator $H$. It can be proved that any quadratic Hamiltonian whose second-quantised form (by which we mean that the $p$ and $q$ making it up are operators obeying the Heisenberg relations) is a positive operator can be diagonalised to a sum of harmonic oscillators by a canonical transformation. In Exercise 11.65 you are asked to try it out for a single degree of freedom. The case of two degrees of freedom is solved in detail in [Moshinski and Winternitz (1980)]. If a canonical state $\rho = Z^{-1}e^{-\beta H}$ is to be of trace class, clearly $H$ must be bounded below. When written in diagonal form, we apply our result above to show that $\rho$ is quasifree. The symplectic transformation does not change the quadratic nature of the exponent in the generating function, so we conclude that all density operators with quadratic Hamiltonians are quasifree. The converse is also true; the general quasifree state has generating function of the form given in Eq. (11.19), and we can diagonalise the operator $A$ by a unitary transformation, and remove the $f$ by a change in origin, $z \mapsto z - \zeta$, where $\zeta$ is a suitable c-number. Then the generating function has the form of a product over quasiparticle modes, and is therfore a canonical state for some diagonal oscillator Hamiltonian. This state has the maximum entropy of any state with the same mean energy.

We can now find the entropy of a general quasifree state for $N$ degrees of freedom. As usual, we first treat the case $N = 1$. From $\rho = Z^{-1}e^{\beta H}$ we find that for $H = \varepsilon a^* a$ we get

$$
\begin{aligned}
S &= -Tr\,\rho \log \rho = \log Z + \beta\langle \rho, H \rangle \\
&= -\log(1 - e^{-\beta\varepsilon}) + \beta\varepsilon e^{-\beta\varepsilon}/(1 - e^{-\beta\varepsilon}).
\end{aligned}
$$

We can rewrite this in terms of the mean number

$$
\nu = \rho \cdot (a^* a) = \zeta/(1 - \zeta),
$$

where $\zeta = e^{-\beta\varepsilon}$. We find

$$
S(\rho) = -\nu \log \nu + (1 + \nu) \log(1 + \nu). \tag{11.23}
$$

We see that the energy of the mode does not enter directly. It follows that if we have $N$ independent modes, the canonical state is a product of independent modes, and the entropy is additive:

$$
S(\rho) = \sum_{k=1}^{N} \left( -\nu_k \log \nu_k + (\nu_k + 1) \log(\nu_k + 1) \right).
$$

The classical theory uses just the first term, which is correct in the low-density limit, but which is not by itself positive if $\nu$ is not small. For the general quasifree state, we first diagonalise it into modes, and then use Eq. (11.23).

In quantum field theory the momentum of a particle can be moderately sharp only if the state of the field (of which the particle is the quantum) shows phase correlations over a fair distance. The trouble with the stoss map is that it destroys the correlations even over very short distances; one result of this is that there is no flow term $\mathbf{v}.\nabla$ in the resulting dynamics. We therefore are unable to reproduce the hydrodynamic equations if we randomise by the stoss map. The models studied so far exhibit a diffusion term but no convection, which needs collective flow. In classical mechanics the momentum of a small element of fluid is one of the labels of phase-space. In our treatment so far, these elements lose their individual velocity, because the system is thermalised by the $LTE$ map, which assigns all kinetic energy to a communal pot. This is too drastic a simplification if we want to describe convection rather than diffusion of heat. The quasifree states of a boson field offer us a way to simplify the description, and to retain at least the one-body (i.e. the quadratic) correlations over some distance, while retaining the second law of thermodynamics. Instead of the $LTE$ map, or the stoss map, as a way to mix the energy-shells, we have available the *quasifree map*. Let us denote by $Q$ the (non-linear) map on the set of regular states of the Segal-Weyl algebra that we get by replacing a state $\rho$ by the quasi-free state with the same moments up to second order. In the derivation of the general form of the quasifree state (11.19), we use the positivity condition on the moments, but only up to second order. In other words, any state $\rho$ with finite one- and two-point functions does indeed define a unique quasifree state $Q\rho$ having the same first and second moments. The importance of this is in the crucial remark, made in [Wichmann (1963)], that the entropy of $Q\rho$ is not less than that of $\rho$. This follows from Theorem 9.5, which, combined with the Kullback lemma, also shows that the increase in entropy under $Q$ is strict; see also [Streater (1987)]. The argument goes as follows. Let $\rho$ be any state with finite two-point functions, and let $\rho_\beta = Q\rho$. Then there exists a self-adjoint operator $H$, quadratic in the creators and annihilators, such that $\rho_\beta$ is the canonical state of $H$. Since $\langle \rho, H \rangle = \langle \rho_\beta, H \rangle$ we can apply the theorem. The proof only applies to systems with a finite number of degrees of freedom, because otherwise we must consider $H$ to be an infinite sum of quadratic expressions, and we must discuss convergence.

A similar result was proved for some systems with an infinite number of degrees of freedom in [Lanford and Robinson (1972)]. They show that the state of a free, infinite system of bosons (or fermions), initially in any

entropy. One might argue that an element of fluid, small
on the time-scale of every-day objects is well approximated
the thermalising forces, by an infinite volume of fluid obse
nite time. This, together with [Lanford and Robinson (197
justification of the quasifree map. Wichmann [Wichmann (
Bayesian view (previously advocated in [Jaynes (1957)], an
den (1992)]); he argues that if we have information only a
values of certain observables, then we should assign the st
imises the entropy subject to these mean values. If the obs
quadratic expressions in the creation and annihilation ope
densities and correlations, then the state of maximum ent
mean values is the quasifree one. This type of argument wa
Krylov, who derides it as not being physics: the physicist's
to devise good theories for deriving the state at time $t$, gi
time 0; his job is not to guess what the state at time zero i
it that Krylov would not have been happy with the book
(1967)] (apart from the fact that it has no references whate
his book on the ideas of Jaynes. The principle of maxim
is too subjective; our knowledge or lack of it does not mak
to the way a system evolves, says Krylov. In considering t
to mix the energy shell (the stoss map, the $LTE$-map, the
we must think of these as giving us different models of t
always in theoretical physics, we must try to test the pre
experiment, and decide which is the best. All are worthy
they obey the laws of thermodynamics. We have seen th
strict Lyapunov function for the quasifree map. In a mo
are boson fields representing electromagnetic radiation, sa
or phonons, and we wish to retain some of the coherence o
one time-step of the model, then the quasifree map can b

of the $LTE$ map. We can also describe atoms and molecules as bosons, if
the effects of the hard core can be neglected. Thus, if $x \in \Lambda$ represents
a region that can contain an enormous number of atoms, it is convenient
to choose the Hilbert space of the chemical to be $F(\mathbf{C}^{|\Lambda|})$. Then the map
$Q$ in place of the $LTE$-map will retain the coherence of the one-particle
wave-functions over the whole space $\Lambda$, and will allow collective motion to
persist, in spite of the thermalising nature of the dynamics.

## 11.3   Fermions and Ions with a Hard Core

Not all particles are bosons; particles with spin equal to half-odd-integer in
units of $\hbar$ satisfy Fermi-Dirac statistics, which is the algebraic statement of
the Pauli exclusion principle. The requirement that the wave-function of
$n$ indistinguishable particles should be antisymmetric under permutations
of the position variables ensures that the wave-function is zero if any two
particles sit at the same point. For electrons, which have two states of
spin, the wave-function is antisymmetric under the combined exchange of
the positions and the spin states of any pair of particles. The same holds
for any other particle of spin $1/2$, such as proton or neutron. This ensures
that the Pauli principle takes the form 'at most two electrons can occupy
the same position in space'; the same holds if we substitute proton or
neutron for electron in the sentence. The antisymmetry can be ensured if we
use creation and annihilation operators obeying anticommutation relations,
instead of the commutation relations of the boson oscillators. Suppose that
the one-particle space is the complex Hilbert space $\mathcal{H}_1$; for example, for a
particle without spin, $\mathcal{H}_1$ might be $L^2(\mathbf{R}^3)$, but such a particle would be
a boson, by the spin-statistics theorem [Streater and Wightman (1964)].
This theorem holds for particles described by a relativistic quantum field
theory, and does not mathematically prevent this choice for $\mathcal{H}_1$ in our
nonrelativistic discussion. However, such a choice would not describe real
particles. For a particle with spin $\frac{1}{2}$ we would take $\mathcal{H}_1 = L^2(\mathbf{R}^3) \otimes \mathbf{C}^2$,
and for a particle of spin $s$, half the odd integer $2s$, we would choose $\mathcal{H}_1 =
L^2(\mathbf{R}^3) \otimes \mathbf{C}^{2s+1}$. For any choice of $\mathcal{H}_1$, let $\psi \mapsto a^*(\psi)$ be a linear map
from $\mathcal{H}_1$ to $B(\mathcal{K})$, the bounded operators on a Hilbert space $\mathcal{K}$, such that,
putting $a(\psi) := (a^*(\psi))^*$, which is antilinear in the 'test-function' $\psi$, we

A representation of the $CAR$ over a Hilbert space $\mathcal{H}_1$ is
$\pi : \psi \mapsto \pi(\psi)$, possibly different from that given by $a$ in the
obeys the same equations, (11.24), (11.25). Two such rep
and $\pi_2$ on Hilbert spaces $\mathcal{K}_1$ and $\mathcal{K}_2$ are said to be unitar
there exists a unitary operator $U : \mathcal{K}_2 \to \mathcal{K}_2$ such that $U\pi_1($
for all $\psi \in \mathcal{H}_1$. If $\dim \mathcal{H}_1 < \infty$, then there exists only
representation of these relations up to unitary equivalen
Wigner (1928)], where it is also shown that the algebra ge
representation of the $CAR$ is isomorphic to the tensor pro
copies of $\mathbf{M}_2$. We take this algebra to be the $C^*$-algebra o
From (11.25) we see that $a^*(\psi)^2 = 0$, which expresses that w
two particles in the state $\psi$ without getting zero. In any re
whether reducible or not, the operators $\pi^*(\psi)$ are all boun
obey $\|\pi^*(\psi)\| \le \|\psi\|$ for all $\psi \in \mathcal{H}_1$. If $\dim \mathcal{H}_1 = \infty$, there
$C^*$-algebra, given abstractly, which is obtained by taking w
*inductive limit* $\mathcal{A}$ of the algebras of all finite-dimensional s
We then get a mapping, call it $a^*$ again, from $\mathcal{H}_1$ into the a
$\mathcal{A}$, obeying (11.24), (11.25). However, in the case of infin
there are infinitely many inequivalent irreducible representa
'Fock' representation, in finite as well as infinite dimension
one containing a vector $\Psi_0 \in \mathcal{K}$, the Fock vacuum, which
by all the operators $a(\psi)$, thus: $a(\psi)\Psi_0 = 0$ for all $\psi \in$
representation can be given in an explicit manner on the
Fock space over $\mathcal{H}_1$:

$$\mathcal{K} = \wedge \mathcal{H}_1 := \mathbf{C} \oplus \mathcal{H}_1 \oplus \overset{2}{\wedge} \mathcal{H}_1 \oplus \overset{3}{\wedge} \mathcal{H}_1 \oplus \ldots$$

where $\mathbf{C}$ is the space of complex numbers, a one-dimension
spanned by the Fock vacuum $\Psi_0$, and for $n = 2, 3, \ldots$, $\wedge$
tisymmetric part of the $n$-fold tensor product $\mathcal{H}_1 \otimes \ldots \mathcal{H}_1$

$$a^*(\psi)\varphi_1 \wedge \cdots \wedge \varphi_n \;=\; \psi \wedge \varphi_1 \wedge \cdots \wedge \varphi_n$$

In words, $a^*(\psi)$ creates the particle with wave-function $\psi$
adjoint $a(\psi)$ is then computable, from the definition of
in $\wedge \mathcal{H}_1$ as the sum of the scalar products in the summa
see and do Exercise 11.67. If $\dim \mathcal{H}_1 = n < \infty$, then th
at the 'top form' $\wedge^n \mathcal{H}_1$: there cannot be more than $n$ pa
which is Pauli's exclusion principle. The dimension of
is then $2^n$. Any non-zero element of the top form is so
*anti-Fock vacuum*; it is annihilated by the action of $a^*($
The fermionic Fock space has a functorial property sin
Fock space:

$$\wedge (\mathcal{H}_1 \oplus \mathcal{H}_2) = \wedge \mathcal{H}_1 \otimes \wedge \mathcal{H}_2.$$

This is the property implicitly used by Heisenberg w
isotopic spin and thereafter treated the proton and ne
of the same particle.

A state $\rho$ on the $CAR$-algebra $\mathcal{A}$ is determined by the

$$\rho \cdot (a^*(\psi_1) \ldots a^*(\psi_m) a(\phi_1) \ldots a(\phi_{n-m}$$

as $\psi_i$ and $\phi_j$ run over $\mathcal{H}_1$. Because of the anticommutati
we need only give the expectation values of all the Wick
the creators are to the left and the annihilators are to t
expectation values of a product in any order is determ
the Fock vacuum is determined by the vanishing of all t

The $CAR$-algebra $\mathcal{A}$ is not taken as the algebra
quantum theory; this role is taken to be a subalgebra c
important subalgebras, the gauge-invariant subalgebra
algebra. The $CAR$, (11.24) and (11.25), are invariant
transformation of $\mathcal{H}_1$. That is, if $\psi \mapsto a(\psi)$ and $\psi \mapsto a$

and (11.25), then so do the maps $\psi \mapsto a(U\psi)$ and $\psi \mapsto a^*(U\psi)$ for any unitary operator $U$ on $\mathcal{H}_1$. Now, the $CAR$-algebra is unique up to isomorphism; the sets of operators $\{a(\psi) : \psi \in \mathcal{H}_1\}$ and $\{a(U\psi) : \psi \in \mathcal{H}_1\}$ are exactly the same, so the uniqueness means that there is an automorphism $\gamma_U$ of $\mathcal{A}$ such that $\gamma_U(a(\psi)) = a(U\psi)$ for all $\psi$. Now, it can be proved that if $\dim \mathcal{H}_1 < \infty$, any automorphism of $\mathcal{A}$ is spatial; that is, it is implemented by a unitary operator; thus, in the case at hand, $\gamma_U$ can be implemented by a unitary operator, $\Gamma_U$, called the second quantisation of $U$. It has the explicit form on the $n$-particle state

$$\Gamma_U(\psi_1 \wedge \psi_2 \wedge \ldots \wedge \psi_n) = U\psi_1 \wedge U\psi_2 \ldots \wedge U\psi_n.$$

When $U$ is the multiplication of $\psi$ by the phase $e^{i\theta}$, where $\theta \in \mathbf{R}$, the corresponding automorphism $\gamma_\theta$, say, is called a gauge transformation (of the first kind). Such automorphisms form the *gauge group*, $\mathcal{G}$ say. One can argue that all observables are gauge invariant, and that any observable of a theory with fermions should be the gauge-invariant part:

$$\mathcal{A}_\mathcal{G} := \{A \in \mathcal{A} : \gamma_\theta(A) = A\}$$

for all gauge transformations $g \in \mathcal{G}$. This is obviously a subalgebra of $\mathcal{A}$. A state $\rho$ is said to be invariant under the automorphism $\gamma$ if the dual action to $\gamma$, acting on the states, which can be regarded as the right action $\rho \mapsto \rho\gamma$, leaves $\rho$ unchanged. It is often said that only gauge-invariant states are physical. However, the modern point of view is dual to this: the only observables are gauge invariant, and two states are physically the same if they reduce to the same state on the gauge-invariant subalgebra. From this point of view, a non-gauge-invariant state always defines a state on $\mathcal{A}_\mathcal{G}$. By integrating the state over the circle group we get a more mixed state of $\mathcal{A}$ that is gauge invariant, and which is physically the same as original one. Thus the non-gauge-invariance of a state does not mean that gauge invariance does not hold in the corresponding cyclic representation. If a state is gauge invariant, then the $n$-point functions are zero unless the number of creators, $m$ in (11.30), is the same as the number of annihilators, $n - m$.

The other important subalgebra of $\mathcal{A}$ is the even subalgebra. This is the algebra which is invariant under the single gauge transformation $\psi \mapsto -\psi$. A state is said to be even if it is invariant under the dual map induced on the states. An even state has the property that the expectation value of the product of an odd number of creators and annihilators vanishes. For an even state $\rho$ we can usefully define the concept of the correlation functions (cumulants or truncated functions are two other names). Let $b$ denote $a$ or $a^*$ (which one, $a$ or $a^*$, being fixed by the wave-function $\psi_j$); then let $\rho_\tau$ denote the correlation function, defined inductively by

$$\rho\left(b(\psi_1)\ldots b(\psi_n)\right) = \rho_\tau\left(b(\psi_1)\ldots b(\psi_n)\right)$$
$$+ \sum_I P(I)\rho_\tau(I_1)\rho_\tau(I_2)\ldots\rho_\tau(I_r). \qquad (11.31)$$

Here, $I$ runs over all partitions $I = I_1 \cup I_2 \cup \ldots \cup I_r$ of the set $\{b(\psi_1),\ldots,b(\psi_n)\}$ into $r$ parts, and then $r$ runs from 2 to $n$. Because the state $\rho$ is even, only parts with an even number of elements enter. In each partition, the symbol $I_m$ stands for the product of the $b(\psi_j)$ in increasing order in $j$, and $\psi_j \in I_m$. Finally, $P(I)$ is the parity of the permutation taking the order $I_1 I_2 \ldots I_r$ to the natural order $1, 2, \ldots, n$. Because each part contains an even number of terms, this parity does not depend on the choice (arbitrary) of the order in which we write the parts. The two-point correlation functions are the same as the two-point functions, since the expectation of a single power of $b$ is zero. In general the $2n$-point correlation function measures the mutual quantum correlation of the $2n$ operators in the state $\rho$.

Balslev and Verbeure [Balslev and Verbeure (1968)] define a quasifree state of the $CAR$-algebra to be an even state for which the correlation functions vanish for $n \geq 3$. For gauge-invariant states this is equivalent to the earlier definition of Shale and Stinespring [Shale and Steinspring (1964)]. From the definition (11.31) we see that for a gauge-invariant quasifree state, the $n$-point function is determined by the two-point functions by the formula

$$\rho\left(a^*(\psi_1)\ldots a^*(\psi_n)a(\phi_1)\ldots a(\phi_n)\right) = \det\left[\rho\left(a^*(\psi_i)a(\phi_j)\right)\right]. \qquad (11.32)$$

The general two-point function of a gauge-invariant state on the $CAR$-algebra is determined by the operator $A$ such that

$$\rho\left(a^*(\psi)a(\phi)\right) = \langle \phi, A\psi \rangle_{\mathcal{H}_1}. \qquad (11.33)$$

Since this is positive-definite, $A$ must be a positive operator on the one-particle space $\mathcal{H}_1$. More, since $a(\psi)$ is bounded by $\|\phi\|$, the operator $A$ must be bounded by the unit operator. Conversely, any $A$ obeying $0 \leq A \leq 1$ defines the two-point function (11.33) of a quasifree state. It follows that we may define the (non-linear) quasifree map $Q$ on any state $\rho$ as taking $\rho$ to the quasifree state with the same two-point function. It can be shown that the quasifree state defined by the operator $A$ is pure (as a state on $\mathcal{A}$) if and only if $A$ is a projection. If $A$ is a projection and $\dim A < \infty$, then the subspace of $\mathcal{H}_1$ defined by $A$ represents those states in $\mathcal{H}_1$ that are occupied, or in Dirac's language, 'filled' by particles. The representation given by the state is then equivalent to the Fock representation. If $\dim A = \infty$ then the state is not a vector in Fock space, because infinitely many levels are occupied. The one-particle states in the subspace defined by $A$ fill a 'sea', and the representation is non-Fock.

We shall be interested in a generalisation of this concept of gauge group, because we need to describe a process in which a fermion decays into a state with several fermions, so that the quantum number, total number of fermions, is not conserved. This can occur in nuclear physics, atomic physics, as well as in chemistry and molecular physics. Of course, the total charge $q$ is conserved, as is the total number, $N_B$ say, of baryons. It turns out that the lepton number $\ell_e$ is also conserved. This is the number of electrons, minus the number of positrons, plus the number of neutrinos, minus the number of antineutrinos. Here, the neutrino and its antiparticle are of the type associated with the electron. It is thought that there are also two other conserved lepton numbers, $\ell_\mu$ and $\ell_\tau$, associated in a similar way with the muon and the tau. Each of these quantum numbers is believed to generate a superselection rule. This means the following. Each type of particle, say the $i$-th particle, has an integer amount of charge, $q_i$; similarly, baryon number, and lepton number of the three kinds are integers. These integers should then be eigenvalues of the operators representing charge, baryon number etc. To set this up if all the particles are fermions, the one-particle space $\mathcal{H}_1$ should be the direct sum of spaces of one particle of each type, thus:

$$\mathcal{H}_1 = \oplus_i \mathcal{H}_i.$$

On the subspace $\mathcal{H}_i$, $q$ has the value $q_i$, and on $\mathcal{H}_1$, $q$ is defined by linearity. Similarly, the other conserved numbers are defined on $\mathcal{H}_1$. Thus we define five unitary gauge groups acting on $\mathcal{H}_1$; for example, $U_q(\theta) = \exp\{iq\theta\}$ is the electromagnetic gauge group. The second quantisation of these unitary groups give us five commuting automorphism groups of the $CAR$-algebra over $\mathcal{H}_1$. Let us denote any of these by $\alpha(\theta)$, acting on the left of the algebra. Then the concept of superselection rule can be summarised by saying that the *observables* of the theory lie in the subalgebra of operators that are invariant under these groups: $\mathcal{A}_0 = \{A \in \mathcal{A} : \alpha(\theta)A = A\}$ holds for all $\theta \in \mathbf{R}$. This replaces the gauge-invariant subalgebra $\mathcal{A}_\mathcal{G}$ in models where the particles have not got the same charge, baryon number and lepton numbers. As before, this means that any state can be chosen to be invariant under the gauge groups. Recall that given any state, $\rho$, gauge invariant or not, then its gauge-group average, $\int \rho \alpha(\theta)\, d\theta$, gives the same expectation value as does $\rho$ for each gauge-invariant operator. Indeed, this is exactly what is meant by saying that when there is a superselection rule, there can be no coherent superposition of states with different values of the charge or other quantum number defining the group. It should be said that our gauge groups are of the first kind; gauge invariance of the second kind involves using a different element of the gauge group at each point of space or space-time, coupled with a compensating change in the quantum field of the gauge boson. We have not been talking about this.

The $CAR$ over $\mathcal{H}$ can be related to the Clifford algebra over $\mathcal{H}$, regarded as a *real* vector space with scalar product $(\psi, \phi) = Re\langle \psi, \phi \rangle_{\mathcal{H}}$. That is, we forget the complex structure of $\mathcal{H}$, leaving just the orthogonal structure defined by $(\cdot, \cdot)$. We then seek a real-linear map $A : \mathcal{H} \to B(\mathcal{K})$, where $\mathcal{K}$ is a complex Hilbert space, such that

(1) $A(\psi)A(\phi) + A(\phi)A(\psi) = 2(\psi, \phi)_{\mathcal{H}} I_{\mathcal{K}}$;
(2) $A(\psi)^* = A(\psi)$

for all $\psi$ and $\phi$ in $\mathcal{H}$. Such a mapping $A$ is said to be a representation of the Clifford algebra over the real Hilbert space $\mathcal{H}$. If $A$ is a representation, then so is $A \circ T$, where $T$ is an orthogonal operator on $\mathcal{H}$; see Exercise 11.68. To get a representation of the $CAR$ from $A$ we must choose a *complex structure* $J$ on $\mathcal{H}$; that is, $J$ is an orthogonal operator on $\mathcal{H}$ such that $J^2 = -1$. This gives a definition of multiplication by making the structure $(\mathcal{H}, J)$ into a complex Hilbert space. Indeed, using $J$, we define the multiplication of any vector $\psi \in \mathcal{H}$ by the complex number $\alpha + i\beta$ to be $\alpha\psi + \beta J\psi$. The scalar product in $\mathcal{H}$, regarded as a complex Hilbert space, needs an imaginary part; it is easy to show that

$$\langle \psi, \phi \rangle_J = (\psi, \phi) - i(\psi, J\phi) \tag{11.34}$$

makes $\mathcal{H}$ into a complex Hilbert space; see Exercise 11.69. Then we can define maps $a$ and $a^*$ from $\mathcal{H}$ to $B(\mathcal{K})$ by

$$a(\psi) = \frac{1}{2}(A(\psi) + iA(J\psi)); \quad a^*(\psi) = \frac{1}{2}(A(\psi) - iA(J\psi)). \tag{11.35}$$

These then obey the $CAR$; see Exercise 11.70.

In addition to the choice of $J$, we may specify a conjugation on $\mathcal{H}$; this is a $J$-antilinear map on $\mathcal{H}$, say $\psi \mapsto \overline{\psi}$, whose square is the identity. This enables us to define the *real* vectors in $\mathcal{H}$, namely, those for which $\overline{\psi} = \psi$. Choosing the conjugation is like making the choice of a Lagrangian submanifold in classical mechanics; we already met this idea in the theory of bosons. It enables us to define the fermion field $\Psi(\psi) = 2^{-1/2}(a^*(\psi)+a(\overline{\psi}))$, which is **C**-linear in the vector $\psi$. This is what is needed in quantum field theory, where $\psi$ is a test-function [Streater and Wightman (1964)]. Physically, the conjugation is related to charge conjugation.

The orthogonal group of $\mathcal{H}$ is larger than the unitary group of $\mathcal{H}_1$; the latter leaves both the real and imaginary parts of the complex scalar product (11.34) unchanged, whereas the orthogonal group leaves only the real part unchanged. Recall that for bosons, it is exactly the opposite: the canonical transformations, which mix $p$ and $q$, are the symplectic transformations, namely, those that leave the *imaginary* part of the scalar product invariant. For fermions, a useful theorem states that any positive Hamiltonian, quadratic in the creators and annihilators, can be diagonalised by an orthogonal transformation. This is the analogue of the theorem for bosons, which involves a symplectic transformation instead of an orthogonal one. In either case, fermionic or bosonic, maps preserving the *CAR* or the *CCR* are said to be *canonical*. If the maps are generated by a real linear map on $\mathcal{H}$, then they are termed *Bogoliubov* transformations. They generate automorphisms of the corresponding **C**\*-algebra. These automorphisms can be thrown onto the states by duality; they do not leave the Fock vacuum invariant, unless the one-particle map is unitary. It can be shown that any pure quasifree of the *CAR* or the *CCR* is a Bogoliubov transform of the Fock vacuum [Balslev and Verbeure (1968)]. In various models, an approximate ground state of the interacting Hamiltonian is constructed, which is pure and quasifree; it is not annihilated by the original operators $a(\psi)$, but there is a Bogoliubov transformation to new variables $a'(\psi)$ which do kill it. The states created by $a'^*$ are called 'quasi-particles'. They are collective modes of the original quanta, and occur in several important models such as the *BCS* (Bardeen, Cooper and Schrieffer) model of superconductivity. The quasifree states can be completely characterised:

**Theorem 11.7.** *Let $\mathcal{H}$ be a real separable Hilbert space with scalar product $(\bullet, \bullet)$, and let $\psi \mapsto A(\psi)$ define the Clifford algebra over $\mathcal{H}$. Let $\rho$ be a quasifree state. Then $\rho$ determines a bounded operator $T$ on $\mathcal{H}$ by the equation*

$$\rho \cdot (A(\psi)A(\phi)) = (\psi, \phi) + i(\psi, T\phi), \qquad \psi, \phi \in \mathcal{H} \tag{11.36}$$

*and $T$ satisfies $T^d = -T$ and $\|T\| \leq 1$. Moreover, $\rho$ is pure if and only if $T^2 = -1$, in which case $T$ is a conjugation, and the representation is the*

*Fock representation with this $T$ as the $J$ of (11.34). Conversely, any such $T$ determines a quasifree state.*

Here, $T^d$ is the adjoint of $T$ relative to the real scalar product $(\bullet, \bullet)$.

**Proof.**   We shall prove the necessity of the conditions; you are referred to [Balslev and Verbeure (1968)] for the sufficiency.

   Suppose then that $\rho$ is any state on the Clifford algebra. From the anticommutation relations, it is easy to see that

$$\rho \cdot (A(\psi)A(\phi)) - (\psi, \phi) = i\sigma$$

defines a real, antisymmetric bilinear form $\sigma$. A necessary condition for positivity is

$$\rho \cdot ((A(\psi) + iA(\phi)(A(\psi) - iA(\phi))) \geq 0,$$

since by (2), $A(\psi)$ is self-adjoint. This yields $|\sigma| \leq 1$, so by the Riesz duality theorem, $\sigma$ defines a bounded operator $T$ such that $\sigma = (\psi, T\phi)$. The antisymmetry of $\sigma$ gives that of $T$. Finally, if $T^2 = -1$, it is a conjugation, and the corresponding creators and annihilators, (11.35), obey $\rho \cdot (a^*a) = 0$; thus the representation is the Fock representation, known to be irreducible, so that $\rho$ is pure.                                            □

   In the proof of the converse (that any operator $T$ obeying the conditions of the theorem defines a quasifree state), Balslev, Manuceau and Verbeure give a useful formula for the representation obtained from $\rho$ by the Gelfand-Naimark-Segal construction. The case when $\dim_{\mathbf{R}} \mathcal{H}$ is even or infinite is simpler, and is also more important, since only in that can $\mathcal{H}$ be complexified. So let $J$ be a complex structure on $\mathcal{H}$, and define $j = J \oplus -J$; this is a complex structure on $\mathcal{H}' := \oplus \mathcal{H}$. Let $\pi_j$ be the Fock representation of the $CAR$ over $\mathcal{H}'$. This takes place in the tensor product of the $J$-Fock and the $-J$-Fock spaces, by the functorial property (11.29) of these spaces. Let $T'$ be the map from $\mathcal{H}$ into $\mathcal{H}'$ given by $\psi \mapsto 2^{-\frac{1}{2}} \left( (1 + |T|^{\frac{1}{2}})\psi \oplus (1 - |T|^{\frac{1}{2}})\psi \right)$. Then the representation of the Clifford algebra defined by $\rho$ is equivalent to $\pi_j \circ T'$, applied to the cyclic subspace generated from the vacuum. We can see a similarity with the tensor structure of the representation defined by a $KMS$ state; in fact, this is exactly what we have, for the *even* subalgebra. All this was already understood by Araki and Wyss in [Araki and Wyss (1964)]. The representation of the fermion operators $a(\psi), a^*(\psi)$ is not quite a tensor product, since the parts in the two factors anticommute rather than commute. The special case $T = 0$ corresponds to the tracial state (the microcanonical state, the state with $\beta = 0$) in the case where $\dim \mathcal{H} < \infty$. In the infinite case the vacuum defines a central state, which is one obeying $\rho(AB) = \rho(BA)$ for all elements $A, B$ of the algebra. This shows that the representation generates (by weak closure) a $W^*$-algebra of type $II_1$ in the classification of Murray and von Neumann. In general, $0 < \|T\| < 1$ and the representation is of type $III$, which was a big surprise to mathematicians.

It follows from Theorem 11.7, that we can define the *quasifree map*, denoted by $\rho \mapsto \rho Q$; this maps any state to the quasifree state with the same two-point functions: such a quasifree state always exists.

While quasifree states have a simple description, they carry no correlations between three or more particles. It might be thought that a more detailed description would be possible, by allowing for example that the four-point correlations to be non-zero, but taking for simplicity all higher than four to be zero. Unfortunately, there are no such states; as with the classical case, and the bosonic case, any state of the $CAR$ with correlation functions zero beyond some $n > 2$ must be quasifree. Thus unless we include all the correlations, however remote, we may include none beyond those of second order, without violating the condition of positivity of the state. This has hampered the construction of non-quasifree theories. The use of higher-order temperatures [Ingarden (1992)] is a way out. This will be explained in Chapter 15.

Quasifree states are quite tractable: the entropy of a quasifree state has an explicit expression, and the canonical and the grand canonical states of an oscillator are quasifree. To find the entropy of a quasifree state, we first deal with the case with only one fermion, in which case $\mathcal{H}_1 = \mathbf{C}$. The fermionic Fock space is then just $\mathcal{F} = \mathbf{C} \oplus \mathbf{C}$. A quasifree state is determined by the operator $A = \lambda$, where $\lambda$ is just a real number, $0 \leq \lambda \leq 1$, which is the probability of occupation. The density matrix of the state in the basis determined by the direct sum $\mathcal{F}$ is $\operatorname{diag}(1 - \lambda, \lambda)$, and its entropy is $S = -\lambda \log \lambda - (1 - \lambda) \log(1 - \lambda)$. Suppose now that we have more than one mode, so that $\dim \mathcal{H}_1 > 1$. When $A$ has finite rank, we may diagonalise it into modes and can take $\mathcal{H}_1$ to be the non-zero, finite-dimensional, part. Write $A = \sum \lambda_j P_j$, its spectral resolution. The relevant part of Fock space is, by the functorial property (11.29), the tensor product of copies of the above two-dimensional space, $\mathcal{F}$. The state is also a product state, so the entropy of the quasifree state determined by $A$ is the sum of the entropy of the factors:

$$S = -\sum_j \{\lambda_j \log \lambda_j + (1 - \lambda_j) \log(1 - \lambda_j)\}$$

$$= -Tr\{A \log A + (1 - A) \log(1 - A)\}. \tag{11.37}$$

Even if $\dim \mathcal{H}_1 = \infty$, (11.37) may make sense by converging; then we say that the state has finite entropy. It is interesting to compare the fermion entropy with the entropy of a quasifree boson, (11.23); we see that as there it is determined by the mean particle density, and that we have a similar formula except for some signs. Again, the first term is the same as the Boltzmann entropy.

We shall find the canonical state if $\dim \mathcal{H}_1 < \infty$ for any Hamiltonian which is the second quantisation of a one-particle operator. These are sometimes called quadratic Hamiltonians. They lead to linear equations of motion for the creators and annihilators, $a, a^*$. We shall be interested in the case where the Hamiltonian $H$ has discrete spectrum with finite multiplicity; we first diagonalise $H$ by a canonical transformation. So suppose that $\{\psi_j\}$ are the normalised eigenstates of $H$, with eigenvalues $\{\mathcal{E}_j\}$. The second quantisation of $H$ can then be written

$$d\Gamma H = \sum_j \mathcal{E}_j a^*(\psi_j) a(\psi_j).$$

Let us start with the simple case where $\dim \mathcal{H}_1 = 1$ with energy $\mathcal{E}$. The canonical state $\rho$, with inverse temperature $\beta$, is determined by its generating function

$$\mathcal{F}(\lambda, \mu) = \rho(\exp\{\lambda a^*\} \exp\{\mu a\})$$

$$= Z^{-1} Tr \left( \exp\{-\beta \mathcal{E} a^* a\} \exp\{\lambda a^*\} \exp\{\mu a\} \right).$$

It will be quasifree if all the $n$-point functions are given by (11.32). For one mode we can use the representation

$$a^* = \begin{pmatrix} 0 & 1 \\ 0 & 0 \end{pmatrix} \qquad a = \begin{pmatrix} 0 & 0 \\ 1 & 0 \end{pmatrix} \qquad a^* a = \begin{pmatrix} 1 & 0 \\ 0 & 0 \end{pmatrix}. \tag{11.38}$$

It is elementary to find the exponentials of these matrices, since $a^2 = 0$. We get, in Exercise 11.71,

$$\mathcal{F}(\lambda, \mu) = 1 + \frac{e^{-\beta \mathcal{E}}}{1 + e^{-\beta \mathcal{E}}} \lambda \mu. \tag{11.39}$$

We see that the partition function is $Z = 1 + e^{-\beta \mathcal{E}}$ and that the two-point function is

$$\rho(a^* a) = \frac{e^{-\beta \mathcal{E}}}{1 + e^{-\beta \mathcal{E}}}.$$

We get the mean density with several mutually orthogonal modes $\psi_j$ by summing over $j$; the mean energy is then

$$\overline{E} = \sum_j \mathcal{E}_j \frac{e^{-\beta \mathcal{E}_j}}{1 + e^{-\beta \mathcal{E}_j}}.$$

This is the famous Fermi-Dirac distribution. We note that the generating function (11.39) contains only a constant term, 1, and the quadratic term proportional to $\lambda \mu$. The higher powers of $\lambda$ and $\mu$ are zero, which comes from the fact that $a^2 = 0$. Similarly the generating function for the case of several modes, defined to be

$$\mathcal{F}(\lambda_i, \mu_j) = \rho \left( \exp \left\{ \sum_i \lambda_i a_i^* \right\} \exp \left\{ \sum_j \mu_j a_j \right\} \right)$$

will contain a product $\prod_i \lambda_i \mu_i$ with no repetitions. So the $n$-point functions in Eq. (11.32) are all zero unless all the creators refer to different modes, and the same for the annihilators. Then only one term survives, in which each creator is contracted with its own mode, and we arrive at the determinant of Eq. (11.32) with only one non-zero term. So the canonical state is quasifree.

Now we can easily see that the quasifree map is entropy increasing. For, to maximise the entropy among all states, subject to finitely many quadratic conditions, leads via Lagrange multipliers to a canonical state with a quadratic Hamiltonian. But this is quasifree. The proof is easier than the original one [Lanford and Robinson (1972)], which however also applies to the infinite-dimensional case.

We can describe a gas of fermions, moving on a lattice $\Lambda$, by the $CAR$-algebra over $\mathcal{H}_1 = \oplus_x \mathcal{H}_x$, where we have a finite-dimensional Hilbert space $\mathcal{H}_x$ at each point. The states in $\mathcal{H}_x$ label the spin states of the fermion, and also the different types of fermion, such as neutron, proton, heavy hydrogen etc. To get a simple model, we choose the Hamiltonian to be that of the free particles; thus if the chemical energy per molecule, of the $j$-th type of particle is $\varepsilon_j$, then the Hamiltonian is

$$H_c = \sum_{x \in \Lambda} \sum_j \varepsilon_j a_j^* a_j.$$

This is the simplest choice, in which there is no kinetic energy assigned to the molecules. Usually in quantum mechanics the kinetic energy enters through gradient terms in the quadratic part; on a lattice the corresponding expression should be a linking of nearest neighbours. This is simplest if $\Lambda$ is a cubic lattice, with periodic boundary conditions. Let us illustrate this in one dimension, and choose $\Lambda = \mathbf{Z}_N$, a ring of $N$ atoms. The finite translation group $\mathbf{Z}_N$ acts on itself, and a translation-invariant kinetic energy can be chosen to be

$$H_1 = \frac{1}{2m} \sum_x \left( (a_{x+1}^* - a_x^*)(a_{x+1} - a_x) \right)$$

for each atom, whose mass is $m$. In the term $x = N$, we interpret $x + 1$ as 1, as we have a ring. Clearly, $H_1$ is invariant under translations of the lattice, which leads to the conservation of momentum. There are $N$ eigenstates of momentum, labelled by the dual group (here also $\mathbf{Z}_N$). The total energy is taken to be $H = H_c + H_1$, with spectral projections $P_j$ say. Let $\mathcal{A}$ be the $CAR$-algebra of all atoms at all sites. The energy-shells are the subalgebras $\mathcal{A}_{ij} = P_i \mathcal{A} P_j$. We get a model by choosing a bistochastic completely positive map $T$ on the $\mathcal{A}$, which maps each $\mathcal{A}_{ij}$ into itself. This is designed to give a non-zero transition probability for each process we want to describe by the model, and at the same time, is to map to themselves each 'number-shell' of any conserved quantity (like charge, baryon number). The number-shell is defined similarly to the energy-shell. Because the kinetic energy $H_1$ links different sites, it is no longer true that the stoss map conserves the mean energy. But we have available the quasi-free map $Q$, which conserves the mean (in the current state $\rho$) of all quadratic expressions, including each term in the sum for the kinetic energy. Since $Q$ also increases the entropy, we have a viable theory if we take one time-step of the dynamics of $\mathcal{A}$ to be

$$\rho \mapsto \rho T Q.$$

This will obey both laws of thermodynamics. In the next section we show that the quantum Boltzmann equation (for discrete time) is obtained if the map $T$ is a suitable unitary conjugation commuting with $H$.

We can add a heat-particle to the above model, to allow for the energy located in the elecromagnetic field. It is natural to localise the heat-particle on the bond between the sites, rather than at each site. If we omit the kinetic energy-term, we can use the heat-particle also to model the kinetic energy. In that case we can use the stoss map to add more mixing. This simpler model is like the classical models discussed in part I: there is no mechanism for collective flow, as the correlations between distant parts (and even neighbours) is destroyed. The advantage of the quasifree map over the stoss map is that it increases the entropy without destroying the correlations between distant particles. It is this correlation that gives rise to convection.

In a theory with a single type of fermion the exclusion principle acts as a repulsion between the particles, which behave as if they had a hard core. When there are several fermions, then each site can be occupied by more than one particle, and it is no longer true that the theory seems to describe particles with a hard core. Actual molecules very often exhibit a strong repulsion, and a large kinetic energy is then needed to bring them together, before a reaction can occur. This energy is called the threshold by physicists and the activation energy by chemists. We can model $n$ atoms with a hard core by choosing at each site $x \in \Lambda$ the Hilbert space $\mathcal{H}_x = \mathbf{C}^{n+1}$. Let $\{\psi_j\}_{j=0,\ldots,n}$ be an orthonormal basis in $\mathcal{H}_x$; one vector, say $\psi_0$, will be taken to represent the empty state (no particles present) and the remaining $n$ basis vectors will represent the presence of just one of the atoms. Let $\mathcal{A}_x := B(\mathcal{H}_x)$; then the algebra $\mathcal{A} = \bigotimes_x \mathcal{A}_x$ will be called the *hard-core* algebra. It generalises the fermion model, in that if $n = 1$ the algebra reduces to the $CAR$-algebra of $|\Lambda|$ fermions, without spin. The $n$ excited states of an excitable atom are well described by the hard-core model. For some chemicals, the hard-core model might not be a good one; a large organic molecule might occupy more than the parts out of which it is made. But for some gases and dilute solutions, Guy Lussac's law is nearly true. This states that the size of each molecule is approximately the same. Let us choose the lattice of points in space, $\Lambda$, so that one point can accommodate at most one molecule. When two molecules combine to form a third, this can occupy just one site, unlike the two molecules of which it is made. As a result, the total number of occupied sites in a given volume changes when a reaction takes place, and this means that the pressure changes as well. In this respect, the hard-core model gives predictions quite different from those of fermions. The usual theory [Sewell (1986)], which chooses the same class of algebras, has no scope to describe unoccupied sites; all the states of $\mathcal{H}_x$ represent states of a particle.

We can easily find the algebraic properties of the hard-core algebra. We first represent the operators of $\mathbf{M}_{n+1}$ by matrices with rows and columns labelled $\{0, 1, \ldots, n\}$, and act on it by left-multiplication. The states are then row-vectors with $(n + 1)$ components; the operators act on these vectors by right action. The one-dimensional projection operators onto the states $\psi_j$, $j = 1, \ldots, n$ are the number operators $N_j$ for the $n$ molecules; further, the creation operator $a_j^*$ for the $j$-th particle moves $\psi_0$ to $\psi_j$, and so is represented by the matrix with zeros everywhere except in the zeroth row, $j$-th column, where the entry is 1. Let us call this the *standard* representation of the hard-core algebra. In the standard representation the number operator $N_j = a_j^* a_j$ is the matrix of zeros except on the diagonal in the $j$-th place, and $a_j^* a_k = E_{j,k}$, the matrix unit with zeros except in the $j$-th row and the $k$-th column. In the standard representation $N_0$, the projection onto the vacuum, has a 1 in the zeroth row and column. We have

$$a_j a_j^* = N_0 \text{ for } j = 1, 2, \ldots, n \; ; \qquad a_j a_k = 0 \; ;$$
$$a_j a_k^* = 0 \text{ if } j \neq k \; ; \qquad N_0 a_j = a_j \text{ for all } j \; . \qquad (11.40)$$

For fermions, $n = 1$ and the relations reduce to the $CAR$. A representation of the hard core algebra is then a mapping $a$ from $\{1, 2, \ldots, n\}$ into a $C^*$-algebra obeying the relations (11.40). These relations characterise the algebra up to isomorphism. That is, we have the following theorem:

**Theorem 11.8.** *Let $\{a_1, \ldots, a_n\}$ be an irreducible set of operators obeying (11.40) such that $N_0 \neq 0$. Then $a_1, \ldots, a_n$ generate $\mathbf{M}_{n+1}$ and are equivalent to the standard representation.*

**Proof.** We see that $N_0^2 = N_0 a_j a_j^* = a_j a_j^* = N_0$, so $N_0$ has eigenvalues 0 and 1. If $N_0 \neq 0$ then there is a vector $\psi_0$ in the representation space such that $\|\psi_0\| = 1$ and for all $j$, $a_j a_j^* \psi_0 = \psi_0$. Consider $\psi_j = a_j^* \psi_0$; we find that these are normalised and mutually orthogonal. They are also orthogonal to $\psi_0$, since $\langle \psi_0, a_k^* \psi_0 \rangle = \langle a_j a_j^* \psi_0, a_k^* \psi_0 \rangle = \langle a_j^*, a_j^* a_k^* \psi_0 \rangle = 0$. By direct calculation we show that Span $\{\psi_0, \psi_1, \ldots, \psi_n\}$ is invariant under the action of $a_j, a_k^*$, and as the algebra is irreducible, this is the whole space. We find the matrix element $\langle \psi_j, a_\ell \psi_k \rangle = 0$ unless $j = 0$ and $\ell = k \neq 0$, when it is 1; the representation in this basis is standard. $\qquad \square$

By using the defining relations (11.40) we can prove that $N_0 + \sum_j a_j^* a_j = 1$. This reduces to the $CAR$ when $n = 1$. We note that the operator $\theta = 1 - 2N_0$ anticommutes with all the $a_j$ and $a_j^*$.

Let us define a quasifree state with given two-point functions

$$\rho(a_j a_k), \; \rho(a_j^* a_k), \; \rho(a_j^* a_k^*)$$

to be the state with greatest entropy having these values. Such a state always exists; it is unique by the strict concavity of the von Neumann entropy on $\mathbf{M}_{n+1}$. By the method of Lagrange multipliers, the density matrices of the quasifree states will be exponentials of quadratic operators in $a_j, a_k^*$.

On a set $\Lambda$, we can introduce one copy of $\mathbf{M}_{n+1}$ at each site, and take the tensor product. In that way the creators at different sites commute. This is isomorphic to the usual algebra of spins. In the fermion algebra (the case with $n=1$) the creators at different sites anticommute instead of commuting as in the algebra of spins. This is achieved by putting a twist in the tensor product, using the operator $\theta$. We get the analogue of the fermion algebra inductively as follows. Let $\Lambda$ be given, and let $\mathcal{A}(\Lambda) = \mathbf{M}_{n+1}^{\otimes|\Lambda|}$. Suppose that $\pi$ is a representation of the hard-core algebra over $\Lambda$, such that the creators at different sites anticommute. Let $\theta_x = 1 - 2N_0(x)$, $x \in \Lambda$; this anticommutes with all the creators in $\mathcal{A}_x$, and commutes with all the others in $\Lambda$. Let $y \notin \Lambda$, and let $a_j(y)$ be the standard representation of the hard-core algebra in $\mathbf{M}_{n+1}(y)$. Then the hard-core algebra over $\Lambda \cup \{y\}$ is represented in $\mathcal{A}(\Lambda) \bigotimes \mathbf{M}_{n+1}$ as follows:

- $a_j(x)$, with $x \in \Lambda$, is represented by the ampliation $\pi\left(a_j(x)\right) \otimes 1$;
- $a_j(y)$ is represented by $\prod_{x \in \Lambda} \theta(x) \otimes a_j(y)$.

The creators at different points anticommute. Starting at one site, we construct the anticommuting hard-core algebra for two sites, and so on, getting an isomorphism between the resulting algebra and the usual algebra of spins. The actual isomorphism depends on the order in which points of $\Lambda$ are included. For $n = 1$ this construction reduces to that for the $CAR$, due to Jordan and Wigner [Jordan and Wigner (1928)].

A typical Hamiltonian for a system of chemicals moving in $\Lambda$ will consist of the chemical energy $\sum_x \sum_j \mathcal{E}_j a_j^* a_j$ and some kinetic energy

$$(2m)^{-1} \sum_{x,y} \sum_j (a_j^*(x) - a_j^*(y))(a_j(x) - a_j(y)).$$

Here, $y$ runs over the neighbours of $x$. We should also add the heat-particle, as a balancing item. This could be located on the sites, or on the bonds between neighbours. The Hamiltonian is then the sum of the chemical, kinetic and heat energy. To get one time-step in the dynamics, we apply a bistochastic completely positive map $T$ on the algebra, which leaves invariant the spectral projections of the Hamiltonian. $T$ is conveniently constructed in terms of creators and annihilators, which are used to make up the operators $X_j$ in Kraus's form for a completely positive map. That is, $X_j$ is a product of creators and annihilators that conserve energy, in that the sum of the energy of the creators balances that of the annihilators. These operators will not be localised at one point $x \in \Lambda$, because of the hard-core property. Several points will be needed to arrive at a non-zero polynomial. For example, we could choose one $X$, a unitary operator $e^{iK}$, where $K$ is a typical Hermitian interaction Hamiltonian $\lambda a_j(x)a_k(y)a_\ell^*(z)$

(plus Hermitian conjugate) which represents the combination of atoms $j$ at $x$ and $k$ at $y$ to produce atom $\ell$ at $z$, and the inverse. To ensure that $K$ commutes with the kinetic energy, we sum over all sites. Let $\tau$ be the induced action on the states, written as right-multiplication.

The next step in statistical dynamics is to choose a suitable randomising map that conserves means of all the quantities we want to include in the first law. These will already have been conserved by $\tau$. It is clear that we need a generalisation of quasifree map; if the atoms have different quantum numbers such as baryon number most of the two-point functions will be zero, and we lose a lot of quantum coherence that we might want to keep. We shall return to this in Chapter 15 in the context of generalised temperatures, where a more general method will be introduced.

## 11.4 The Quantum Boltzmann Equation

The quantum Boltzmann equation was written down by Uehling and Uhlenbeck in 1933 [Ueling and Uhlenbeck (1933)]. For the bosonic process

$$A \rightleftharpoons B + \gamma$$

it takes the form

$$\frac{dn_A}{dt} = -\frac{dn_B}{dt} \tag{11.41}$$

$$= -\frac{dn_\gamma}{dt} \tag{11.42}$$

$$= \kappa \left( -n_A (1 + n_B)(1 + n_\gamma) + n_B n_\gamma (1 + n_A) \right). \tag{11.43}$$

The equation for the same process, if $A$ and $B$ are fermions, is

$$\frac{dn_A}{dt} = -\frac{dn_B}{dt} \tag{11.44}$$

$$= \frac{dn_\gamma}{dt} \tag{11.45}$$

$$= \kappa \left( -n_A (1 - n_B)(1 + n_\gamma) + n_B n_\gamma (1 - n_A) \right). \tag{11.46}$$

In these equations, $n$ is the mean number of the particle labelled in the suffix. The parameter $\kappa$ determines the rate at which the process occurs. If there are several modes, the rate for each is a sum over all processes; the original version [Ueling and Uhlenbeck (1933)] had a continuum of modes, and the right-hand side was an integral over these modes; then $\kappa$ was related to the cross-section of the process. We must decide whether we wish to regard the Boltzmann equation as giving an approximation to a Hamiltonian dynamics, or to regard it as defining a theory in its own right. Let us first adopt the Hamiltonian point of view, such as that in [Balescu (1975)]. We introduce creators $a_A^*$, $a_B^*$ and $a_\gamma$ for the $A$, $B$ and heat-particle respectively. Then a justification in the simplest terms of (11.43) is to choose an interaction Hamiltonian

$$H_I = e\left(a_A^* a_B a_\gamma + a_B^* a_\gamma^* a_A\right). \qquad (11.47)$$

This is to be added to the free Hamiltonian

$$H = H_0 + H_I \quad \text{where } H_0 = \mathcal{E}_A a_A^* a_A + \mathcal{E}_B a_B^* a_B + \mathcal{E}_\gamma a_\gamma^* a_\gamma. \qquad (11.48)$$

We see that $H_I$ commutes with $H_0$ if and only if we have a balance of energy:

$$\mathcal{E}_A = \mathcal{E}_B + \mathcal{E}_\gamma. \qquad (11.49)$$

For, by the commutation relations for creators and annihilators,

$$[aa^*, a^*] = a^* \text{ and } [aa^*, a] = -a \text{ for each species.}$$

It follows that $[\mathcal{N}_A, a_A a_B^* a_\gamma^*] = -a_A a_B^* a_\gamma^*$, and so

$$[H_0, a_A a_B^* a_\gamma^*] = (-\mathcal{E}_A + \mathcal{E}_B + \mathcal{E}_\gamma)\, a_A a_B^* a_\gamma^* = 0 \ .$$

Taking Hermitian conjugates and adding, gives $[H_0, H_I] = 0$ if and only if the energy balances. In the interaction picture, the unitary time-evolution is determined by

$$U_I(t) = e^{-iH_0 t} e^{iHt}$$

which causes the transitions of (11.43) to occur. The usual derivation of the Boltzmann equation starts with a computation of the matrix element $\langle \psi_1, H_I \psi_2 \rangle = M$. Then according to the Fermi golden rule, $|M|^2$ is proportional to the transition rate from $\psi_1$ to $\psi_2$. Let $a^*$ be a bosonic creator, and denote the normalised vacuum by $\psi_0$. Then $\|a^{*n}\psi_0\| = \sqrt{(n!)}$; this is elementary: see Exercise 11.72. Let us denote the state $a_A^{*n}\psi_0$ by $\psi_A^n$ etc., and let us omit the tensor product symbol from the product of such states. Then the transition matrix for the forward process is

$$\langle \psi_A^{n_A} \psi_B^{n_B} \psi_\gamma^{n_\gamma},\ H_I \psi_A^{n_A-1} \psi_B^{n_B+1} \psi_\gamma^{n_\gamma+1} \rangle.$$

Here, $n_A = \langle \rho, \mathcal{N}_A \rangle$ etc., where $\rho$ is the current state. Only the term $e a_A a_B^* a_\gamma^*$ contributes, and we find

$$|M|^2 = e^2 n_A (n_B + 1)(n_\gamma + 1).$$

There are some extra factors arising from the golden rule, which is obscure in this simple model. We shall explain these below. The transition probability for the inverse is computed in the same way. We see that the rate of the forward process contains the factors $(1+n_B)$ and $(1+n_\gamma)$. This means that the more $B$ particles or heat-particles that the final state contains the faster the initial state decays to it. The part proportional to $n$ is the celebrated 'stimulated emission' predicted with uncanny insight by Einstein in his seminal paper [Einstein (1917)]. The remaining part, proportional to 1, he called the spontaneous part. Quantum field theory gives both parts together in a natural way. Since the right-hand side of the equation is the transition rate, we put it equal to the rate of *loss* of $n_A$. At this point we have ensured that the dynamics is irreversible. By using the transition probability $|M|^2$ instead of a complex amplitude, we increase randomness. It can be pictured as a measurement process, in which we measure $n_A$, $n_B$ and $n_\gamma$. This step, in modern terms, should be implemented by a completely positive map.

This whole argument has been criticised by Krylov in his brilliant polemic [Krylov (1979)]. In the first place, the eigenvalues and eigenvectors of $H_0$ are not the same as those of $H$. If we take $H$ as being the true energy, then the time-evolution $e^{iHt}$ cannot cause any transitions between the eigenstates. Why should we be interested in the transitions between the eigenstates of $H_0$, which is not the true energy? The answer we get cannot be a good approximation to the rate of transition between the eigenstates of $H$, since the latter is zero. The dynamics using the full Hamiltonian cannot lead to (11.43), since the latter converges to a mixed state when we start from a pure state. In addition to these points made by Krylov, the use of Fermi's golden rule for this model is dubious, as the rule is derived in a theory with space-translation invariance in which an asymptotic condition holds. In fact, in the special case when $H_0$ and $H_I$ commute we have $U_I(t) = e^{iH_I t} = 1 + iH_I t + \ldots$ and we get an expansion for the transition amplitude:

$$\langle \psi_1, U_I(t)\psi_2 \rangle = \langle \psi_1, \psi_2 \rangle + it\langle \psi_1, H_I \psi_2 \rangle + \ldots .$$

Taking $\psi_1$ and $\psi_2$ to be orthogonal, the first term is zero, the term of lowest order is proportional to $t$, so the transition probability has a factor $t^2$ multiplying $|M|^2$. The transition *rate* is then proportional to $t$, which goes to zero as $t \to 0$. This is what is behind the 'Zeno paradox' of continuous measurement, sometimes called 'the paradox of the watched pot' (which never boils). This has often been cited as one of the puzzles of quantum theory. It seems that if we compute the instantaneous rate of the process, then this is zero. We therefore need a more careful study of the problem. A serious attempt to arrive at the Boltzmann equation from a purely Hamiltonian theory must involve some sort of limit. The most successful idea is that of van Hove [van Hove (1955)]: the parameter $t$ must go to infinity, to represent the fact that the macroscopic time we are using for one time-step in the Boltzmann equation is very large on the microscopic scale. We then use one of the time-factors in $|M|^2$ to provide the factor $e^2 t = \tau$, which is held fixed by taking the coupling $e$ to zero. Then $\tau$ is taken to be one unit of macroscopic time. The result is the weak coupling limit. It has been implemented rigorously for a certain class of models [Ho, Landau and Wilkins (1993); Landau (1994)]. The other factor $t$ multiplying $|M^2|$ is removed in a more involved way, which we now explain.

The equation of motion in the interaction picture is

$$\frac{dU_I(t)}{dt} = iH_I(t)U_I(t)$$

where $H_I(t) = e^{-iH_0 t}H_I e^{iH_0 t}$. We impose the boundary value $U_I(t_1) = 1$. This is equivalent to the integral equation

$$U_I(t) = 1 + \int_{t_1}^{t_2} H_I(s)U_I(s)\,ds.$$

Scattering takes place from time $t_1 = -\infty$ to time $t_2 = +\infty$, and so we are interested in the transition amplitude from $\psi_1$ at time $-t$ to $\psi_2$ at time $t$, as $t$ becomes large. The Heisenberg dynamics must be compared with the free dynamics in the same period of $2t$; that is, the free dynamics is to be cancelled out by using $U_I(t)$ instead of the full dynamics $U(t)$. Using perturbation theory, we can solve by the Picard iteration method, to get

$$U_I(2t) = 1 + \int_{-t}^{t} H_I(s)\,ds + \dots.$$

Then if $\psi_1$ and $\psi_2$ are orthogonal, we get

$$\langle \psi_1, U_I(2t)\psi_2 \rangle = \langle \psi_1, \int_{-t}^{t} \exp\{-iH_0 s\}H_I \exp\{iH_0 s\}\psi_2\rangle$$

$$= M \int_{-t}^{t} \exp\{is(\mathcal{E}_2 - \mathcal{E}_1)\}ds.$$

Here, $\mathcal{E}_j$ is the energy of $\psi_j$, taken to be an eigenstate of $H_0$. For this to make sense, we must regard $H_0$ as the energy of the free ingoing and outgoing states, as in scattering theory. In the model we have here, the asymptotic condition does not hold. But in a more realistic theory, with quantised fields in an infinite space and a local interaction $H_I$, we can expect that the particles become asymptotically free, at least for a dilute gas. The transition probability is obtained by squaring the amplitude, so we get for this

$$|M|^2 \int_{-t}^{t} \int_{-t}^{t} \exp\{i(s - s')(\mathcal{E}_2 - \mathcal{E}_1)\}\, ds\, ds'$$

$$= |M|^2 (2t) \int_{-t}^{t} \exp\{is(\mathcal{E}_1 - \mathcal{E}_2)\}\, ds \to 2t|M|^2 2\pi\delta(\mathcal{E}_2 - \mathcal{E}_1)$$

as $t \to \infty$. This has the factor $2t$ over the period of time $2t$. The rate is therefore $2\pi|M|^2$, which is Fermi's golden rule. The other factor $t$ in the rate has gone out, as promised, to provide us with an overall energy-conserving $\delta$-function. In our theory with a discrete set of energy-levels, this factor is infinite. We must imagine that our discrete levels are obtained by starting with a continuum theory in an infinite volume, for which scattering theory holds, and then forming the discrete model by integrating over energy-bands. The $\delta$-function might then be replaced by a Kronecker $\delta$. Something like this has been achieved by Kiegerl and Schürrer [Kiegerl and Schürrer (1990)] by starting with the classical Boltzmann equation and discretising it. One of the difficulties with the van Hove limit is that each of the terms of higher order in the perturbation series become infinite. Thus, for example if energy balance (11.49) holds, the term of second order in the Dyson expansion is

$$e^2 t^2 /(2!)\langle \psi_1, H_I^2 \psi_I \rangle$$

whose square has the factor $e^4 t^4 = \tau^2 t^2$. So the rate has the factor $\tau^2 t$, which has no limit as $t \to \infty$.

Our point of view is not to derive the Boltzmann equation from a purely Hamiltonian theory, but to show that it comes from a bistochastic map $T$ followed by a randomising procedure which increases entropy. This will be enough to show that it leads to dynamics obeying both laws of thermodynamics. For comparison with the above 'derivation', our total energy is analogous to the asymptotic energy $H_0$, and the bistochastic map $T$, acting on the right, plays the role of the scattering operator (or a random mixture of such), which in the limit $t \to \infty$ commutes with the energy operator of the free ingoing and outgoing particles. In a theory with discrete energy levels, the energy-conserving $\delta$-function requires that $T$ maps the energy-shells to themselves, which has been our requirement all along.

The fixed points of (11.43) obviously obey

$$\frac{n_A}{1+n_A} = \frac{n_B}{1+n_B}\frac{n_\gamma}{1+n_\gamma} .\tag{11.50}$$

Suppose that each particle is in the grand canonical state at the same temperature; that is, the density matrix of the system is the tensor product of the states

$$\rho_A = Z_A^{-1}\exp\{-\beta(\mathcal{E}_A - \mu_A)a_A^* a_A\} ,\tag{11.51}$$

$$\rho_B = Z_B^{-1}\exp\{-\beta(\mathcal{E}_B - \mu_B)a_B^* a_B\} ,\tag{11.52}$$

$$\rho_\gamma = Z_\gamma^{-1}\exp\{-\beta\mathcal{E}_\gamma a_\gamma^* a_\gamma\} .\tag{11.53}$$

Then the mean values of the number operators are related to the intensive parameters $\beta$, $\mu_A$ and $\mu_B$ by

$$n_A = \frac{\lambda_A}{1 - \lambda_A};\ n_B = \frac{\lambda_B}{1 - \lambda_B};\ n_\gamma = \frac{\lambda_\gamma}{1 - \lambda_\gamma}.$$

Here, $\lambda_A$ etc. are the activities:

$$\lambda_A = \exp\{-\beta(\mathcal{E}_A - \mu_A)\}$$
$$\lambda_B = \exp\{-\beta(\mathcal{E}_B - \mu_B)\}$$
$$\lambda_\gamma = \exp\{-\beta\mathcal{E}_\gamma\}.$$

Then the condition for equilibrium is simply $\lambda_A = \lambda_B\lambda_\gamma$, which we have met as the condition for equilibrium in chemistry (because $a$ is used as an annihilation operator, we use $\lambda$ for the activity in this section). In view of $\mathcal{E}_A = \mathcal{E}_B + \mathcal{E}_\gamma$, this condition becomes simply $\mu_A = \mu_B$, and the equilibrium state is the grand canonical state of the whole system with a certain temperature and a common chemical potential for the chemicals. The dynamics can be regarded as a flow through the space of the extensive parameters $n_A$, $n_B$, $n_\gamma$, or alternatively through the space of intensive parameters $\mu_A$, $\mu_B$, $\beta$. The ratio of the backward to the forward rate contains the factor $n_\gamma/(n_\gamma + 1) = \lambda_\gamma$, which we recognise as the factor required by detailed balance. It is very likely that in getting his kinetic law for stimulated emission, Einstein worked backwards from the equilibrium, known to be the Planck distribution, using detailed balance. It is not possible to arrive at the law of stimulated emission uniquely by this method, since we have seen that the *activity-led* chemical laws

$$\frac{dn_A}{dt} = \kappa(-\lambda_A + \lambda_B\lambda_\gamma)$$

have the same fixed point and obey detailed balance. These seem to be good laws for reactions in dense liquids; they are not good for photons.

Einstein may have been guided to the right law by the following argument (or was he simply trying to avoid fractions?). The rate of the inverse process $B + \gamma \rightarrow A$ must be proportional to the number of $B$ present. It must also be proportional to the number of $\gamma$ present. So the direct process $A \rightarrow B + \gamma$, which ends up with $(n_B + 1)$ $B$'s and $(n_\gamma + 1)$ heat particles, must be proportional to $(n_B + 1)(n_\gamma + 1)$, by microscopic reversibility. It is also proportional to $n_A$. This assumes that the particles do not get in each other's way. We must not confuse the hopping probability rate from an occupied level with exactly $n$ particles in it with the rate of change of the mean number of particles. The Boltzmann equation uses the means. I shall now concoct a *classical* model which uses this intuition to choose a bistochastic map $T$ leading to the quantum Boltzmann equation with discrete time. This is adapted from [Koseki (1993)] and appeared in [Streater (1993b)].

**Example 11.9.** Consider the model consisting of two chemicals, $A$ and $B$, and one heat particle, $\gamma$. The sample space is $\Omega = \Omega_A \times \Omega_B \times \Omega_\gamma$, where each factor is $\ell^2 = \{0, 1, 2, \dots\}$, the classical Fock space. A typical sample point is $\omega = (i, j, k)$ where $i, j, k = 0, 1, 2, \dots$. The algebra of observables is $\mathcal{A} = \ell^\infty(\Omega)$, the abelian algebra of all bounded random variables on $\Omega$. The dynamics will be given by a bistochastic energy-conserving map $T$ on $\mathcal{A}$, which is linear, followed by a non-linear randomising map to mix the energy-shells. For this, we are going to use the $LTE$ map, which was called $Q$, to provide the increase in entropy. Then the state $p$ at time $t$ will be in $LTE$, and so can be written $p = p_A \otimes p_B \otimes p_\gamma$ where

$$p_A(i) = Z_A^{-1} \lambda_A^i; \qquad p_B(j) = Z_B^{-1} \lambda_B^j; \qquad p_\gamma(k) = Z_\gamma^{-1} \lambda_\gamma^k.$$

Any random variable on $\Omega$ is a function $f(\omega) = f(i, j, k)$; it can be written as a sum of the projections $P_{i,j,k}$:

$$P_{i',j',k'}(\omega) = \begin{cases} 1 & \text{if } i - i', \ j = j', \ k = k'; \\ 0 & \text{otherwise.} \end{cases} \qquad (11.54)$$

In (11.54), $\omega$ is the point $(i, j, k)$. Then we can write

$$f(\omega) = \sum_{\omega'} f(\omega') P_{\omega'}(\omega) .$$

In particular, $\mathcal{N}_A$ is given by

$$\mathcal{N}_A = \sum_{\omega'} i' P_{\omega'}.$$

A state is also a function on $\Omega$. Thus the state $p$ in $LTE$ has the form

$$p(\omega) = Z_A^{-1} Z_B^{-1} Z_\gamma^{-1} \sum_{i',j',k'=0}^{\infty} \lambda_A^{i'} \lambda_B^{j'} \lambda_\gamma^{k'} P_{i',j',k'}(\omega).$$

Because of the map $Q$, in one time-step we move from one $LTE$-state, say $p = (n_A, n_B, n_\gamma)$ to another, say $p' = (n'_A, n'_B, n'_\gamma)$. Since $Q$ conserves $\mathcal{N}_A$, $\mathcal{N}_B$ and $\mathcal{N}_\gamma$, any change in $n_A$ in the time-step is caused by bistochastic map $T$:

$$\langle p', \mathcal{N}_A \rangle = \sum_\omega p T Q(\omega) \mathcal{N}_A(\omega)$$
$$= \langle pT, \mathcal{N}_A \rangle. \tag{11.55}$$

The action of $T$ on $p$ or a random variable is given by its action on all the $P_{i,j,k}$. We shall choose $T$ to be symmetric in the Hilbert-Schmidt scalar product, so that left and right actions are the same. This is the simplest expression of microscopic reversibility. Following Einstein's intuition, $T$ can be split into a forward part, $T_+$, associated with $A \to B + \gamma$, and a backward part, $T_-$, associated with the reverse. We shall define these, and then put $T = T_+ + T_-$. We now define

$$T_+ P_{i,j,k} = (1 - i(j+1)(k+1)\kappa) P_{i,j,k}$$
$$+ i(j+1)(k+1)\kappa P_{i-1,j+1,k+1} \,,$$
$$T_- P_{i,j,k} = (1 - (i+1)jk\kappa) P_{i,j,k} + (i+1)jk\kappa P_{i+1,j-1,k-1}. \tag{11.56}$$

In order for the discrete map $T$ to be stochastic, for a given $i, j, k$ we must have

$$(i(j+1)(k+1) + (i+1)jk)\kappa \le 2.$$

Thus $\kappa$ must be taken smaller, the greater the number of particles there are in the state. As $\kappa$ is proportional to the time-step, we must reduce the time-step when we apply $T$ to states of more and more particles. Thus, there is no stochastic map on the algebra which works for all elements. This is one of the features of bosonic quantum fields − the rate of reactions gets large when there are many particles. The differential equation with continuous time, however, seems to show no pathology. Taking a small enough $\kappa$, the change of $n_A$ is then got by inserting $T$ in (11.55):

$$\langle T_+^\dagger p, \mathcal{N}_A \rangle = \sum_{\omega,\omega'} i' p(i,j,k) \langle \{P_{i,j,k}(1 - \kappa i(j+1)(k+1))$$
$$+ P_{i-1,j+1,k+1}\kappa i(j+1)(k+1)\}, P_{i',j',k'} \rangle$$

which reduces to

$$n'_A = n_A - \kappa n_A (n_B + 1)(n_\gamma + 1)$$

giving the forward part of the discrete version of (11.43) except for a factor 2 in the definition of $\kappa$. Similarly, $T_-$ gives the backward part. The limit as the time step goes to zero gives us (11.43).

To derive (11.43) in the spirit of *quantum* statistical dynamics, let us assume energy balance, (11.49), and put $S = \exp\{iH_I\}$. This involves questions of rigour, since to exponentiate an operator it must be essentially self-adjoint. One way this might be achieved in the bosonic case is to add a high power of a conserved positive operator such as the number operator, and to show that $H_I$ is Kato-small [Kato (1966)]. This problem does not arise for fermions in a discrete space. The operator $S$, once defined, is thus expressed as a function of the free creators and annihilators, in the spirit of the Haag expansion [Haag (1955)]. Then conjugation of the operators with $S$ is a completely positive bistochastic map leaving the spectral projections of $H_0$ invariant. We regard this as the scattering operator in one large time-step. We imagine that the scattering is over and done with, even during a small fraction $t$ of this time-step. Thus we can consider $e^{iH_I t}$ as the scattering that takes place in $t$ steps, the ingoing states for one time step being the outgoing states of the previous step. The time-evolution of the number operator under this unitary group is

$$\mathcal{N}_A(t) = \mathcal{N}_A(0) + it[\mathcal{N}_A, H_I] + (it)^2/2![[\mathcal{N}_A, H_I], H_I] + \dots. \qquad (11.57)$$

Let us follow the scattering over one time-step of size $t$ not by another scattering automorphism, but by the quasifree map. This acknowledges that after the time $t$, which is large on the microscopic scale, the particles are far from their starting points, and meet new particles. The higher correlations among them are lost. This is the modern form of the Fermi golden rule. We start with an even state, $\rho$, which is quasifree, and take the expectation of (11.57). Since $H_I$ is an odd power of the fields (it is cubic) the term of first degree in $t$ has zero expectation value. The second term gives

$$-t^2/2[[\mathcal{N}_A, H_I], H_I] = e^2 t^2 \{(\mathcal{N}_A + 1)\mathcal{N}_B\mathcal{N}_\gamma - \mathcal{N}_A(\mathcal{N}_B + 1)(\mathcal{N}_\gamma + 1)\}.$$

Suppose that $\rho$ is gauge invariant under the separate $U(1)$ groups generated by the number operators, and put $\rho(\mathcal{N}_A) = n_A$ etc.; then the expectation of the second-order term is

$$e^2 t^2 \{(n_A + 1)n_B n_\gamma - n_A(n_B + 1)(n_\gamma + 1)\}.$$

If we put $e^2 t = \kappa$, and divide by $t$, and take the limit as $t \to 0$, we get the Boltzmann equation (11.43). The higher-order terms vanish in this limit. After one time-step, we construct the quasifree state with the same two-point functions and start again with the scattering automorphism.

We can get the fermion version of the Boltzmann equation, (11.46) in the same way by choosing the $A$ and $B$ to be fermions, and taking the same form for $H_I$.

We have mentioned that one of the limitations of using the $CCR$ and $CAR$ is that there are no states in which only a finite number of the many-body correlations are not zero except the quasifree states. Such states might omit correlations that might have physical importance. Even worse, if the interaction energy is a quartic expression in the creators and annihilators then the quasifree map will not conserve the mean energy. In Chapter 15 we shall use the idea of higher order temperatures to suggest a wider class of states and an entropy-increasing map which is more convenient than the quasifree map for interacting systems.

## 11.5   Exercises

**Exercise 11.58.** Let $\mathcal{A}_x = \mathcal{M}_2$ for all $x \in \Lambda$, where $|\Lambda| < \infty$. Let $H_x = \sigma_3$ for all $x$. Let $\mathcal{A} = \bigotimes \mathcal{A}_x$, and $H = \sum H_x$. Show that the only state that is both uniform on the energy-shell and a product state is a canonical state.

**Exercise 11.59.** Show that the permutation group $S_n$ acts in a unitary way on $\otimes^n \mathcal{H}$ by linear extension of the map

$$\pi \left( \psi_1 \otimes \psi_1 \ldots \psi_n \right) = \psi_{\pi(1)} \otimes \psi_{\pi(2)} \otimes \psi_{\pi(n)} \qquad \text{where } j \mapsto \pi(j) \text{ is in } S_n.$$

Show that the space of symmetric tensors is spanned by product vectors $\psi \otimes \psi \otimes \ldots \otimes \psi$, with $\psi \in \mathcal{H}$.

**Exercise 11.60.** Show that the Segal-Weyl operators obey the relation

$$\langle W(\psi)e^\chi, W(\psi)e^\phi \rangle = \langle e^\chi, e^\phi \rangle$$

for all coherent states $e^\chi$ and $e^\phi$.

**Exercise 11.61.** Show that $W(\psi)$ obeys Eq. (11.15).

**Exercise 11.62.** Show that

$$\langle e^\phi, W(\chi)e^\psi \rangle = \exp \left( -\|\chi\|^2/2 - \langle \chi, \psi \rangle + \langle \phi, \chi \rangle + \langle \phi, \psi \rangle \right).$$

Hence obtain Eq. (11.20).

Exercise 11.63. Show that the canonical state of 2 bosonic oscillators, with given mean values for all the first and second moments, is quasifree.

Exercise 11.64. Suppose that $z = \{q_1, q_2, \ldots, q_N, p_1, p_2, \ldots, p_N\}$ and $z' = Sz$, where $q_i$ and $p_j$ obey Heisenberg's commutation relations, and $S \in GL(2N, \mathbf{R})$. Show that $[z'_j, z'_k] = i\hbar G_{jk}$ if and only if $SGS^t = G$, where $G$ is the symplectic metric.

Exercise 11.65. Put $z = q + ip \in \mathbf{C}^N$, and the same with dashed symbols. Show that

$$Im\langle z', z \rangle = z'Gz.$$

Exercise 11.66. Show that in any representation $\pi$ of the $CAR$, we have

$$\|\pi(\psi)\| = \|\pi^*(\psi)\| = \|\psi\|_{\mathcal{H}}.$$

Exercise 11.67. Show that the fermion creation and annihilation operators defined on the antisymmetric Fock space by Eq. (11.28) obey the CAR relations Eqs. (11.24), (11.25).

Exercise 11.68. Suppose that $A$ is a representation of the Clifford algebra over the real Hilbert space $\mathcal{H}$. Let $T$ be an orthogonal operator on $\mathcal{H}$, that is, an $\mathbf{R}$-linear map obeying $(T\psi, T\phi) = (\psi, \phi)$ for all $\psi$ and $\phi$ in $\mathcal{H}$. Show that $A \circ T$ is a representation of the Clifford algebra.

Exercise 11.69. Show that $\langle \bullet, \bullet \rangle_J$ defined in Eq. (11.34) is a scalar product; that is, it is positive definite, $\mathbf{C}$-linear in the second variable, and Hermitian: $\langle \psi, \phi \rangle_J^* = \langle \phi, \psi \rangle_J$.

Exercise 11.70. Let $A$ be a representation of the Clifford algebra over the real Hilbert space $\mathcal{H}$, and let $J$ be a complex structure on $\mathcal{H}$. Let $a$ and $a^*$ be given by Eq. (11.35). Show that they provide a representation of the $CAR$ over the complex Hilbert space $(\mathcal{H}, J)$.

Exercise 11.71. Find the generating function for a single fermion mode

$$\rho \left( \exp\{\lambda a^*\} \exp\{\mu a\} \right)$$

where $\rho = Z^{-1} \exp(-\beta \mathcal{E} a^* a)$.

Exercise 11.72. Let $\psi_0$ be the Fock vacuum of a bosonic oscillator; show that the norm of $a^{*n}\psi_0$ is $\sqrt{n!}$

# Chapter 12

# Isothermal and Driven Systems

## 12.1 Isothermal Quantum Dynamics

Consider a quantum system, described by a $C^*$-algebra $\mathcal{A}_c$, coupled to a heat-bath at beta $\beta$, described by the algebra $\mathcal{A}_\gamma$. The combined system is described by $\mathcal{A} = \mathcal{A}_c \otimes \mathcal{A}_\gamma$; we shall use $\rho$ for a state on $\mathcal{A}$. In the simplest models, the energy is taken to be the sum $H = H_c \otimes 1 + 1 \otimes H_\gamma$. One time-step of the isolated dynamics of the combined system is given by an energy-conserving bistochastic map $\tau$, followed by the stoss map. Thus one time-step in the isolated dynamics is the map

$$\rho \mapsto \rho\tau \mapsto (\rho\tau)\,\mathcal{M}_c \otimes (\rho\tau)\,\mathcal{M}_\gamma.$$

Even if the state $\rho$ at time $t$ is the product of a state $\rho_c$ with a thermal state $\rho_{\gamma\beta}$, the interaction $\tau$ will spoil this form. In the isothermal dynamics we assume that the chemical dynamics is so slow that within the time-step it is possible to restore the state of the heat bath to the same state $\rho_{\gamma\beta}$ as before the action of $\tau$. Thus one time step in the isothermal dynamics is given by the map $T$ defined by

$$\rho_c T = ((\rho_c \otimes \rho_{\gamma\beta})\tau)\,\mathcal{M}_c \tag{12.1}$$

with $\rho_{\gamma\beta}$ being constant in time. Unlike the isolated dynamics with stoss map, this is a linear map in the state $\rho_c$. Further, this dynamics does not conserve the mean value of the energy, either of the chemical system or the combined system; physically, this is due to the assumed heat flows needed to restore the state $((\rho_c \otimes \rho_{\gamma\beta})\tau)\,\mathcal{M}_\gamma$ back to $\rho_{\gamma\beta}$. In the case where $\tau$ is unitary conjugation (non-random dynamics), this map has been called the 'reduced Heisenberg dynamics' in [Gorini, Kossakowski and Sudarshan (1976)].

The map $\tau$ is assumed to be energy-conserving; thus, if $H = \sum_j \mathcal{E}_j P_j$ is the spectral resolution of the total energy then we assume that for each $j$, we have $P_j \tau = P_j$. Moreover, we require that $\tau^*$, the adjoint of $\tau$ on the Hilbert space of Hilbert-Schmidt operators, also leaves each $P_j$ invariant. As a consequence, the canonical state $\rho_{c\beta}$ of $H_c$ is a fixed point of $T$. Thus

$$\rho_{c\beta} T = ((\rho_{c\beta} \otimes \rho_{\gamma\beta})\tau) \, \mathcal{M}_c = (\rho_{c\beta} \otimes \rho_{\gamma\beta}) \, \mathcal{M}_c = \rho_{c\beta}. \tag{12.2}$$

Here, we have used the fact that each $P_j$ projects to a finite dimensional space and so is of Hilbert-Schmidt class.

Just as in the classical theory, in isothermal dynamics the second law is expressed as the free-energy theorem. This states that the thermodynamic function $\Psi(\rho) = S(\rho) - \beta\rho \cdot H_c$, called the Massieu function, does not decrease under a completely positive stochastic map. This is effectively contained in a result of Lindblad [Lindblad (1975)] of which a simple form is

**Theorem 12.1.** *Let $T$ be a completely positive stochastic map on $\mathbf{C}^n$; denote the density matrix of a state $\rho$ by $\Delta\rho$. Then*

$$Tr \left( \Delta\rho(\log \Delta\rho - \log \Delta\sigma) \right) \geq Tr \left( \Delta\rho T(\log \Delta\rho T - \log \Delta\sigma T) \right). \tag{12.3}$$

The point is that if $\sigma$ is the canonical state for the Hamiltonian $H_c$ and is a right fixed point of $T$, then $\log \Delta\sigma = -\beta H_c - Z_\beta$ and the inequality becomes

$$S(\rho T) - \beta\rho T \cdot H_c \geq S(\rho) - \beta\rho \cdot H_c \tag{12.4}$$

as the terms in $\log Z$ cancel. In the classical case all positive maps are completely positive, and the result reduces to a theorem on relative entropy contained in [Penrose (1970)].

We shall prove a version of this result for isothermal dynamics; it does not use that the map is completely positive, and this leads one to ponder whether complete positivity automatically holds for any isothermal dynamics. This is true, as was first proved by Kraus [Kraus (1971)] for the reduced Heisenberg dynamics, in which $\tau$ is a unitary conjugation, and so by convexity, it is true for random reversible dynamics. For the general bistochastic case, it is not known. Our more general result [Streater (1992)] is

**Theorem 12.2 (The Quantum Free-energy Theorem).** *Any isothermal dynamics $T$ obeys*

$$S\left( \rho_c T \right) - \beta\rho_c T \cdot H_c \geq S(\rho_c) - \beta\rho_c T \cdot H_c. \tag{12.5}$$

**Proof.** Let $\tau$ be the energy-conserving bistochastic map which defines the isolated dynamics by

$$\rho_c \otimes \rho_\gamma \mapsto \rho'_c \otimes \rho'_\gamma = ((\rho_c \otimes \rho_\gamma)\tau)\,\mathcal{M}_c \otimes ((\rho_c \otimes \rho_\gamma)\tau)\,\mathcal{M}_c.$$

The quantum $H$-theorem for isolated systems says that

$$S(\rho_c) + S(\rho_\gamma) \leq S(\rho'_c) + S(\rho'_\gamma).$$

Apply this when $\Delta\rho_\gamma = Z^{-1}\exp -\beta H_\gamma$, in which case $\log \Delta(\rho_\gamma) = -\beta H_\gamma - \log Z$. We get, using Klein's inequality on the way

$$
\begin{aligned}
S(\rho_c) - \rho_\gamma \cdot (-\beta H_\gamma - \log Z) &= S(\rho_c) + S(\rho_\gamma) \\
&\leq S(\rho'_c) - \rho'_\gamma \cdot \log \Delta\rho'_\gamma \\
&= S(\rho') - \rho'_\gamma \cdot \log \Delta\rho_\gamma + \rho'_\gamma \cdot (\log \Delta\rho_\gamma - \log \Delta\rho'_\gamma) \\
&\leq S(\rho') - \rho'_\gamma \cdot \log \Delta\rho_\gamma \\
&= S(\rho')\rho'_\gamma \cdot (\beta H_\gamma - \log Z).
\end{aligned}
$$

As in the classical case and Lindblad's theorem, the term in $\log Z$ cancels, and we are left with

$$S(\rho_c) + \beta\rho_\gamma \cdot H_\gamma \leq S(\rho'_c) + \beta\rho'_\gamma \cdot H_\gamma. \qquad (12.6)$$

But the isolated dynamics preserves the mean total energy, say $E$, so

$$E = \rho_c \cdot H_c + \rho_\gamma \cdot H_\gamma = \rho'_c \cdot H_c + \rho'_\gamma \cdot H_\gamma$$

which substituted in (12.6) gives

$$S(\rho_c) - \beta\rho_c \cdot H_c \leq S(\rho'_c) - \beta\rho'_c \cdot H_c$$

which is the F-theorem. □

If, in the proof, we use the quantum Kullback inequality instead of the Klein inequality, we see that there is an increase of at least $\varepsilon = 1/2\|\rho'_\gamma - \rho_\gamma\|_1^2$ which is positive unless $\rho'_\gamma = \rho_\beta$; that is, there is a strict increase in the Massieu function $\Psi$ unless the state of the heat-particle is not disturbed by the isolated dynamics. As in the classical theory, the gain in entropy in one time-step is at least $\beta(\rho' \cdot H_c - \rho \cdot H_c) = dQ/T$ when the heat flow (needed for the readjustment of the state of the heat particle) is $dQ$, in addition to $\varepsilon$ coming from the sharp form of the Kullback lemma.

Since the isothermal dynamics $T$ is an affine map on the state space $\Sigma(\mathcal{A}_c)$, it can be extended to a linear map on Span $\Sigma_c$ and so is the dual of a linear map on $\mathcal{A}_c$. Let us give some of its properties.

**Theorem 12.3.** *Any isothermal dynamics $T$, coming from a bistochastic map $\tau$, obeys*

(1) $T$ *is the right action of a stochastic map and has* $\rho_{c\beta}$ *as fixed point.*
*The relation*

$$\rho_c T \cdot A = (\rho_c \otimes \rho_{\gamma\beta}) \cdot \tau(A \otimes 1_\gamma) \qquad (12.7)$$

*holds, where* $1_\gamma$ *is the unit operator in* $\mathcal{A}_\gamma$.
(2) *If* $\tau$ *is completely positive, then* $T$ *is completely positive.*

**Proof.** (1) Since $T$ is a bistochastic map followed by tracing over the heat-particle, it maps $\Sigma(\mathcal{A}_c)$ to itself, and so is normalised and positive. It is therefore the dual to a stochastic map. We have already shown in (12.2) that $\rho_{c\beta}$ is a fixed point. For the last part, for any $A \in \mathcal{A}_c$,

$$\begin{aligned}
\rho_c \cdot TA = \rho_c T \cdot A &= Tr_1(\Delta \rho_c TA) \\
&= Tr_1 A Tr_2 \Delta(\rho_c \otimes \rho_{\gamma\beta}\tau) \\
&= Tr_1 Tr_2 \left(\Delta(\rho_c \otimes \rho_{\gamma\beta}\tau)(A \otimes 1_\gamma)\right) \\
&= Tr_1 Tr_2(A \otimes 1_\gamma)\tau(\rho_c \otimes \rho_{\gamma\beta}) = \rho_c \otimes \rho_{\gamma\beta} \cdot \tau(A \otimes 1_\gamma).
\end{aligned}$$

(2) You are referred to [Kraus (1971)] for the case of reduced Heisenberg dynamics. The general case is similar.                                                   □

Note that [Gorini, Kossakowski and Sudarshan (1976)], Appendix, has an error, since it assumes that the convex hull of products of pure states is dense in the state-space of a tensor product. This cannot be true, since it would exclude the existence of pure entangled states, such as those used in Bell's theorem.

Recall that a random reversible dynamics is a convex mixture of unitary conjugations, each of which leaves the spectral projections of the total Hamiltonian invariant. This is a special class of bistochastic energy-conserving maps, since not every bistochastic map is a mixture of automorphisms. If $\tau$ is a random reversible dynamics, then the corresponding isothermal dynamics has some further important properties. One of these is *detailed balance*; this is a quantum version of Definition 5.2. We give it in the general form suggested in [Streater (1993c)].

**Definition 12.4.** Let $\mathcal{A}_c$ be a $C^*$-algebra with identity and let $T$ be a stochastic map on $\mathcal{A}_c$. Let $\rho$ be a faithful state on $\mathcal{A}_c$, invariant under $T$, and let $\{\pi_\rho, \mathcal{H}_\rho, \psi_\rho\}$ be the representation, Hilbert space and cyclic vector given by the Gelfand-Naimark-Segal construction. Let $\pi_\rho(T)$ denote the operator on $\mathcal{H}_\rho$

$$\pi_\rho(T)\pi_\rho(A)\psi_\rho = \pi_\rho(TA)\psi_\rho, \qquad (A \in \mathcal{A}) \qquad (12.8)$$

defined on the dense domain $D = \pi_\rho(\mathcal{A}_c)\psi_\rho$. We say that $T$ obeys *detailed balance relative to* $\rho$ if there exists a stochastic map $T^{(\rho)}$ such that $\pi_\rho(T)^* = \pi_\rho(T^{(\rho)})$ on $D$.

Note that $T^{(\rho)}(A) = T(A\Delta\rho)(\Delta\rho)^{-1}$, as in Exercise 12.16. In the following theorem, the *modular automorphism* is defined to be conjugation by $e^{iH_c t}$, that is, it is the reversible Heisenberg dynamics of the chemical algebra.

**Theorem 12.5.** *Let $T$ be the isothermal dynamics defined by $\rho_c \mapsto \rho_c T = (\rho_c \otimes \rho_{\gamma\beta})\tau\mathcal{M}_c$, where $\tau$ is a random reversible dynamics. Then*

*(1) $T$ obeys detailed balance relative to $\rho_{c\beta}$ in the general form of Definition 12.4.*

*(2) $T$ commutes with the modular automorphism.*

*(3) $T$ is a contraction on $\mathcal{A}_c$ in the norm $\|A\|_\beta = |\rho_{c\beta} \cdot A^* A|^{1/2}$.*

*(4) $T$ is a right contraction on $\Sigma(\mathcal{A}_c)$ in the norm (on density matrices) $\| \bullet \|_{-\beta}$.*

**Proof.** (1) Let $\tau A = \sum_i \lambda_i U_i A U_i^*$ be the random dynamics, with each $U_i$ commuting with the total Hamiltonian: $H := H_c \otimes 1 + 1 \otimes H_\gamma$. Here, $A \in \mathcal{A} = \mathcal{A}_c \otimes \mathcal{A}_\gamma$ and $U_i \in \mathcal{A}$. Let $\rho_\beta = \rho_{c\beta} \otimes \rho_{\gamma\beta}$ be the canonical state and denote $T^{(\rho_{c\beta})}$ by $T^{(\beta)}$. Then

$$\tau(A\Delta\rho_\beta) = \sum_i \lambda_i U_i^* A \rho_\beta U_i = \sum_i \lambda_i U_i^* A U_i U_i^* \rho_\beta U_i$$
$$= \tau(A)\rho_\beta \tag{12.9}$$

since each $U_i$ commutes with $\rho_\beta$. It follows that for any state $\rho_c$, we have

$$\rho_c \cdot T^{(\beta)}(A) = \rho_c \cdot \left(T(A\rho_{c\beta})\rho_{c\beta}^{-1}\right) = \rho_c \rho_{c\beta}^{-1} \cdot T(A\rho_{c\beta})$$
$$= \left(\rho_c \rho_{c\beta}^{-1}\right) T \cdot A\rho_{c\beta} = \left(\rho_c \rho_{c\beta}^{-1} \otimes 1\right)\tau \cdot A\rho_{c\beta} \otimes \rho_{\gamma\beta}$$
$$= \rho_c \rho_{c\beta}^{-1} \otimes 1 \cdot \tau\left(A\rho_{c\beta} \otimes \rho_{\gamma\beta}\right)$$
$$= \rho_c \rho_{c\beta}^{-1} \cdot \tau(A \otimes 1)(\rho_{c\beta} \otimes \rho_{\gamma\beta}) \qquad \text{by (12.9)}$$
$$= \rho_c \otimes \rho_{\gamma\beta} \cdot \tau(A \otimes 1).$$

It follows that $T^{(\beta)}$ is the isothermal dynamics determined by $\tau$, by Exercise 12.7. It follows from Theorem 12.3 that $T^{(\beta)}$ is stochastic. Note that it is the positivity that is the hard part, since $T^{(\rho_c)}1 = 1$ is easy (Exercise 12.16). Note further that the condition of detailed balance follows from bistochasticity and the one condition (12.9), and does not require that the dynamics be random conservative. Thus the theorem is rather more general than stated.

(2) To show that $T$ commutes with the modular automorphism, take any $\rho_c$; then we have

$$\rho_c \cdot \left(\rho_{c\beta} T(A) \rho_{c\beta}^{-1}\right) = Tr_c \left(\rho_c \rho_{c\beta} T(A) \rho_{c\beta}^{-1}\right) = Tr_c \rho_{c\beta}^{-1} \rho_c \rho_{c\beta} T(A)$$

$$= Tr \left(\rho_{c\beta}^{-1} \rho_c \rho_{c\beta} \otimes \rho_{\gamma\beta} \tau(A \otimes 1)\right)$$

$$= (\text{writing } \rho_c \text{ instead of } \Delta\rho_c \text{ and so on})$$

$$= Tr \left(\rho_\beta^{-1}(\rho_c \otimes \rho_{\gamma\beta})\rho_\beta \tau(A \otimes 1)\right)$$

$$= Tr \left((\rho_c \otimes \rho_{\gamma\beta})\tau[\rho_\beta(A \otimes 1)\rho_\beta^{-1}]\right)$$

$$= \rho_c \otimes \rho_{\gamma\beta} \cdot \tau[\rho_{c\beta} A \rho_{c\beta}^{-1} \otimes 1] = \rho_c \cdot T(\rho_{c\beta} A \rho_{c\beta}^{-1}).$$

Since $\rho_c$ was an arbitrary state, we conclude that $T(A)$ commutes with conjugation by $\rho_{c\beta}$. Hence it commutes with all powers of $\rho_{c\beta}$ and hence with $\log \rho_{c\beta}$ and so with the modular automorphism group, namely, conjugation with $(\log \rho_{c\beta})^{it} = e^{-i\beta H_c t}$.

(3) Let us follow [Majewski and Streater (1998)]. In fact, we show that the contractivity of $T$ is a consequence of detailed balance (which by (1) is a consequence of the construction of $T$ from a random reversible dynamics). So, choose the faithful state $\rho$ mentioned in (12.4) to be $\rho = \rho_{c\beta}$. We then have that $\pi_\rho(T)^*$ is induced by the stochastic map $T^{(\rho)}$. By Størmer's theorem, [Størmer (1963)] both $T$ and $T^{(\rho)}$ are norm contractions on $\mathcal{A}_{\text{J}}$. Then

$$\|\pi_\rho(T)\pi_\rho(A)\psi_\rho\|^2 = \rho(T(A^*)T(A)) = \rho\left(A^* T^{(\rho)}(T(A))\right)$$

$$\leq (\rho(A^*A))^{1/2}[\rho\left(T^{(\rho)}(TA)^* T^{(\rho)}(TA)\right)]^{1/2}$$

$$= [\rho(A^*A)]^{1/2}[\rho\left(A^* T^{(\rho)}(T(T^{(\rho)}(TA)))\right)]^{1/2}$$

$$\leq [\rho(A^*A)]^{3/4}[\rho\left(T^{(\rho)}(T(T^{(\rho)}(TA^*)))T^{(\rho)}(T(T^{(\rho)}(TA)))\right)]^{1/4} \cdots$$

$$\leq [\rho(A^*A)]^{(1-2^{-n})}[\rho\left(T^{(\rho)}(T \ldots (TA^*)T^{(\rho)}(T \ldots T(A) \ldots))\right)]^{2^{-n}}$$

$$\to C\rho(A^*A)$$

as $n \to \infty$, where $C \leq 1$. In the last step we used that $T$ and $T^{(\rho)}$ are contractions in norm, and so any iteration of the map $T^{(\rho)}T(\bullet)$ is a contraction.

(4) This follows from (3) by duality. $\qquad\square$

The first results along the lines of (2) and (3) were obtained by Majewski [Majewski (1984)] who used a version of the condition of detailed balance based on the physical idea of time-reversal invariance of an underlying dynamics. This was a generalisation of the property of microscopic reversibility. Majewski's version implies the version given here, but the converse has only been established in special cases [Majewski and Streater (1998)]. We see that detailed balance is a strong and subtle condition, whose full implications have not yet been fully explored.

It is easy to show that $\rho_{c\beta}$ is the state with the smallest norm $\| \bullet \|_{-\beta}$. See Exercise 12.17. So the dynamics, if it converges to equilibrium, is a contraction from the initial state to the state of smallest norm. In the next section we give sufficient conditions which ensure this, at least if $\dim \mathcal{A}_c < \infty$.

## 12.2 Convergence to Equilibrium

The long-time behaviour of a dynamical system can be studied by Lyapunov's direct method, which starts with the construction of a Lyapunov function. In the case of isothermal dynamics, we have such a function, the free energy or its relation $\Psi$, both in the classical and the quantum versions. The general theorem of Lyapunov can be applied if three conditions hold: the state space is compact, the Lyapunov function is strict, and there is a unique fixed point. We are not free to choose the sense in which the state-space is compact, since the topology must be such that both the Lyapunov function and the dynamics are continuous, at least in the elementary version given in Theorem 3.17. There is no problem in this connection, if $\dim \mathcal{A}_c < \infty$. However, the theorem leaves something to be desired, and in the spirit of Krylov [Krylov (1979)] we should ask for a stronger result: we need a strong (=large) norm in which the convergence takes place, and also we should get an estimate of the relaxation time. In good cases too there should be exponential approach to equilibrium. Finally, the theory should have something to say if $\dim \mathcal{A}_c = \infty$. We can achieve some of these objectives if we postulate strong enough ergodic conditions on the dynamics $\tau$. Here we are guided by what happens in the case of random reversible dynamics.

Let $\mathcal{A} = \mathcal{A}_c \bigotimes \mathcal{A}_\gamma$ be the algebra of all bounded operators on $\mathcal{H} = \mathcal{H}_c \otimes \mathcal{H}_\gamma$, with Hamiltonian $H = H_c \otimes 1 + 1 \otimes H_\gamma$, and let $\tau = \sum \lambda_i U_i$ be a mixture of unitary conjugations of $\mathcal{A}$, each $U_i$ commuting with $H$. Let $H = \sum \mathcal{E}_j P_j$ be the spectral resolution of $H$, with $\dim P_j < \infty$ for all $j$. The state space $\Sigma(\mathcal{A})$ spans a complex vector space, known as $\ell^1(\mathcal{A})$ by analogy with the commutative case; this, and the analogues of all the $\ell^p$ spaces, were introduced and studied by von Neumann and Schatten. The span of the normal states can be identified with a subalgebra $\mathcal{A}_1$ of $\mathcal{A}$ by the map $\Delta$, which assigns the density operator $\Delta\rho$ to the normal state $\rho$. Then $\mathcal{A}_1$ is the direct sum of algebras

$$\mathcal{A}_1 = \overline{\bigoplus}_{ij} P_i \mathcal{A}_1 P_j. \tag{12.10}$$

The meaning of the bar is that the algebraic direct sum must be closed in the trace-norm. The diagonal algebras $P_i \mathcal{A} P_i$ contain the microcanonical states $P_i$, which act as units, and the others, $P_i \mathcal{A} P_j$, with $i \neq j$, are nilpotent. Each of these algebras is invariant under $\tau$. It is therefore reasonable to say that a bistochastic map, $\tau$, is energy-conserving if each $\mathcal{A}_{ij}$ is invariant under $\tau$. Let us say that $\tau$ is ergodic relative to $H$ if for each $j$, multiples of $P_j$ are the only invariant elements of $P_j \mathcal{A} P_j$, and zero is the only invariant element of $P_j \mathcal{A} P_k$, $k \neq j$. We shall also need a further property, which ensures that the energy-shells link all the energy levels of $H_c$. For example if the energies were incommensurate, then only the identity map $\tau = Id$ would conserve energy. Let us say that two energy-levels $\mathcal{E}_{cj}$ and $\mathcal{E}_{ck}$ are linked if there exist two energy-levels $\mathcal{E}_{\gamma 1}$ and $\mathcal{E}_{\gamma 2}$ such that

$$\mathcal{E}_{cj} + \mathcal{E}_{\gamma 1} = \mathcal{E}_{ck} + \mathcal{E}_{\gamma 2}.$$

We can regard being linked as a relation like being nearest neighbours in a graph. To get mixing, we shall need to assume that each energy-level $\mathcal{E}_{cj}$ is linked to any other by a finite chain. A similar assumption was needed in the classical case (the irreducibility of the Markov chain). We express this by saying that the energy-shells are connected. Then we have

**Theorem 12.6.** *Let $\tau$ be energy-conserving and ergodic relative to $H$. Suppose that $\dim \mathcal{A}_c < \infty$ and that the energy shells of $H = H_c \otimes 1 + 1 \otimes H_\gamma$ are connected. Then the iterated isothermal dynamics converges to equilibrium, for any initial state $\rho_c$.*

**Proof.** Consider the action of $\tau$. Any invariant state $\rho = \rho_c \otimes \rho_{\gamma\beta}$ can be written as the orthogonal sum (in the Hilbert-Schmidt sense) of its parts in the spectral subspaces of $H_c = \sum \mathcal{E}_c P_{cj}$ and $H_\gamma = \sum \mathcal{E}_\ell P_{\gamma\ell}$:

$$\rho = \sum_{jk\ell} (P_{cj} \otimes P_{\gamma\ell})(\rho_c \otimes Z_{\gamma\beta}^{-1} e^{-\beta \mathcal{E}_{\gamma\ell}} 1)(P_{ck} \otimes P_{\gamma\ell}). \tag{12.11}$$

Let us choose energy-shells (of $H$) labelled by $\mathcal{E}_1$ and $\mathcal{E}_2$, and consider the partial sum over all $k, \ell$ such that $\mathcal{E}_{cj} + \mathcal{E}_{\gamma\ell} = \mathcal{E}_1$ and $\mathcal{E}_{ck} + \mathcal{E}_{\gamma\ell} = \mathcal{E}_2$. This is an element of $P_1 A P_2$, which is mapped to itself by $\tau$. By the uniqueness of the Hilbert-Schmidt orthogonal sum, this element must be invariant, since $\rho$ is invariant, and so by ergodicity is zero if $\mathcal{E}_1 \neq \mathcal{E}_2$, and is a multiple of $P_1$ if $\mathcal{E}_2 = \mathcal{E}_2$. We note that

$$H = H_c \otimes 1 + 1 \otimes H_\gamma = \sum_{j\ell} \left( P_{cj}\mathcal{E}_{cj} \otimes P_{\gamma\ell} + P_{cj} \otimes P_{\gamma\ell}\mathcal{E}_{\gamma\ell} \right)$$

$$= \sum_{\mathcal{E}} \sum_{j\ell:\mathcal{E}_{cj}+\mathcal{E}_{\gamma\ell}=\mathcal{E}} \mathcal{E} \left( P_{cj} \otimes P_{\gamma\ell} \right).$$

It follows that

$$P_1 = \sum_{j\ell:\mathcal{E}_{cj}+\mathcal{E}_{\gamma\ell}} P_{cj} \otimes P_{\gamma\ell}$$

holds, and that the sum is an orthogonal one. Since we have just proved that the part of $\rho$, (12.11) on the energy shell $\mathcal{E}_1$ is a multiple of $P_1$, we can equate the orthogonal components, to get

$$\left( P_{cj} \otimes P_{\gamma\ell} \right)\left( \rho_c \otimes Z_{\gamma\beta}^{-1} e^{-\beta\mathcal{E}_{\gamma\ell}} 1 \right)\left( P_{cj} \otimes P_{\gamma\ell} \right) = \mu P_{cj} \otimes P_{\gamma\ell}$$

for all $j$ and $\ell$ leading to the same energy-shell $\mathcal{E}_1$. The factor $\mu$ depends on $\mathcal{E}_1$ but not on $j$ and $\ell$ separately. So the part of $\rho_c$ in the algebra $P_{cj} A_c P_{cj}$ must be proportional to $P_{cj}$, with a factor $e^{-\beta\mathcal{E}_{cj}}$ to balance the factor $e^{-\beta\mathcal{E}_{\gamma\ell}}$. The overall factor is not determined by looking at one energy-shell, but we do get the Boltzmann factors

$$P_{cj}\rho_c P_{cj} = \lambda_j P_{cj} \text{ with } \lambda_j = e^{-\beta(\mathcal{E}_{cj}-\mathcal{E}_{ck})}\lambda_k$$

for all $j$ participating in the energy-shell $\mathcal{E}_1$. But since we have assumed that the energy-shells are connected, we get a family of such relations including all levels, and we conclude that $\rho_c$ is the canonical state with beta $\beta$. Thus $\tau$ has a unique fixed point. By the free-energy theorem, $\Psi$ is a strict Lyapunov function, and by the compactness of $\Sigma(A_c)$, we can apply Lyapunov's theorem, 3.17, to show that the isothermal dynamics converges.

$\square$

If the isothermal dynamics satisfies detailed balance then it is a contraction in the norm $\| \bullet \|_{-\beta}$, and we get convergence by contraction from the original state $\rho_c$ to the state of smallest norm. This is because in finite dimensions all norms are equivalent. If $\dim A_c = \infty$ we have seen that detailed balance is enough to ensure that the isothermal dynamics is a contraction in this norm, but we must make some assumptions about the growth in multiplicity of the energy-eigenvalues in order to get convergence [Streater (1985)]; it remains an open question whether we get convergence in the norm $\| \bullet \|_{-\beta}$ in the infinite case.

## 12.3   Driven Quantum Systems

Lebowitz and Spohn [Spohn and Lebowitz (1977)] formulate quantum dynamics in which various points are kept at different temperatures, by being coupled to heat-baths. They work in the limit of weak coupling, and impose the condition of detailed balance (at the local beta) on the local generator of a completely positive semi-group. The full dynamics is a convex sum of all these. We now develop a general scheme for driven systems, which includes a version of this construction as a special case. In the description of isothermal systems we start with one particle, the heat-particle, in a canonical state, and after one time-step, we inject heat, or remove it, to restore the heat-particle to its previous state. In a similar way, we can construct isopotential dynamics, for any chemical potential. Thus, suppose that the full system is described by the algebra $\mathcal{A} = \mathcal{A}_1 \bigotimes \mathcal{A}_2$, where $\mathcal{A}_2$ concerns the chemicals whose potentials are to be kept fixed. Let $\tau$ be a bistochastic map on $\mathcal{A}$. We assume that the initial state is $\rho_1 \otimes \rho_2$, where $\rho_2$ is a grand canonical state of all the chemicals described by $\mathcal{A}_2$. Then the isopotential dynamics is the map

$$\rho_1 \otimes \rho_2 \mapsto ((\rho_1 \otimes \rho_2)\tau)\,\mathcal{M}_1 \otimes \rho_2.$$

The only difference between an isopotential map and a driven dynamics is that in the isopotential dynamics all the chemicals in $\mathcal{A}_2$ are conserved quantities, whereas in a driven system they are not. In the first case, we can use the conservation laws to establish a free-energy theorem, so that isopotential systems converge to equilibrium. Driven systems, in which some non-conserved quantities are fed from outside, might or might not have a Lyapunov function, and in general do not. In the models with several temperatures [Spohn and Lebowitz (1977)], the local energy is not conserved, so they are driven, not isopotential systems. We illustrate the general theory with the Fröhlich pumped phonon model [Fröhlich (1968); Sewell (1986)]. The model is 'formulated at the kinetic level', which means that the correct Boltzmann equation is written down by instinct. We shall show that it fits into statistical dynamics.

**Example 12.7.** The model consists of a system, which has $n$ bosons, called here $A$-particles, which are modes of a phonon field. So the Hilbert space of this is $\mathcal{F}(\mathbf{C}^n)$. Let $a_j$, $a_j^*$ be the annihilators and creators, and let the energy of the system be $H_c = \sum_j \mathcal{E}_j a_j^* a_j$; we assume that $\mathcal{E}_k > \mathcal{E}_j$ if $k > j$. Thus $\mathcal{E}_1$ is the A-particle with the smallest energy. The A-particles can be absorbed directly by a heat-bath, and can be emitted by the bath. They can also be converted, one into the other, by absorbing or emitting a suitable quantum from the heat-bath. In addition, the population of phonons is fed by a source, which increases the number of phonons of type $j$ at the rate $s_j$. Thus the system is clearly a driven one, but one in which the driving mechanism cannot modelled by an external chemical potential kept constant in time, as we have used earlier. The driving mechanism must be such that the actual input of quanta is at the predetermined rate $s_j$ (which could depend on time in a more elaborate version). It is rather easy to model this; we just add $s_j$ to the rate of change in the mean number of quanta $j$.

To treat this example, we first set up the theory of an isolated system obeying both laws of thermodynamics. To this end, we introduce a suitable collection of heat-particles, which act as the balancing item. We then move to the isothermal system as usual by keeping the heat-particles at their canonical state, by feeding in heat or removing it. We then add the changes in the mean number due to the sources. Finally, we apply the quasifree map to find the quasifree state after one time-step.

The energy-levels $\mathcal{E}_j$ are not assumed to be rationally related to each other, yet they make transitions among themselves; so we need one heat-particle $\gamma_{jk}$, of energy $\mathcal{E}_{\gamma jk}$ for each pair of phonon modes, such that

$$\mathcal{E}_k - \mathcal{E}_j = \mathcal{E}_{\gamma jk}, \text{ where } k > j. \qquad (12.12)$$

This allows the process

$$A_k \rightleftharpoons A_j + \gamma_{jk}$$

to proceed with the conservation of energy. Let $b_{jk}$ denote the annihilator of this heat-particle. The model also allows a phonon, say $A_j$, to be absorbed by the heat-bath, or created by it; this means that it is converted directly into a heat-particle, so we need a heat-particle $\gamma_j$ with energy exactly $\mathcal{E}_j$. Let $b_j$ denote its annihilator. The absorption and its inverse is therefore the process

$$A_j \rightleftharpoons \gamma_j.$$

The energy of the heat-particles is thus

$$H_\gamma = \sum_j \mathcal{E}_j b_j^* b_j + \sum_{j<k} \mathcal{E}_{jk} b_{jk}^* b_{jk}.$$

A simple operator implementing the desired reactions is

$$H_I = \sum_j C_j (a_j^* b_j + b_j^* a_j) + \sum_{j<k} C_{jk} \left( a_j^* a_k b_{jk}^* + a_j a_k^* b_{jk} \right).$$

We only need to consider $C_{jk}$ with $j < k$, since otherwise the reaction is forbidden by energy-conservation. We note that $H_I$ has a quadratic term and a cubic term. Let $U_I = e^{iH_I}$. Then the Boltzmann equation for the isolated dynamics of any observable will be chosen to be conjugation by $U_I$; we throw this dynamics onto the states, by duality, and follow one time-step by the quasifree map. We shall also let the time-interval go to zero, and the couplings large, so that only the second-order term survives (the anti-van Hove limit). Suppose that at time $t = 0$ the state (which can be assumed to be quasifree, from the previous time-step) is a tensor product $\rho = \rho_c \otimes \rho_\gamma$. We then find the change in the state in the time-step to $t = 1$. The quasifree map does not change the mean numbers of any type of boson, so the change of for example $n_j = \rho(a_j^* a_j)$ is given in one time step by the change due to the conjugation with $U_I$:

$$n_j' = \rho(U_I a_j^* a_j U_I^{-1})$$
$$= \rho\left(a_j^* a_j + i[H_I, a_j^* a_j] + i^2/2!\, [H_I, [H_I, a_j^* a_j]]\right) . \quad (12.13)$$

The operator of first order is (12.18)

$$i[H_I, \mathcal{N}_j] = iC_j(b_j^* a_j - a_j^* b_j)$$
$$+ i \sum_{k:k>j} C_{jk}\left(a_k^* b_{jk} a_j - a_k b_{jk}^* a_j^*\right)$$
$$+ i \sum_{\ell:\ell<j} C_{\ell j}\left(a_j a_\ell^* b_{\ell j}^* - a_\ell a_j^* b_{\ell j}\right) . \quad (12.14)$$

This has a quadratic term, and a cubic term under the sum; each is linear in the heat-variables, and as we assume that the heat-particles are (in the isothermal dynamics) in the thermal state, the expectation value of (12.14) in $\rho$ is zero. So the dynamics comes from the second-order term. If $\rho_c$ is an even state, then the only parts which survive in the second-order term $i^2/2[H_I, [H_I, \mathcal{N}_j]]$ in Eq. (12.13) are the commutator of the quadratic terms, and the commutator of the cubic terms. These are, respectively,

$$i^2/2 \sum_m C_m C_j \left[a_m^* b_m + a_m b_m^*, b_j^* a_j - b_j a_j^*\right]$$

and

$$\sum_{j'<k',k>j,\ell<j} \frac{i^2}{2} C_{j'k'} \left[a_{j'}^* a_{k'} b_{j'k'}^* + a_{j'} a_{k'}^* b_{j'k'},\right.$$
$$\left. C_{jk}(a_k^* a_j b_{jk} - a_k a_j^* b_{jk}^*) + C_{\ell j}(a_j a_\ell^* b_{\ell j}^* - a_\ell a_j^* b_{\ell j})\right] .$$

These can be evaluated in the state $\rho$, which is quasifree. We get terms involving the two-point functions $\rho(a_j^* a_k)$, $\rho(a_j^* b_j)$ etc. These express correlations between the different modes of the phonon field, and between the phonons and the heat-particles. To get a closed system of equations we would need to find equations for all these two-point functions. In the spirit of statistical dynamics, we can use the stoss map to replace the state $\rho'$ by the product state of the system and the heat-particles. This step is part of the change-over to isothermal dynamics from isolated dynamics. The heat-bath, by definition, always remains independent of the system. The stoss map is entropy increasing and conserves mean energy, so we remain with a viable theory if we assume that $\rho = \rho_c \otimes \rho_\gamma$, and maintain this independence after one time-step. As for the correlations between the different phonon modes, we get a tractable theory if the number of modes is small, say 2. The original model [Fröhlich (1968)] did not derive the equations in this way, and ignored the phonon-phonon correlations. That is alright; we can apply a further randomising map, also entropy-increasing, putting these correlations zero after each step. Then $\rho_c$ is a product state $\otimes \rho_{cj}$, one factor for each phonon, at each time. Evaluating the two expressions contributing to the second-order term, we find for the first,

$$n'_j = n_j - C_j(n_j - n_{\gamma j})$$

and defining $C_{jk}$ for $j > k$ to be $C_{kj}$ for convenience, the second is

$$\sum_{k \neq j} C_{jk}^2 \left( n_k(n_j + 1)(n_{\gamma jk} + 1) - n_j(n_k + 1)(n_{\gamma kj} + 1) \right).$$

We have assumed that the first moments of the phonon field are zero. If this is true at time zero, then it is true for all time. There is no freedom to put them zero as part of an entropy-increasing map, and if the initial state is a coherent state of a mode of the phonon, with non-zero one-point function, we must take account of the dynamics of this parameter. We would expect such a state to lose its coherence because of the use of entropy-increasing maps. Thus far, we have derived the isolated dynamics.

The move to isothermal dynamics consists in putting the Planck distribution at the same beta for all the $n_\gamma$ occurring in the dynamics. We have already used the map which puts to zero the correlations between heat-particles and the system. Finally, we drive the system by adding the source $s_j$ to the rate of change of $n_j$, the mean number of quanta in the phonon mode $j$; we get the coupled system of equations

$$\frac{dn_j}{dt} = s_j + C_j(n_{\gamma j} - n_j)$$
$$+ \sum C_{jk}^2 \left( n_k(n_j + 1)(n_{\gamma jk} + 1) - n_j(n_k + 1)n_{\gamma kj} \right). \qquad (12.15)$$

This partially agrees with [Sewell (1986)], p. 210, Eq. (115); we can relate our constants $C_j$ and $C_{jk}$ to Sewell's constants $a_j$ and $b_{jk}$, which however become temperature-dependent.

If $s_j = 0$ then there is an equilibrium point when $n_j = n_{\gamma j}$, which is thermal equilibrium. The principle of detailed balance holds: both terms are separately zero at the fixed point; this holds by virtue of the relation (12.12). If $s_j > 0$, then there is a fixed point in which the state of the phonon $A_j$ is a grand canonical state with some chemical potential related to the source strength [Fröhlich (1968); Sewell (1986)]. Duffield [Duffield (1988)] has shown that independently of any derivation of the Fröhlich equations, they do define a one-parameter group of completely positive maps.

## 12.4  Exercises

Exercise 12.16. Show that $T^{(\rho)}(A) = ((A\rho)T)\rho^{-1}$; hence show that

$$T^{(\rho_{c\beta})} 1 = 1 \ .$$

Exercise 12.17. Let $\rho$ be a faithful normal state on $B(\mathcal{H})$. Show that the unique normal state $\sigma$ with the smallest value of $Tr\left(\rho^{-1}\sigma^2\right)$ is $\sigma = \rho$.

Exercise 12.18. As in the pumped phonon model, let

$$H_I = \sum_j C_j \left(a_j^* b_j + a_j b_j^*\right) + \sum_{j<k} C_{jk} \left(a_j^* a_k b_{jk}^* + a_j a_k^* b_{jk}\right)$$

and $\mathcal{N}_j = a_j^* a_j$. Prove (12.14).

# Chapter 13

# Infinite Systems

In the theory of equilibrium states in statistical mechanics the important question as to whether a system undergoes a phase transition is best addressed in the thermodynamic limit. This is the limit in which the size of the system becomes infinite. Although any actual system is of finite size, this device is a very useful idealisation. Almost any sensible model of finite size shows no phase transition at all; all the thermodynamic variables are analytic functions of the intensive variables such as beta. A phase transition is recognised as a discontinuity in an interesting variable, usually as a function of beta and an external field. For example, a system of particles of spin $1/2$ on a finite lattice $\Lambda$ in an external magnetic field $\boldsymbol{B}$ may be described by the algebra $\mathcal{A} = \bigotimes_{x \in \Lambda} \mathbf{M}_2(x)$, and the Hamiltonian $H = H_0 + \boldsymbol{B}.\boldsymbol{\sigma}$, where $H_0$ is the Hamiltonian in the absence of the magnetic field, and $\boldsymbol{\sigma}$ are the Pauli matrices. The magnetisation as a function of $\boldsymbol{B}$ is

$$m = \rho_{\beta,\boldsymbol{B}} \cdot \boldsymbol{\sigma}.$$

Here, $\rho_{\beta,\boldsymbol{B}}$ is the canonical state with Hamiltonian $H(\boldsymbol{B})$. We define the *spontaneous magnetisation* to be the limit of $m$ as $\boldsymbol{B}$ converges to zero. It can be shown that if $|\Lambda| < \infty$, then the spontaneous magnetisation is zero at all beta. On the other hand, the celebrated work of Peierls [Peierls (1936); Griffiths (1964)] shows that for the Ising model, the limit of the magnetisation per site converges, as $|\Lambda| \to \infty$ to a discontinuous function of beta, being zero for small beta, and non-zero above a critical beta, $\beta_c$, called the *Curie point*. The behaviour for a large but not infinite system is presumably near to that of the infinite system in some numerical sense, but there is no developed theory of functions that are continuous but nearly discontinuous. The infinite system not only shows the desired effect, but is simpler to describe. One reason for the simplicity is translation invariance, which holds for the infinite system. The strategy of studying the system in infinite volume is called 'taking the thermodynamic limit'. There are some advantages to adopting the same strategy in non-equilibrium statistical mechanics; this is particularly true in the fundamental problem of deriving irreversible behaviour from Hamiltonian dynamics. The weak limit of a sequence of unitary operators need not be unitary, but could be a contraction. So the dynamics of a limit of theories with reversible dynamics might be irreversible, just as in equilibrium theory the limit of functions continuous in the parameters $\beta, B$ might not be continuous. The idea, then, is an attempt at a more fundamental theory, in which irreversibility is not put in, but is derived. This is only partly achieved, as the infinite-volume approximation has its own limitations. In this chapter we shall discuss some of the easier methods and results for an infinite system, though it is outside our self-imposed framework of dealing with only finite systems. We begin with the definition of the spin algebra for an infinite lattice. This uses the idea of an inductive limit. We get a definition of the hard-core algebra for an infinite system, which does not seem to have been defined before. In discussing time-evolution for infinite systems it is easier to choose the time to be continuous; then we are able to show that with an interaction of finite range, the time evolution is a strongly continuous group of automorphisms of the algebra, following [Streater (1967)]. In the next section, we analyse the possible one-parameter groups of completely positive semigroups of contractions. This is the heart of the subject, and was set up by Kraus, Gorini, Sudarshan, Kossakowski and Lindblad among others. Here we follow the elementary treatment described in [Landau and Streater (1993)], which assumes that the system is of finite dimension. But then, applying the same estimates as in the case of automorphisms, we can extend the semigroup to the inductive limit. In previous treatments it was thought that it was necessary to impose the condition of detailed balance on the dynamics. From our point of view, detailed balance (in the sense of Glauber) is a consequence of moving to the isothermal dynamics; we construct the isolated dynamics, in which the dissipative part is determined by a bistochastic map. We can then move to the isothermal version after this. This makes the construction of the theories easier than imposing detailed balance from the start.

## 13.1 The Algebra of an Infinite System

Let $\Lambda$ be an infinite set, and let $\Lambda_0$ be a typical finite subset. As usual we define the algebra of observables $\mathcal{A}(\Lambda_0)$ to be the tensor product of $\mathbf{M}_n$ at each site $x \in \Lambda_0$. Each algebra $\mathcal{A}(x)$ is naturally embedded in each $\mathcal{A}(\Lambda_0)$ for which $x \in \Lambda_0$ and each operator in $\mathcal{A}(\Lambda_0)$ can be identified with a unique element (its ampliation) in $\mathcal{A}(\Lambda_1)$ for any finite $\Lambda_1$ containing $\Lambda_0$. This identification defines an injective $C^*$-homomorhpism,

$$\phi(\Lambda_0, \Lambda_1): \quad \mathcal{A}(\Lambda_0) \to \mathcal{A}(\Lambda_1)$$

which is transitive in the sense that $\phi(\Lambda_2, \Lambda_1)\phi(\Lambda_1, \Lambda_0) = \phi(\Lambda_2, \Lambda_0)$. This allows us to put an equivalence relation on the union of all $\mathcal{A}(\Lambda_0)$ over all finite subsets $\Lambda_0$, namely, two elements $A_1 \in \Lambda_1$ and $A_2 \in \Lambda_2$ are equivalent if there exists $\Lambda_0$ containing both $\Lambda_1$ and $\Lambda_2$ such that $\phi(\Lambda_0, \Lambda_1)A_1 = \phi(\Lambda_0, \Lambda_2)A_2$. Since the ampliation is norm-preserving, we see that equivalent elements have the same norm. We can multiply equivalence classes by multiplying representatives; the result is independent of the choice of representative. We can add elements in the same way, and form the Hermitian conjugate. The $C^*$-identity $\|A^*A\| = \|A\|^2$ also holds. In this way, the set of equivalence classes becomes a *-algebra, which except for completeness, obeys all the axioms of a $C^*$-algebra. Let us call this algebra $\mathcal{A}_0$, the algebra of local observables. Its completion is therefore a $C^*$-algebra, $\mathcal{A}$, the algebra of quasilocal observables. As always when we complete a space, the original one is dense in the completion. So any quasilocal observable is the norm limit of a sequence of local observables. We adopt the convention of calling the algebra the observable algebra even though it contains non-Hermitian elements. Each $\mathcal{A}(\Lambda_0)$ can be identified with a unique subalgebra of $\mathcal{A}$. This construction of $\mathcal{A}$ from the local algebras is called the inductive limit. It is not hard to show that $\mathcal{A}$ is separable; it is one of the classes of $C^*$-algebras introduced and studied by Glimm, the uniformly hyperfinite class. This algebra has a special state, the trace. We define the trace on the matrix algebra $\mathcal{A}(\Lambda_0)$ as

$$\rho_0(A) = d^{-1}\mathrm{Tr}\, A \qquad \text{where } d = n^{|\Lambda_0|}$$

is the dimension of $\mathcal{A}$. This is unchanged on ampliation, and so defines a positive linear functional on the space of equivalence classes. Since $\rho_0(1) = 1$, it is bounded, and so can be extended to the completion. This state obeys $\rho_0(AB) = \rho_0(BA)$. It corresponds to the state at infinite temperature, that is, zero beta. This state is a fixed point of all bistochastic maps on

$\mathcal{A}$. We defined the hard-core algebra for any finite $\Lambda_0$. We could choose
the operators so that different modes commute, by injecting the standard
representation into $\mathcal{A}(\Lambda_0)$. Or, we could choose them so that different
modes anticommute. Recall that this is achieved by ordering the modes
and forming the tensor product of the algebras containing the modes one
by one. Anticommutativity is achieved by inserting the factor $\theta$ which
anticommutes with all modes used up to then. The hard-core algebra is
defined as the mapping of the symbols obeying the hard-core relations into
the $C^*$-algebra $\mathcal{A}$. The fermion algebra is the special case where $n = 2$.

## 13.2    The Reversible Dynamics

If $\Lambda = \mathbf{Z}^s$, where $s$ is the space-dimension, then the translation group
$\{\sigma_x\}_{x\in\Lambda}$ acts naturally on $\mathcal{A}$ as automorphisms. Thus, if $A \in \mathcal{A}(x)$, then
$\sigma_y A$ is the same element $A \in \mathbf{M}_n$ ampliated to be an element of $\mathcal{A}(x + y)$.
The map $\sigma_x$ is bounded and can be extended to an automorphism of the
quasilocal algebra. The map $\sigma :\quad x \mapsto \sigma_x$ is a homomorphism from the
group $\Lambda$ into the group $AUT\mathcal{A}$.

We now show how to define the time-evolution with continuous time
in the simplest case, when the interaction provides a coupling between the
operators at a point $x \in \Lambda$ and its nearest neighbours, and is translation
invariant. We might as well assume that $s = 1$, the case of the spin chain,
in which case there are two nearest neighbours, and the energy can be
assigned to the bond between them. The proof for space of any dimension
is very similar. This and other more general cases were treated in [Streater
(1967)]. Thus let $H(x) \in \mathbf{M}_n(x) \bigotimes \mathbf{M}_n(x + 1)$ represent the energy on the
bond between $x$ and $x + 1$. The operator $H(x)$ is Hermitian, and is taken
to be covariant under the translation group:    $H(x + y) = \sigma_y H(x)$. The
total energy in a finite interval $\Lambda_1 \subseteq \Lambda = \mathbf{Z}$ is taken to be

$$H(\Lambda_1) = \sum_{x\in\Lambda_1} H(x).$$

This is an operator in the interval $\Lambda_1$ extended by one point on each side.
The sum for the energy of the whole system is divergent, and so the definition of the time-evolution needs attention. We denote by $H$ the sum
of indefinitely many terms; as remarked, it does not converge, but gives a
well-defined sequence of operators.

We first define the time-evolution of a local operator, one in $\mathcal{A}_0$ and therefore in some region $\Lambda_0$: $A \in \mathcal{A}(\Lambda_0)$. We start with the series

$$A(t) = A + it[H, A] + (it)^2/2! \, [H, [H, A]] + \dots \, . \tag{13.1}$$

In each bracket of a multiple commutator we choose as many terms in $H$ as is needed so that the inclusion of any more does not change the expression. We shall now prove that this converges in norm to an element of $\mathcal{A}$ if $t$ is small enough. Taking the first commutator, we note that $A$ commutes with $H(y)$ unless $y$ is a neighbour of $\Lambda_0$, and there are $|\Lambda_0| + 2$ such $y$. So this is the number of terms needed. The commutator has two terms, and each is bounded in norm by $\|H(x)\| \, \|A\|$. So we get the estimate for the second term of the series

$$\|it[H, A]\| \leq 2 \, (|\Lambda_0| + 2) \, |t| \, \|H(x)\| \, \|A\|.$$

Let $\Lambda_1$ denote $\Lambda_0$ together with its neighbours. So $|\Lambda_1| = |\Lambda_0| + 2$. Then each of the $2 \, (|\Lambda_0| + 2)$ terms in $[H, A]$ lies in $\Lambda_1$. Now consider the number of terms in $H$ in the outer bracket of third term, the double commutator. Since each of the $2 \, (|\Lambda_0| + 2)$ terms in the inner commutator lies in $\Lambda_1$, we must keep $|\Lambda_1| + 2 = |\Lambda_0| + 4$ terms in $H$. There are four terms in each contribution to the double commutator, each of norm $\leq \|H(x)\|^2 \, \|A\|$. Hence the third term is bounded by

$$\|(it)^2/2[H, [H, A]]\| \leq |t|^2/2.4. \, (|\Lambda_0| + 2) \, (|\Lambda_0| + 4) \, \|H(x)\|^2 \|A\|.$$

The third term, the double commutator, lies in $\Lambda_2$ say, which is $\Lambda_1$ together with its nearest neighbours. This has $|\Lambda_0|+4$ points, so we must keep $|\Lambda_0|+6$ terms in the outermost $H$ in the fourth term, the triple commutator, and it is similarly bounded in norm by

$$(|\Lambda_0| + 2) \, (|\Lambda_0| + 4) \, (|\Lambda_0| + 6) \, 8\|H(x)\|^3 \, \|A\| \, |t|^3/3!.$$

By induction the ratio of the norms of the $n + 1$-th to the $n$-th term in the series is bounded by

$$(|\Lambda_0| + 2n) \, .2. \|H(x)\|.|t|/n \to r < 1$$

as $n \to \infty$ if $t < (4\|H(x)\|)^{-1}$. So for small $t$, the series converges in norm to an element of $\mathcal{A}$, as $\mathcal{A}$ is complete. Let $\tau_t A$ denote the limit. This result is the heart of the construction of the time-evolution. The rest is tidying up.

Now, for fixed $\Lambda_1$, $H(\Lambda_1)$ can be identified with a Hermitian matrix, and it is well known that

$$A + it[H(\Lambda_1), A] + (it)^2/2![H(\Lambda_1), [H(\Lambda_1), A]] + \dots$$

converges in norm to

$$\tau_{1t}A = e^{iH(\Lambda_1)t}Ae^{-iH(\Lambda_1)t}$$

and clearly, $\|\tau_{1t}A\| = \|A\|$. We see that for large $\Lambda_1$ the first terms of the series for $\tau_{1t}A$ coincide with the first terms of the series in (13.1), and we get more terms coinciding as $\Lambda_1$ becomes large. Since the series converge in norm, we have for $A \in \mathcal{A}_0$ and $t$ small,

$$\tau_t A = \lim_{\Lambda_1 \to \mathbf{Z}} \tau_{1t}A.$$

It follows that $\|\tau_t A\| = \|A\|$, and that

$$(\tau_t A)^* = \tau_t A^*; \qquad \tau_t(AB) = \tau_t A \tau_t B; \qquad \text{and that } \tau_t \text{ is linear.}$$

The continuity of $\tau_t$ in time follows from the convergence of the series. Since $\tau_t$ is a bounded map, it can be uniquely extended to an automorphism of $\mathcal{A}$, the completion of $\mathcal{A}_0$. The extension is continuous in time, and obeys the group law for small time; see and do Exercises 13.8, 13.9. We can show that the resulting time-evolution commutes with space-translation; this is first done on $\mathcal{A}_0$ and small time simply by commuting $\tau_t$ and $\sigma_j$ on the series in (13.1) term by term, and using the space translation-invariance of the indefinite sums defining $H$. The commutativity can then be extended to the whole algebra.

We have constructed the time-evolution for a one-dimensional lattice, with an interaction of finite range, and with translation invariance. In [Streater (1967)] we deal with any dimension, and cover the case of interactions of any finite range (which must be independent of position). Further extensions to more general cases are described in [Ruelle (1969); Bratteli and Robinson (1979)].

## 13.3    Return to Equilibrium

The tracial state $\rho_0$ is easily shown to be invariant under $\tau_t$ and $\sigma_t$. In fact, $\rho_0$ is *universally invariant*: it is invariant under all automorphisms and indeed all bistochastic maps. The representation obtained from it using the Gelfand-Naimark-Segal construction leads to a represented algebra which is of type $II_1$. It corresponds to the case $\beta = 0$. In a seminal paper [Haag, Hugenholtz and Winnick (1967)] Haag, Hugenholtz and Winnink show how to define states at any beta. The starting point is the notable remark of Kubo [Kubo (1957)], and of Martin and Schwinger [Martin and Schwinger

(1959)] that the canonical state possesses some analytic properties, now dubbed the $KMS$ condition. This goes as follows.

**Definition 13.1.** Let $\mathcal{A}$ be a $C^*$-algebra, with continuous time-evolution $\tau_t$. Let $A_t$ denote the function of time $\tau_t A$. Then a state $\rho \in \Sigma(\mathcal{A})$ is said to obey the $KMS$ condition relative to the time-evolution $\tau_t$ if

(1) For any elements $A, B \in \mathcal{A}$, the function of time $\rho(A_t B)$ is the boundary value of an analytic function $F(t)$ of the complex variable $t$ in the strip $-\beta < Im\, t < 0$, and is continuous on the boundaries. Also the function $G(t) = \rho(AB_t)$ is the boundary value of an analytic function in the strip $0 < Im\, t < \beta$, and continuous on the boundary.

(2) $F(t) = G(t + i\beta)$ holds for all $t \in \mathbf{R}$.

These certainly hold in finite systems, as can be seen by the cyclicity of the trace:

$$Tr\left(e^{-\beta H} e^{iHt} A e^{-iHt} B\right) = Tr\left(e^{-\beta H} B e^{-\beta H} e^{iHt} A e^{-iHt} e^{\beta H}\right)$$
$$= \rho\left(B e^{iH(t+i\beta)} A e^{-iH(t+i\beta)}\right).$$

We expect the $KMS$ condition to hold if we take the limit to an infinite system, since in the domain of complex $t$ considered, the rising exponential $e^{iHt}$ is beaten by the decaying exponential $e^{-\beta H}$; this relies on the positivity of the microscopic Hamiltonian. In fact, for our system of spins with finite-range interactions, and for a somewhat larger class too, Robinson [Robinson (1968)] has proved that the canonical state for a region $\Lambda_1$, using the Hamiltonian $H(\Lambda_1)$, converges to a state obeying the $KMS$ condition relative to the automorphism $\tau_t$ that we have constructed above.

It was noticed in [Haag, Hugenholtz and Winnick (1967)] that the $KMS$ condition can be expressed in terms solely of the automorphism of the $C^*$-algebra, and does not depend on the Hilbert space in which we might have defined the operators. They are able to construct the equilibrium statistical mechanics of an infinite system entirely in terms of the Gelfand-Naimark-Segal representation defined by a $KMS$ state. One of the basic results of the theory is that a $KMS$-state is always faithful. This reflects the property, obvious for finite classical systems, that the equilibrium measure is everywhere positive, and in finite quantum systems, the canonical density matrix has no zero eigenvalues. Another important theorem is that any $KMS$-state is invariant under the automorphism group $\{\tau_t\}$. This has the important consequence that the time-evolution is given by a unitary group with self-adjoint generator in the representation given by the $KMS$-state.

The generator, called the *modular Hamiltonian*, is never bounded below (unless the algebras $\mathcal{A}$ are of finite dimension). This is quite natural: an infinite system at a positive temperature contains an infinite amount of energy, and by removing some from a state we can get to another state of lower mean energy than we had before. This unboundedness below shows up in finite systems, in that the canonical state is conveniently taken to have zero energy; then states with some heat-particles removed will have negative energy. More recently, the theory has become based on the state itself, especially on its property of being faithful; the time-automorphism then become a derived concept, determined by the state. In the Tomita-Takesaki theory, we start with a $C^*$-algebra and a faithful state $\rho$. We then prove that there exists a unique automorphism group $\tau_t$ on the algebra such that the given faithful state obeys the $KMS$-condition relative to this automorphism. This is called the modular automorphism group defined by $\rho$. The state can be used in the Gelfand-Naimark-Segal construction to find a representation $\pi_\rho$ on a Hilbert space $\mathcal{H}_\rho$. Surprisingly, the representations obtained are always of type $III$ in von Neumann's classification (unless the system is finite). This has been a great stimulus to this branch of mathematics. See [Bratteli and Robinson (1979)] for a complete account, with applications to equilibrium statistical mechanics. Of particular physical interest is the work of Woronowicz and Pusz [Pusz and Woronowicz (1978)]. They introduce the concept of *passive* states, which are very stable under small perturbations. To give a small flavour of the stability question, we now give a typical result in the subject. We shall also see the limitations.

**Theorem 13.2.** *Let $\mathcal{A}$ be a $C^*$-algebra and $\rho$ a faithful state, such that the Gelfand-Naimark-Segal representation $\pi_\rho$ on the Hilbert space $\mathcal{H}_\rho$ is of type $III$. Let $\psi_\rho$ be the cyclic vector of the construction. Then there exists a dense set $\mathcal{D}$ of vectors in $\mathcal{H}_\rho$ such that for any $A \in \mathcal{A}$ and any normalised $\psi \in \mathcal{D}$, the mean of $A_t = \tau_t A$ in the state $\psi$ converges to its equilibrium value:*

$$\langle \psi, A_t \psi \rangle \to \langle \psi_\rho, A \psi_\rho \rangle = \rho(A) \ as \ t \to \infty.$$

**Proof.**   Since $\rho$ is faithful, the vector $\psi_\rho$ is separating, that is, $\pi_\rho(A)\psi_\rho = 0$ implies that $A = 0$. Since $\psi_\rho$ is separating for the algebra $\pi_\rho(\mathcal{A})$, it is cyclic for the commutant. Thus the vectors

$$\mathcal{D} = \{ B\psi_\rho : \ B : \mathcal{H}_\rho \to \mathcal{H}_\rho \text{ and } [B, \pi_\rho(A)] = 0 \text{ for all } A \in \mathcal{A} \}$$

give us a dense set $\mathcal{D}$. We may scale the operator $B$ so that $B\psi_\rho$ is normalised. Let $H_m$ denote the modular Hamiltonian, and $U(t)$ the unitary time-evolution it generates. Then $\pi_\rho(\tau_t A) = U(t)\pi_\rho(A)U(-t)$, and $U(t)\psi_\rho = \psi_\rho$. We get

$$
\begin{aligned}
\langle B\psi_\rho, \pi_\rho(A_t)B\psi_\rho \rangle &= \langle \psi_\rho, \pi_\rho(A_t)B^*B\psi_\rho \rangle \\
&= \langle \psi_\rho, U(t)\pi_\rho(A)U(-t)B^*B\psi_\rho \rangle \\
&= \langle \pi_\rho(A^*)\psi_\rho, U(-t)B^*B\psi_\rho \rangle \\
&\to \langle \pi_\rho(A^*)\psi_\rho, \psi_\rho \rangle \langle \psi_\rho, B^*B\psi_\rho \rangle \text{ as } t \to \infty \\
&= \rho(A) \text{ since } \|B\psi_\rho\| = 1
\end{aligned}
$$

by the Riemann-Lebesgue lemma. Here we use the assumption that the algebra is of type $III$ only inasmuch as this implies that $H_m$ has an absolutely continuous spectrum, except for the zero eigenvalue from the stationary state $\psi_\rho$. Our calculation shows that the vector state $B\psi_\rho$ converges weakly to the equilibrium state $\rho$ for large times. $\qquad\square$

The theorem gives no unique rate of convergence. Can we get the exponential convergence demanded by Krylov? By restricting $B$ we can, but not with a uniform rate. For example, let $h(t)$ be a test-function such that its Fourier transform $\hat{h}$ has compact support. Then replace $B$ in the proof by $B(h) = \int U(t)BU(-t)h(t)\,dt$, which lies in the commutant of the algebra since the modular time-evolution $U(t)$ maps this to itself. Then the convergence given by the proof is exponential, but the rate is determined by the support of $\hat{h}$, not by an intrinsic physical property, as required by Krylov. For this type of model, this lack of relaxation time is quite natural; there are indeed some states which do converge extremely slowly to equilibrium. These might be very hard to find, and might correspond (in this model) to the states that exhibit the *spin-echo effect*. This is the name given to an apparent violation of the second law. In the experiment, energy was sent into a system of ordered spins in the form of spin-waves; after a long time on the atomic scale, the system was a jumble of spins, and it seemed at first sight that the spins had been randomised. However, on reversing the applied magnetic field, the spins retraced their steps, and nearly reproduced the ordered spins of the original sample. The true relaxation time for the system was very much longer than that seen in coarse-grained measurements. The state seen at the half-way stage was one of the complicated states that are deemed to be overwhelmingly improbable. This may be so, but experimentalists are very clever, and did make it, and the theorem we proved allows for it. Another way to put Krylov's requirement that a theory should predict a relaxation time is to ask, how can we make a choice of coarse-graining in a theory such as the infinite-volume spin-system, such

that the chosen variables exhibit a relaxation to equilibrium exponentially
at a minimum rate? We cannot do this in a self-contained Hamiltonian
theory; the point is that it is the *fast* variables, the very variables that we
do *not* use in our description, that cause the relaxation of the variables we
do use.

## 13.4    Irreversible Linear Dynamics

If time is continuous, we can ask that the time-evolution should be a one-
parameter group of completely positive maps; the general form for such
systems was found in Lindblad [Lindblad (1976)], and [Gorini, Kossakowski
and Sudarshan (1976)]. We follow the argument of [Landau and Streater
(1993)]. This analysis does not include the conditions that each time-step be
bistochastic and conserve energy. Recall the notation of Exercise 9.24: if $\rho$ is
a linear functional on the algebra $\mathbf{M}_n$, not necessarily positive, we denoted
by $\Delta(\rho)$ the operator in $\mathbf{M}_n$ such that $\rho(A) = Tr\,\Delta(\rho)A$ for all $A \in \mathbf{M}_n$.
If $\rho$ is a state, then $\Delta(\rho)$ is the corresponding density operator. Recall too
the correspondence $T \mapsto L_T$ between a linear map $T : \mathbf{M}_n \to \mathbf{M}_n$ and the
linear functional $L_T : \mathbf{M}_n \times \mathbf{M}_n \to \mathbf{C}$; it is defined on product operators
by

$$L_T(A \otimes B) = \rho_0\left(B^t T(A)\right) = n^{-1}Tr\left(B^t T(A)\right),$$

and by linear extension to the general operator. We shall use the same
notation, $\Delta L$, for the matrix in $\mathbf{M}_n \bigotimes \mathbf{M}_n$ which gives the linear functional
$L$ on $\mathbf{M}_n \bigotimes \mathbf{M}_n$, thus:

$$Tr_{12}\left(\Delta L(A \otimes B)\right) = L(A \otimes B).$$

Here, $Tr_{12}$ means the trace over $\mathbf{M}_n \bigotimes \mathbf{M}_n$. We shall also use the realisation of $\mathbf{M}_n \bigotimes \mathbf{M}_n$ as the set of operators on the Hilbert space $\mathbf{M}_n$, which
is furnished with the normalised Hilbert-Schmidt norm. This is the faithful
representation of $\mathbf{M}_n \bigotimes \mathbf{M}_n$, given for product operators by

$$(A \otimes B)C = ACB^t \qquad\qquad \text{for } A,\, B,\, C \in \mathbf{M}_n \qquad (13.2)$$

and by linearity for the general element, which is of the form $\sum A \otimes B$.
In this realisation, $\Delta L$ denotes an operator on $\mathbf{M}_n$. This has sometimes
been called a superoperator in the Physics literature, being an operator
on matrices in $\mathbf{M}_n$. Naturally, the word is less used now, to avoid confusion with other uses of 'super'. Recall too the linear functional $L_{C,D}$ on

$\mathbf{M}_n \bigotimes \mathbf{M}_n = B(\mathbf{M}_n)$ defined for each pair of elements $C$ and $D$ in $\mathbf{M}_n$ by $L_{C,D}(M) = \langle C, MD \rangle$. In this, $\langle \bullet, \bullet \rangle$ is the Hilbert-Schmidt scalar product

$$\langle A, B \rangle = Tr(A^*B).$$

We shall need the relationship:

**Lemma 13.3.** *The linear functional $L_{C,D}$ corresponds to the operator $T_{C,D} = C^* \bullet D$ on $\mathbf{M}_n$.*

**Proof.** Let $T$ denote $T_{C,D}$; then

$$
\begin{aligned}
L_T(A \otimes B) &= n^{-1}Tr(B^t T(A)) \\
&= n^{-1}Tr(B^t C^* AD) = n^{-1}Tr(C^* ADB^t) \\
&= \langle C, (A \otimes B)D \rangle = L_{C,D}(A \otimes B).
\end{aligned}
$$

Since this is true for all $A$ and $B$, we have $L_T = L_{C,D}$. $\qquad\square$

Let $P_1 \in B(\mathbf{M}_n)$ be the orthogonal projection onto the unit vector $1 \in \mathbf{M}_n$. This is the density matrix of the vector state on $B(\mathbf{M}_n)$ given by the unit vector $1 \in \mathbf{M}_n$, that is, the functional $M \mapsto \langle 1, M1 \rangle = L_{1,1}(M)$. In symbols, $\Delta L_{1,1} = P_1$. Furthermore, the linear map $T_{1,1}$ corresponding to $L_{1,1}$ given by Lemma 13.3 is the identity map of $\mathbf{M}_n$ onto itself. The following relations are elementary but very useful:

**Lemma 13.4.**

$$P_1 \Delta L P_1 = \langle 1, L1 \rangle \Delta L_{1,1} \qquad\qquad (13.3)$$

$$\Delta L P_1 = \Delta L_{1,\Delta L1} \qquad\qquad (13.4)$$

$$P_1 \Delta L = \Delta L_{\Delta L^* 1,1}. \qquad\qquad (13.5)$$

**Proof.** The first just says that if $P_1 = \psi \otimes \psi$ is a one-dimensional projection, and $M$ is any operator, then $P_1 M P_1 = \langle \psi, M\psi \rangle P_1$. For the second,

$$
\begin{aligned}
Tr(\Delta L P_1(A \otimes B)) &= Tr(P_1(A \otimes B)\Delta L) \\
&= \langle 1, (A \otimes B)\Delta L1 \rangle \\
&= L_{1,\Delta L1}(A \otimes B).
\end{aligned}
$$

Since this is true for all $A \otimes B$, we get the result. The third equation, in which $*$ denotes the adjoint on $B(\mathcal{M}_n)$, is similar. $\qquad\square$

We can now define the tilde operation used in [Landau and Streater (1993)]. Let $T$ be a linear map of $\mathbf{M}_n$ to itself, and $L_T$ the corresponding linear functional on $\mathbf{M}_m \otimes \mathbf{M}_n$. Let $P^\perp = 1 - P_1$ be the projection onto the orthogonal complement of the vector $1 \in \mathbf{M}_n$. Then we have

$$\Delta L_T = P_1^\perp \Delta L_T P_1^\perp + P_1 \Delta L_T P_1^\perp + P_1^\perp \Delta L_T + P_1 \Delta L_T P_1$$
$$= P_1^\perp \Delta L_T P_1^\perp + P_1 \Delta L_T + \Delta L_T P_1 - P_1 \Delta L_T P_1$$
$$= \widetilde{\Delta L_T} + \delta L_{\Delta L^* 1, 1} + \delta L_{1, \Delta L 1} - \langle 1, \Delta L_T 1 \rangle L_{1,1}$$

by (13.3), (13.4), (13.5). We have put

$$\widetilde{\Delta L_T} = P_1^\perp \Delta L_T P_1^\perp.$$

Now associate half of the negative term $-P_1 \Delta L_T P_1$ with each of the other two terms, and define

$$S = \Delta L_T 1 - \frac{1}{2} \langle 1, \Delta L_T 1 \rangle 1 \in \mathbf{M}_n,$$

$$S' = \Delta L_T^* 1 - \frac{1}{2} \langle 1, \Delta L_T^* 1 \rangle 1 \in \mathbf{M}_n.$$

Then we get

$$\Delta L_T = \widetilde{\Delta L_T} + \Delta L_{1,S} + \Delta L_{S',1}. \tag{13.6}$$

We correspondingly obtain, from (13.3)

$$T = \tilde{T} + S'^* \bullet + \bullet S. \tag{13.7}$$

The tilde operation is defined as the transformation $T \mapsto \tilde{T}$. We see that if $T$ is Hermitian, then $L_T$ is real and $\Delta L_T = \Delta L_T^*$. Hence $S = S'$. Moreover, if $T$ is completely positive, then $L_T$ and $\Delta L_T$ are positive, and so is $\Delta L_{\tilde{T}}$, as it is a positive operator sandwiched between projections $P_1^\perp$. Hence $\tilde{T}$ is completely positive.

Now suppose that $T(t)$ is a smooth semigroup of completely positive maps. By continuity, $\tilde{T}(t)$ and $S(t)$, defined for each $t$ by (13.6), are smooth functions of $t$. Further, since the identity map corresponds to $P_1$ it follows that $\widetilde{Id} = 0$. Thus

$$\mathcal{J} = \frac{d}{dt} \widetilde{T(t)}|_{t=0} = \lim_{t \to 0} t^{-1} \widetilde{T(t)}$$

is the limit of completely positive maps, and so is completely positive. Then with the definition $K = \dot{S}_{t=0}$, we have proved the 'only if' part of the structure theorem of [Gorini, Kossakowski and Sudarshan (1976)] and [Lindblad (1976)]:

**Theorem 13.5.** *The map* $\Gamma : \mathbf{M}_n \to \mathbf{M}_n$ *generates a completely positive semigroup on* $\mathbf{M}_n$ *if and only if* $\Gamma = \mathcal{J} + \mathcal{K}$, *where* $\mathcal{J}$ *is completely positive, and* $\mathcal{K} = K^* \bullet + \bullet K$ *for some* $K \in \mathbf{M}_n$.

The converse follows easily by exponentiating, and using the Lie-product formula.

Suppose that the anti-hermitian part of $K$ is $H_1$. This part of the generator gives a contribution to $\Gamma$ of the form $i[H_1, \bullet]$. This generates a unitary dynamics, which does not itself contribute to dissipation. If the Hermitian part of $K$, or the completely positive part $\mathcal{J}$, does not commute with $H_1$, then $H_1$ does contribute indirectly to dissipation by helping to mix the system. It might also contribute to mixing if it does not commute with the non-linear part of the dynamics, like the quasifree map or stoss map. Some authors feel that $H_1$ should be identified with the Hamiltonian of the theory. This is not necessary in statistical dynamics. What we do need is that the semigroup as a whole leaves the spectral projections of the Hamiltonian invariant (the first law of thermodynamics). We shall also need to impose that the maps $T(t)$ are bistochastic (second law of thermodynamics). We are far from a general solution to either problem. We can find some interesting special cases as follows. Suppose that $K$ is antihermitian, and that $iK$ is the Heisenberg Hamiltonian $H$. For the semigroup to be stochastic, it is then sufficient that $\mathcal{J}1 = 0$. The rotation group acts on $\mathbf{M}_2$: to $R \in SO(3)$, conjugate with the unitary operator $U \in SU(2)$ corresponding to $R$. This representation decomposes into the scalar and vector parts. The projection onto the vector part (in the Hilbert-Schmidt space) is $A \mapsto A - 1/2 \operatorname{Tr} A\, 1$. This can be taken to map $\mathcal{A}_x$ to itself, and by tensoring with unit maps, ampliated to an operator on the whole local algebra. Summing over $x \in \Lambda$ then gives $\mathcal{J}$, which commutes with $H$ and annihilates 1.

## 13.5 Exercises

Exercise 13.8. Let $\mathcal{A}$ be a $C^*$-algebra with dense sub$^*$-algebra $\mathcal{A}_0$. Suppose that for small $t$, $\tau_t : \mathcal{A} \to \mathcal{A}$ is a $^*$-algebra homomorphism preserving the norm. Show that $\tau_t$ can for small $t$ be extended to an automorphism of $\mathcal{A}$.

Exercise 13.9. If in addition to the properties in Exercise 13.8, $\tau_t$ is norm continuous ($A\tau_t \to A\tau_s$ as $t \to s$) on $\mathcal{A}$, show that its extension is also norm continuous.

# Chapter 14

# Proof of the Second Law

## 14.1 von Neumann Entropy

von Neumann defined the entropy $S(\rho)$ of a quantum system in the state [=density operator] $\rho$ to be

$$S(\rho) := -k_B \, Tr[\rho \log \rho]. \tag{14.1}$$

He knew that, when $\rho \mapsto \rho(t)$ is the dynamics given by unitary transformations of $\rho$, then $S(t) := S(\rho(t))$ is independent of time. The same thing happens in classical dynamics for the differential entropy defined as

$$S_c(\rho) = -k_B \int_\Gamma \rho(q,p) \log \rho(q,p) d^n p d^n q. \tag{14.2}$$

Here, $\Gamma$ is the phase space of $n$ degrees of freedom, and $\rho$ is a classical probability on $\Gamma$; the dynamics $\rho \mapsto \rho(t)$ is given by Hamilton's equations. It is Liouville's theorem, that $d^n p d^n q$ is invariant under Hamiltonian motion, that causes $S_c$ to remain invariant under time-evolution.

Lebowitz argued as follows; macroscopically we measure very few variables. Since entropy is associated with uncertainty, the entropy we detect omits all the uncertainty associated with the variables we have omitted to measure: thermodynamic entropy is much less than $S$, and could increase with time. However, it seems to me that, in that case, we lose the calculation from first principles, involving $k_B$, that is so successful in the theory of equilibrium.

In his book, von Neumann suggested that thermodynamic entropy requires a different formula, related to the measurements made on the system. Such an entropy, say $s(t)$, might increase in time, due to the disturbances of the system, that occur in quantum measurement. The trouble is, how large an increase then seems to depend on what is being measured. This is taken up in [De Roeck, Jacobs, Maes and Netočný (2003)]. The authors define a quantum version of the Kac spin model and a modified entropy, and show that the new entropy, $s(t)$, increases with time under the dynamics, with probability going to 1 as the volume of the system increases.

We show here that the increase in $s(t)$ under the Schrödinger dynamics, obtained by Maes et al. can be regarded as the increase in von Neumann's original entropy, $S$, under an approximate dynamics, up to an overall factor $N$, the number of spins.

Let $\mathcal{A} = \mathcal{M}^n$, the algebra of complex $n \times n$ matrices. Suppose that the dynamics, with discrete time, is given by a group of automorphisms

$$A \mapsto A(t), \qquad t \in \mathbf{Z}, A \in \mathcal{A}. \qquad (14.3)$$

This dynamics can be transferred to the states by duality; a state $\rho$ is a positive normalised linear form on $\mathcal{A}$, and its time evolution $\rho(t)$, is defined by

$$\rho(t) \cdot A := \rho \cdot A(-t). \qquad (14.4)$$

Then it is well known that von Neumann's definition of entropy Eq. (14.1) is constant in time.

The trouble is, thermodynamics involves taking the thermodynamic limit, Volume, $V \longrightarrow \infty$; in the limit, the von Neumann entropy $S$ diverges to infinity, and only the entropy-density $S/V$ remains finite. We take the view that $V$ should remain finite, but is large, and that $S$ *is* indeed the correct thermodynamic entropy, but is evaluated for an approximate dynamics, $\rho \mapsto \rho T(t)$, which gives the same evolution as the true dynamics for a small set of operators, the macrovariables, designated to be the observables of the macroscopic system. The dynamics $T(t)$ are highly non-linear, and increase the von Neumann entropy of the state.

In the paper on the quantum Kac model, the authors do exactly this. $T(t)$ involves changing only the macrovariables of the state, and is the limit (in probability) to infinite $V$ of the true dynamics of the model.

## 14.2   Entropy Increase in Quantum Mechanics

Let us use the word *thermodynamics* to refer to the limit dynamics of the macrovariables, or slow variables, of the model. Each value of the macrovariables defined an approximate state of the model, which is the state of greatest entropy having these values for the macrovariables. This approximate state can hardly be separated from the true state $\rho(t)$, if we limit ourselves to the measurement of only the macrovariables, and the volume is large. We show that (in the quantum Kac model) the von Neumann entropy of Eq. (14.1) of this approximate dynamics increases in time.

Recall that, acting on observables, a linear map $T$ is said to be *stochastic* if

(1) $T$ is a positive map,
(2) $TI = I$.

A stochastic map does not necessarily increase the entropy; its iteration does not always converge. If we impose *detailed balance* on a stochastic map $T$, then its iteration converges to the Gibbs state determined by the temperature arising in the detailed balance. We saw in Chapter 5 and in 5.12 that this situation arises in isothermal theory, when a system is placed in a heat-bath. The free energy decreases, but the von Neumann entropy may increase or decrease. We also saw that to increase (or leave the same) the entropy of a system requires a *bistochastic map*: recall that a stochastic map $T$ is bistochastic if it preserves the trace:

$$Tr(TA) = Tr(A) \qquad \text{for all } A \text{ of trace class.} \qquad (14.5)$$

We throw the action of $T$ onto the states by duality; then

**Theorem 14.1.** *If $T$ is bistochastic, then*

$$S(\rho T) \geq S(\rho). \qquad (14.6)$$

In the classical case we may assume that $\mathcal{A}$ is abelian; then Birkhoff's theorem says that every bistochastic map is a weighted mixture of permutations. In the quantum case one might think that any bistochastic map is a weighted mixture of automorphisms; this is not the case, but one gets a useful class of bistochastic maps by this method; for example, the finite sum

$$TA := \sum_\alpha \lambda_\alpha U_\alpha A U_\alpha^{-1} \qquad (14.7)$$

where $0 < \lambda_\alpha < 1$ and $\sum_\alpha \lambda_\alpha = 1$ hold. In the Kac model, this arises as the mixture of two terms, one the identity automorphism and the other, one time-step of the dynamics.

Non-linear dynamics can also increase entropy. Consider a system of two particles of spin $\frac{1}{2}$. The algebra of observables is then

$$\mathcal{A} = \mathcal{M}^4 = \mathcal{M}^2 \otimes \mathcal{M}^2. \qquad (14.8)$$

The general state $\rho$ of $\mathcal{A}$ will have entanglement between the spins. These
are determined by the *marginal states*

$$\rho_1 := Tr_2\rho \text{ and } \rho_2 := Tr_1\rho. \tag{14.9}$$

The product state

$$\rho Q := \rho_1 \otimes \rho_2 \tag{14.10}$$

maximises the von Neumann entropy $S(\rho)$, subject to these values,
Eq. (14.9) of its marginals. The map $\rho \mapsto \rho Q$ is non-linear in $\rho$; it puts to
zero any correlations between the two particles.

For their quantum Kac model, Maes et al. show that the macroscopic
dynamics is, in the limit $V \to \infty$ probably given by the bistochastic dynamics of the spin

$$T\sigma := (1 - \mu)\sigma + U\sigma U^{-1} \tag{14.11}$$

followed by the tensor product $T\sigma \otimes \ldots \otimes T\sigma$ over the points of $V$.

## 14.3    The Quantum Kac Model

Kac introduced his ring model [Kac (1955, 1959)] to show how reversible
dynamics can appear to show convergence to equilibrium for most initial
states. A quantum version was constructed in [De Roeck, Jacobs, Maes
and Netočný (2003)].

The system consists of $N$ atoms, each of spin $\frac{1}{2}$, situated on a ring. The
Hilbert space is

$$\mathcal{H}_N = \otimes_{j=1}^{N} \mathbf{C}_j^2. \tag{14.12}$$

Fix a hermitian matrix $H = \boldsymbol{h} \cdot \boldsymbol{\sigma} \in \mathcal{M}^2$, the single-site Hamiltonian, and
put $U = e^{iH}$. The energy $E_N$ is given by

$$E_N := \sum_{i=1}^{N} I \otimes \ldots I \otimes H \otimes I \ldots \otimes I \tag{14.13}$$

where the factor $H$ is in the $i$-th position. The points $1, \ldots, N$ are placed
on a ring; on the segment $(j, j+1)$ is a 'scattering variable' $\epsilon_j$, equal to 0
or 1. The configuration $\epsilon = \{\epsilon_1, \ldots, \epsilon_N\}$ does not change with time. We
define the unitary matrix on $\mathbf{C}_j^2$ by

$$U_j^{\epsilon_j} := \exp i\epsilon_j H_j = \begin{cases} 1 & \text{for } \epsilon_j = 0 \\ U & \text{for } \epsilon_j = 1. \end{cases} \tag{14.14}$$

The ring is rotated by the unitary operator $R$ on $\mathcal{H}$, defined by extending the isometric action on product vectors

$$R\left(\eta_1 \otimes \eta_2 \otimes \ldots \otimes \eta_N\right) = \left(\eta_N \otimes \eta_1 \otimes \ldots \otimes \eta_{N-1}\right). \qquad (14.15)$$

One time-step in the dynamics is given by the unitary operator $U_N$ on $\mathcal{H}_N$:

$$U_N := R \otimes_{j=1}^{N} U_j^{\epsilon_j}. \qquad (14.16)$$

This commutes with the energy-operator, which is therefore conserved.

The *slow variables* of the model are those of the form of a sum over sites

$$A_N := \sum_{j=1}^{N} I \otimes \ldots \otimes A \otimes \ldots \otimes I \qquad (14.17)$$

for any $A \in \mathcal{M}^2$. When $A = \sigma_\alpha$, $(\alpha = x, y, z)$, then $A_N$ is the total magnetisation; together with the identity, these are the thermodynamic variables.

## 14.4  Equilibrium

The energy $E_N$ is conserved by the dynamics and the equilibrium is for example the microcanonical distribution. Let $P_{\pm}$ be the projectors onto the eigenvalues $e_{\pm}$ of the one-body Hamiltonian $H$, and let $e \in [e_-, e_+]$. Let

$$\rho_e^N := \frac{1}{Z_e^N} \sum_{e_{\pm}} P_{e_1} \otimes \ldots \otimes P_{e_N}, \qquad (14.18)$$

where the sum runs over all $e_{\pm}$ satisfying

$$Ne \le \sum_j e_j < e_+ - e_- + Ne. \qquad (14.19)$$

The average

$$\langle A_N \rangle_e^N := Tr(\rho_e^N A_N) \qquad (14.20)$$

gives the mean of $A_N$ in equilibrium; for example

$$\mathbf{m}_e = \lim_{N \to \infty} \frac{1}{N} \langle \mathbf{M}_N \rangle_e^N \qquad (14.21)$$

is the mean magnetisation. The limit exists and equals

$$\mathbf{m}_e = e \frac{\mathbf{h}}{\|\mathbf{h}\|^2}. \qquad (14.22)$$

## 14.5 The $\epsilon$-Limit

The results of the paper are for unexceptional choices of $\epsilon$. Let $\mu \in (0,1)$ be chosen and put

$$\Omega_\mu^N := \left\{ \epsilon \in \{0,1\} \mid \sum_1^N \epsilon_j = [\mu N] \right\} \tag{14.23}$$

where $[\mu N]$ is the integer part of $\mu N + 1$. $\Omega_\mu^N$ is thus the set of $\epsilon$ so that the fraction of active scatterers (those with $\epsilon_j = 1$) is about $\mu$. The authors furnish $\Omega_\mu^N$ with the uniform distribution, so that, as $C_{[N\mu]}^N$ is the number of points, we get

$$\mathbf{P}_\mu^N(\epsilon) = \left( C_{[N\mu]}^N \right)^{-1}. \tag{14.24}$$

Limits $N \to \infty$ will often involve this probability; let us say that a sequence of functions $G_N$ on $\Omega_\mu^N$ typically converges to the value $g_\mu$, written as

$$\epsilon - \lim_{N \to \infty} G_N = g_\mu \tag{14.25}$$

if and only if

$$\lim_{N \to \infty} \mathbf{E}_{\mathbf{P}_\mu^N} \left[ (G_N - gg_\mu)^p \right] = 0, \qquad p = 1, 2. \tag{14.26}$$

In fact, $\epsilon$ is fixed, and is not random; but for a given mean value $\mu$, if we can prove that $G_N$ typically converges to $g_\mu$, then the $\epsilon$ for which it does not are very rare. The authors show that $M_N^\alpha(t)$ is typically determined by $M_N^\alpha(t-1)$: the slow variables obey an autonomous dynamical law. The dynamics does depend on $\epsilon$, but for most choices of $\epsilon$ of mean $\mu$, these equations depend only on $\mu$.

## 14.6 The Marginals and Entropy

The time-dynamics $U_N$ is unitary, and does not change the von Neumann entropy. The authors define a new entropy $S_N$,

$$S_N(\rho^N) := -N \, Tr[\nu^N \log \nu^N] \tag{14.27}$$

where

$$\nu^N = \frac{1}{N} \sum_1^N Tr_j[\rho^N]; \tag{14.28}$$

here, $Tr_j$ is the reduced density matrix, a $2 \times 2$ matrix. This changes in time under the dynamics as

$$\nu^N(t) = \frac{1}{N} \sum_1^N Tr_j[\rho^N(t)] \tag{14.29}$$

$$S_N(\rho^N(t)) = -N \, Tr[\nu^N(t) \log \nu^N(t)], \tag{14.30}$$

and if all goes well, which they show to be true for some technical conditions on the initial state:

$$s(t) = \epsilon \lim_{N \uparrow \infty} \frac{1}{N} S_N(\rho^N(t)) \tag{14.31}$$

is an increasing entropy.

## 14.7   The Results

Let $\rho_0^N$ obey some technical conditions. Then $\rho^N(t)$ obeys, for any continuous function $f$:

$$\epsilon \lim_{N \uparrow \infty} Tr\left[ f\left(\frac{M_N^\alpha}{N}\right) \rho^N(t) \right] = f(m^\alpha(t)) \tag{14.32}$$

with $m^\alpha(t)$ obeying the autonomous equation

$$\mathbf{m}_t = \mathbf{m}_e + \mathrm{Re}\left[ \left( \mathbf{m}_o - \mathbf{m}_e + i\frac{\mathbf{h} \times \mathbf{m}_e}{\|\mathbf{h}\|} \right) \left( 1 - \mu + \mu e^{i(e_+ - e_-)} \right)^t \right]. \tag{14.33}$$

The entropy equals

$$s(t) = -\frac{1 + \|\mathbf{m}_t\|}{2} \log\left( \frac{1 + \|\mathbf{m}_t\|}{2} \right) - \frac{1 - \|\mathbf{m}_t\|}{2} \log\left( \frac{1 - \|\mathbf{m}_t\|}{2} \right) \tag{14.34}$$

and increases in time.

We can interpret $S(t)$ as the von Neumann entropy of the approximate dynamics as follows. Start with $\rho_0$ obeying the technical conditions. Then apply the dynamics $U^N$, and compute the new value of the magnetisation. Change the state to the product of the one-body state with this magnetisation, divided by $N$. This state has a greater entropy than that of $\rho_0$, since it is the state of maximum entropy with the given total magnetisation. Now take *this approximate state* and compute its dynamics under $U^N$. This will be different from what we get from taking the true state at time 1 and applying $U^N$, but it will give rise to the same new value of $\mathbf{M}$, since (as was shown by the authors) the true value is (in the limit of large $N$) the same for all states having the previous value of $\mathbf{M}$. Proceed in this way.

We get that the state we use has the same value of **M** as the true state; and its von Neumann entropy is exactly $Ns(t)$.

Indeed, let $\mu$ be the average value of the 'external field' $\epsilon$. Then in one time step, the change in $s(t+1) - s(t)$ is the same as if the average value of $\epsilon$ had acted at each site. That is

$$T\sigma = (1 - \mu)\sigma + \mu U \sigma U^{-1} \tag{14.35}$$

where $\sigma$ is the one-site state, taken to be independent of time, and $U = e^{iH}$ is the unitary time evolution given by the energy operator. After this stochastic map, we then form the product state $T\sigma \otimes \ldots \otimes T\sigma$; its von Neumann entropy is $Ns(t+1)$. Thus, the von Neumann entropy of the approximate dynamics increases in time.

# Chapter 15

# Information Geometry

In this Chapter we start with the question of how to describe interacting systems in terms of quantum states with a simple description, but one more detailed than a quasifree state; this leads us to the idea of generalised temperature. We make use of a principle of maximum entropy, which is very close to the techniques of Jaynes and Ingarden; their theory is the subject of the next section. We recommend adopting their techniques but not all their philosophy. We go on to discuss estimation theory, which has four parts: finite-dimensional classical probability, which is discussed by Amari [Amari (1985)]; finite-dimensional quantum probability, which is discussed by Ohya and Petz [Petz (2002)]; the infinite-dimensional classical probability, due to Pistone and Sempi [Pistone and Sempi (1995)], and the infinite-dimensional quantum probability, which is our own [Streater (2008b)].

## 15.1    The Jaynes-Ingarden Theory

The justification of the canonical state as that which most systems end up in has been a recurring theme of this book, and many before it. The practical failure of the ergodic programme, after the Kolmogorov-Arnol'd-Moser theorem (KAM), led physicists to think of other reasons why we should choose, sometimes the canonical state, sometimes the grand canonical state (or the great grand canonical state). The success of information theory, in the guise of the principle of minimum information, suggested a non-dynamical, subjective reason why nearly all states are canonical. Identifying entropy as negative information, Jaynes [Jaynes (1957)] proposed that in the absence of any information, we should assume that the state is that of maximum entropy. This makes sense for finite systems, and leads

to the microcanonical state. As information becomes available, we modify
the state accordingly. Thus if the mean energy is known, we maximise the
entropy among all states with the observed mean energy. This gives us the
canonical state. If, in addition, the mean particle-number is observed (and
it was assumed that to be observable on the macroscopic scale, the particle-
number should be conserved in time) then we get the grand canonical state.
This idea is similar in spirit to the Bayesian method. There, we start with
a prior distribution; this is often the uniform distribution (if the sample
space is finite). We then feed in information by conditioning, using Bayes's
theorem. The two procedures are slightly different, since in Bayes's method
we use the result of a single sample, whereas in the Jaynes method we use
the known mean of the population, and do not pick a subsample. Now,
the Bayes method (quantum version) was the subject of Krylov's criticism
[Krylov (1979)] well before Jaynes' proposal. Krylov objected to what he
called von Neumann's measurement theory, in which a state (taken to be
the uniform, tracial state) is filtered by projections onto the subspaces of
observed values of measurements, to get the conditioned state, purer than
the original in general. What Krylov objected to was not the theory of
filtering, but the presumption that the initial state is uniform. He was
objecting to the choice of the state of maximum entropy, even if we have
no information about the state; physics predicts the state at a later time,
from the state at an earlier time, and does not make best guesses about the
initial state; this is the job of a different discipline, prediction theory. The
name, prediction theory, is a misnomer; it should really be 'hedging theory',
since its purpose is not to predict anything specific about what the pure
state is, but to devise a strategy so that one's wrong guesses are not likely
to have serious consequences, on average. In spite of Krylov's objections,
we shall use the principle of maximum entropy, as extended by Ingarden, to
guide us in finding useful non-quasifree states, and a randomisation which
can replace the quasifree map or the stoss map, for example in models with
non-quadratic energy or energy not localised at a single site. Later in the
Chapter we shall justify the method of maximum entropy from standard
probability theory, without assuming any prior distribution.

Ingarden's idea is that when we only know how to measure very crude
things like the mean total energy, then we use the canonical state for the description of the state. But if we can measure more detail, such as the first $n$
cumulants of the energy, then we should use $n$ 'higher-order' temperatures,
being the Lagrange multipliers when we maximise the entropy subject to
given values of all $n$ cumulants of the energy, not just the mean. Haag has

described the increase in entropy as being due to 'dust': it is too small to make any real changes in the means of physical observables. Let us see how one defines the second-order temperature for the harmonic oscillator. Here, the Hilbert space is $\ell^2$ with basis $\psi_0 = (1,0,0,\ldots)$, $\psi_1 = (0,1,0,\ldots)$ and so on; the energy is $\varepsilon\mathcal{N}$, where $\mathcal{N}\psi_n = n\psi_n$. So we must maximise $-Tr\,\rho\log\rho$ subject to

- $Tr\,\rho = 1$
- $Tr\,\varepsilon\mathcal{N}\rho = \overline{E}$
- $Tr\,(\varepsilon\mathcal{N} - \overline{E})^2 = V$.

The answer, Exercise 15.109, is of the form

$$\rho_{\beta,\beta'} = Z^{-1}e^{-\beta H - \beta'(H-\overline{E})^2} \tag{15.1}$$

Ingarden calls $\beta'$, or its inverse, a higher-order temperature. Apart from the implicit relation between the intensive parameters $\beta$, $\beta'$ and the given constraints, the state we get is quite tractable. We even have a formula for it. The state that maximises the entropy subject to given mean values of all operators in a subspace $\mathcal{A}_o$ will be called the corresponding *Jaynes state*. We shall not need to make use of constraints non-linear in the state.

Note that we do not try to require the cumulants beyond 2 to be zero; this might lead to contradictions with positivity. The state maximising entropy, subject to such conditions, not only on the energy, and any other finite number of chosen random variables, certainly exists and is unique, by the concavity of the entropy. The choice of which variables we choose to fix seems subjective. One experimenter might be able to measure more than another. Ingarden answers this by saying that we always have to choose the level of description before constructing any model, and having done so the method gives us a unique procedure, with no further subjectivity. This is so: the level of description is part of the model, like the choice of sample space and Hamiltonian. However, we should not go all the way with Jaynes and Ingarden. First, our state is not intended to be stationary under the dynamics. Secondly, information theory has too little predictive power (like prediction theory in general). For example, suppose that a system is in equilibrium (so that the first objection cannot be made). Ingarden says, that if we can measure the second cumulant of the energy as well as the mean, then we should use two generalised temperatures in the description. In fact we can indeed often measure the second cumulant: it is the fluctuation of the energy. We might then think that we would often, or even sometimes, find that the equilibrium state is not the canonical state, but would need two parameters, as in (15.1). But this rarely happens: when we measure the fluctuations of a system in equilibrium, we nearly always find that they are thermal. This was the remarkable thing about the celebrated discovery of the thermal background radiation by COBE, the Cosmic Background Explorer: the law was almost exactly Planckian. The simple explanation is that it is in a canonical state. When we are in equilibrium, and many high moments of the energy can be measured, Ingarden's theory does not say that the distribution cannot be Planckian; but it does not offer any explanation as to why it is Planckian. Ingarden argues that for a macroscopic system there is very little difference between the two states, and that we would need a mesoscopic or microscopic system to be able to detect the higher temperature. We do not have to agree with this.

In the following, we shall use Ingarden's methods to construct models. The reader might well find it hard to see any difference between our theory and Ingarden's. In our version, the states with generalised temperatures are not in equilibrium; the final state, at large times, will usually be the canonical or grand canonical state, (depending on the mixing properties), and we intend that this occur even for a mesoscopic system, such as a few atoms. There is little difference with Ingarden's theory if the relaxation time from the state with generalised temperatures to the final equilibrium is very long. Although the state at time $t$ is described by a few parameters, we are still able to compute, and so predict, any statistical property of the system at time $t$, since we know the full state. We illustrate the idea by a detailed study of the Ising model.

## 15.2  Non-Linear Ising Dynamics

We now apply the general method of Ingarden to define a dynamics for the Ising model, and in such a way that both laws of thermodynamics are true. The dynamics differs from the Glauber dynamics, in that the temperature does not appear in the dynamics of the isolated system. Our dynamics will also be non-linear, unlike the Glauber dynamics, which defines a stochastic process in the usual sense. For us, isothermal dynamics and detailed balance is a secondary construct, obtained from the isolated model by keeping the beta fixed using external heat-flows.

To illustrate the method, we describe the spin-$1/2$ Ising model on $\mathbf{Z}_N$. Recall that this is a classical model, with sample space $\Omega = \prod_x \Omega_x$, where $\Omega_x = \{-1, 1\}$. Thus a configuration, a point of $\Omega$, is a sequence of $N$

numbers $\omega(x)$, each either $+1$ or $-1$. The translation group $\mathbf{Z}_N$ acts on $\Omega$ by $(\omega\tau(j))(x) = \omega(x + j)$. Recall that the addition rule in $\mathbf{Z}_N$ is addition modulo $N$, and that $x = N + 1$ is identified with 1. The energy is

$$\mathcal{E}(\omega) = -J\sum_x \omega(x)\omega(x + 1).$$

Here as usual, $J > 0$. We shall denote the state by a probability $p$ on $\Omega$. There are two states of minimum energy; the lowest energy, $-NJ$, occurs when all the spins are $+1$, and when all the spins are $-1$. It is usual to introduce random variables $S_x$ to describe the spins, thus: $S_x(\omega) = \omega_x$.

The Ising model is a theory of equilibrium; there is no dynamics associated with the Hamiltonian. We cannot get anything from treating it as a quantum theory; since the Hamiltonian is a random variable, it commutes with all the observables. So the Heisenberg equations of motion lead to the trivial dynamics; each observable is constant in time. Dynamics has to be added, and there is no unique way to do this. In the model dynamics suggested in [Glauber (1963)], transitions between configurations are assumed to take place with certain probabilities. These allow transitions between configurations of different energy, and therefore the model is not of the type we have called 'isolated dynamics'. If the energy-difference between two configurations is $E$, Glauber chooses the ratio of the forward and backward transition probabilities to be $e^{-\beta E}$. This has been called detailed balance in (5.16); we preferred the chemists' term microscopic reversibility. It is clear that Glauber's model is an isothermal one, not an isolated one. The temperature appears in the dynamical law, and to this extent the model is rather far from a microscopic one. Our first task is to suggest another Markov chain, one that obeys the first law of thermodynamics. For simplicity, we shall assume that the stochastic matrix is symmetric (and therefore bistochastic). This is microscopic reversibility. We shall also assume that in one time-step at most one spin is flipped from $+1$ to $-1$ or *vice versa*. Naturally, we only allow flips that conserve energy. We first give a bistochastic map that flips only one point say $x = 2$ and affects its two neighbours. There are eight configurations of $\Omega_1 \times \Omega_2 \times \Omega_3$, namely

$$\{(+ + +), (+ + -), (+ - +), (+ - -),$$
$$(- + +), (- + -), (- - +), (- - -)\}.$$

Only four energy-conserving transitions are possible without involving other sites or flipping more than one spin. These are

$$(+ + -) \rightarrow (+ - -); \qquad (+ - -) \rightarrow (+ + -);$$
$$(- + +) \rightarrow (- - +); \qquad (- - +) \rightarrow (- + +).$$

In general the first two will not be on the same energy-shell as the last two; this depends on the next nearest neighbours. So the simplest possible (but not the only) choice of dynamics is to mix the first two with the matrix

$$T_{x=2} = \begin{pmatrix} 1-\lambda & \lambda \\ \lambda & 1-\lambda \end{pmatrix} \tag{15.2}$$

and to mix the last two with the same matrix. The suffix $x$ denotes the point of $\Lambda$ where the spin-flip is happening. The map is the identity on the other factors in $\Omega$, such as at $\Omega_y$, $y \neq 1, 2, 3$. We see that this map changes the numbers of spins up and down; the map conserves energy, but moves it from one bond to the next. The dynamics will therefore give rise to a diffusion of heat, with rate related to the rate of spin-flip, $\lambda$. We get a translation-invariant bistochastic map $T$ by defining $N$ similar maps $T_x$ with any $x \in \Lambda$ replacing 2, and averaging them: $T = N^{-1} \sum_x T_x$.

We now have the linear part of the dynamics. It cannot mix the energy-shells, and we must add a further randomising map to lead us a dynamics converging to the canonical state. Following our discussion, this should be a loss of information as in the stoss map or the Boltzmann map; we do not want to alter the mean energy. According to Ingarden we should do more; we should fix the level of description; in our model, we choose this to be the $2N$ variables given by the mean energy on each of the $N$ bonds, and the mean spin at each of the $N$ sites. We do not want the further randomising map to alter any of these means, as the flow of heat, and the decay of the spins, should be caused by $T$. That decides it; we define the map $Q$ on the set of states to be the replacement of a state $p$ by that state $pQ$ maximising entropy, subject to the constraints

$$pQ \cdot S_x = p \cdot S_x; \qquad pQ \cdot (S_x S_{x+1}) = p \cdot (S_x S_{x+1}). \tag{15.3}$$

The map $Q$ is by construction entropy-increasing, and conserves the local mean energy, and the particle number (if we interpret $+$ as occupied, and $-$ as unoccupied). It is very non-linear. We can nevertheless make some progress; introduce $N$ Lagrange multipliers, $\beta_x$, to go with the constraint on the energy between $x$ and $x+1$ (including between $N$ and 1) and $N$ Lagrange multipliers $\mu_x$ to go with the constraints on spin-density. The state $pQ$ therefore has the form

$$pQ = Z^{-1} \exp\{\sum_x (J\beta_x S_x S_{x+1} + \mu_x S_x)\}. \tag{15.4}$$

We can interpret $\beta_x$ as the inverse temperature at $x$, or rather, on the bond $(x, x+1)$; it is the Legendre dual to the local energy it multiplies, but the word 'intensive variable' would be a misnomer, as it is not uniform through-out the system. The variable $\mu_x$ is not exactly the chemical potential, as the latter usually appears in the state multiplied by $\beta$. Here, the betas at the two neighbouring bonds could be different; there is no beta at $x$ itself; there is no external field in our model, and so there is no energy at $x$. We must get used to the fact that the vocabulary invented for the theory at equilibrium might need some changes for non-equilibrium dynamics.

One time-step in our dynamics is $p \mapsto pTQ$. We now show that this converges to the canonical state at some overall beta, whose value depends on the mean total energy of the initial state. The map $T$, being a proper mixture of permutations, increases the entropy except at a fixed point. The constraints are linear in $p \in \Sigma$, so the concavity of the entropy ensures that $Q$ is strictly entropy-increasing. Thus, entropy is a strict Lyapunov function for the dynamics, which is a motion through states of the form in (15.4). By the principle of detailed balance, any fixed point must be a fixed point of each $T_x$. Since $T_x$ alters the mean spin at $x$ unless it is zero, any equilibrium state must have mean spin-density equal to zero. Since $T_x$ exchanges energy between the bonds $(x-1, x)$ and $(x, x+1)$, it changes the state unless the betas are all equal. So there is a unique fixed point. The state-space is compact, so, applying Lyapunov's theorem, we see that starting at any state the system converges to equilibrium, which is a canonical state.

We now find the equations of motion. We write the rate of change in the local extensive variables $S_x$ and $\mathcal{E}_x$, or rather their means, in terms of the state at the time. We write the state in terms of the local intensive variables $\beta_x$ and $\mu_x$. Thus the equations will be rather implicit. We saw a similar type of equation for chemical reactions, where the rate of change of density is given in terms of the activity. The form of the dynamics ensures that at time $t > 0$ the state is in the form (15.4) and since $Q$ does not change the mean values of the chosen extensive variables, any change must be due to $T$. To find the change in $p \cdot S_x$ it is enough to work with the marginal probabilities on $\Omega_{x-1} \times \Omega_x \times \Omega_{x+1}$. We need only consider the effect of $T_x$ alone, since the others do not flip at $x$. Let $p'$ denote the state at time $t+1$, that is, $pT$, with a similar notation for the marginal probability on the space $\Omega_{x-1} \times \Omega_x \times \Omega_{x+1}$. Let $p_{+++}$ denote the marginal probability of $p$ at the configuration $S_{x-1} = S_x = S_{x+1} = +1$ etc. The marginal maps are linear, and the map $T$ is linear. So we can work entirely in the marginal variables, and our choice of the map $T$ leads to the following change caused by the linear part of the dynamics:

$$p'_{+++} = p_{+++} , \qquad p'_{++-} = p_{++-} - \lambda p_{++-} + \lambda p_{+--} ,$$
$$p'_{+-+} = p_{+-+} , \qquad p'_{+--} = p_{+--} - \lambda p_{+--} + \lambda p_{++-} ,$$
$$p'_{-+-} = p_{-+-} , \qquad p'_{-++} = p_{-++} - \lambda p_{-++} + \lambda p_{--+} ,$$
$$p'_{---} = p_{---} , \qquad p'_{--+} = p_{--+} - \lambda p_{--+} + \lambda p_{-++} . \qquad (15.5)$$

By using this we can compute the change in

$$p \cdot S_x = p_{+++} + p_{++-} + p_{-++} + p_{-+-} - p_{+-+} - p_{+--} - p_{--+} - p_{---}$$

to be

$$\delta p \cdot S_x := p' \cdot S_x - p \cdot S_x = 2\lambda(p_{+--} - p_{++-} + p_{--+} - p_{-++}). \qquad (15.6)$$

The right-hand side is expressed easily in terms of the variables $\beta$ and $\mu$ at time $t$, but not easily in terms of $S_x$ and $\mathcal{E}_x$. The left-hand side can be expressed in terms of the $\beta$, $\mu$, $\beta'$ and $\mu'$ by finding the mean of $S_x$ in the state (15.4), and again with $\beta$ and $\mu$ replaced by $\beta'$ and $\mu'$. We get one such equation for each $x$, so we have $N$ in all. We get $N$ further equations by finding the changes in the $N$ means $p \cdot \mathcal{E}_x$ due to $T_x$ and $T_{x+1}$. For $\delta p \cdot S_x$ no other $T_y$ for $y \in \Lambda$ contributes, but for $\delta p \cdot \mathcal{E}_x$ we must include the change due to $T_{x+1}$ as well as the change due to $T_x$. We see that the change due to $T_x$ leads to

$$p' \cdot \mathcal{E}_x = J \left( -p'_{+++} + p'_{++-} + p'_{+-+} - p'_{+--} \right.$$
$$\left. -p'_{-++} + p'_{-+-} + p'_{--+} - p'_{---} \right).$$

Using (15.5) again, we find that

$$p' \cdot \mathcal{E}_x = p \cdot \mathcal{E}_x + 2\lambda J(p_{+--} - p_{++-} + p_{-++} - p_{--+}). \qquad (15.7)$$

$T_x$ also causes a change in $p \cdot \mathcal{E}_{x-1}$, which is the opposite to the change in $p \cdot \mathcal{E}_x$. Thus $T_{x+1}$ causes a change in $p \cdot \mathcal{E}_x$ given by the same expression as (15.7), but with opposite signs and where the $p_{\pm\pm\pm}$ are the marginal probabilities translated one step to the right. Both terms are to be added and divided by $N$ to give the change due to $T = N^{-1} \sum_x T_x$. When we write $p \cdot \mathcal{E}_x$ in terms of $\beta$, $\mu$ and $p' \cdot \mathcal{E}_x$ in terms of $\beta'$, $\mu'$ we get $N$ equations. Together with the $N$ equations for $\delta p \cdot S_x$ this gives us $2N$ equations for $\beta'$, $\mu'$ in terms of $\beta$ and $\mu$; then we have made one time-step. These are very difficult equations, but they do give a well-defined procedure. They simplify a bit in continuous time, and give rise to a system of $2N$ first-order differential equations in $t$ for $\beta(x,t)$ and $\mu(x,t)$, $x = 1, \ldots, N$.

In this model the transport of heat can be described as being by convection, where the carriers are the domain walls, that is, points where the direction of the spin changes. One process, the spin-flip, causes both the decay of spin-density and the transport of heat from one bond to the next. In the next section we study the rate of return to equilibrium, when near a canonical state.

## 15.3 Ising Model Close to Equilibrium

It is of interest to study the rate of return to equilibrium, when near a canonical state. For this we use the first method of Lyapunov. We illustrate this in the simplest non-trivial case, which is when $N = 3$. So the model is of a molecule composed of three spin $1/2$ atoms in a ring. We maximise entropy subject to fixing the three mean spin-values $\langle p, S_x \rangle$ and the three mean energies $\langle p, \mathcal{E}_x \rangle = -J \sum_\omega S_x(\omega) S_{x+1}(\omega)$. The answer, by the usual method of Lagrange multipliers is described by six parameters $\beta$ and $\mu$. To keep symmetry, we denote by $\beta_1$ the Lagrange multiplier associated with the bond (23), by $\beta_2$ the Lagrange multiplier associated with the bond (31), and so on cyclically. The result is (Exercise 15.111):

$$p_{+++} = Z^{-1} \exp\{\mu_1 + \mu_2 + \mu_3 + J(\beta_1 + \beta_2 + \beta_3)\}$$
$$p_{++-} = Z^{-1} \exp\{\mu_1 + \mu_2 - \mu_3 + J(-\beta_1 - \beta_2 + \beta_3)\}$$
$$p_{+-+} = Z^{-1} \exp\{\mu_1 - \mu_2 + \mu_3 + J(-\beta_1 + \beta_2 - \beta_3)\}$$
$$p_{+--} = Z^{-1} \exp\{\mu_1 - \mu_2 - \mu_3 + J(\beta_1 - \beta_2 - \beta_3)\}$$
$$p_{-++} = Z^{-1} \exp\{-\mu_1 + \mu_2 + \mu_3 + J(\beta_1 - \beta_2 - \beta_3)\}$$
$$p_{-+-} = Z^{-1} \exp\{-\mu_1 + \mu_2 - \mu_3 + J(-\beta_1 + \beta_2 - \beta_3)\}$$
$$p_{--+} = Z^{-1} \exp\{-\mu_1 - \mu_2 + \mu_3 + J(-\beta_1 - \beta_2 + \beta_3)\}$$
$$p_{---} = Z^{-1} \exp\{-\mu_1 - \mu_2 - \mu_3 + J(\beta_1 + \beta_2 + \beta_3)\}. \tag{15.8}$$

Here $Z$ is not a seventh parameter, since it is fixed as the sum of the exponentials on the right. Since the general state for the three-site model needs seven parameters, and here we have six, there is not much reduction in the level of description, using Ingarden's term. This is not altogether an accurate way to say what we have done, since the remaining parameter needed for a full description of the state is still in the model, and can be predicted.

The dynamics is given by $p \mapsto p' = p(T_1 + T_2 + T_3)Q/3$, where $T_x$ is as in (15.2). It is easy to see that the state, with $\mu = 0$ and all betas equal, is a fixed point of the map. So the Lyapunov theory (first method) will set each $\mu_x$ to be a small quantity, and each $\beta_x$ very close to $\beta$; then we study the linearised dynamics about the fixed point. One type of small perturbation, in which $\mu_x$ remains zero, and all $\beta_x$ move together from $\beta$, and remain equal, will not return to the original state, since we have moved to a new canonical state with a slightly different $\beta$. This will show up in the analysis: one of the eigenvalues of the linearised equation of motion will be zero. The details appeared in [Connaughton (1998)].

Let us now find the equation of motion of $s_x = \rho(S_x)$. It is enough to do one, say $s_2$. We saw that

$$\delta s_2 = 2\lambda(p_{+--} - p_{++-} + p_{--+} - p_{-++}).  \tag{15.9}$$

On substituting (15.8) and keeping only first-order terms in $\mu_x$ and $\beta - \beta_x$, we get (Exercise 15.112)

$$\delta s_2 = -8\lambda_0 Z^{-1}(\beta)e^{-J\beta}\mu_2 ,  \tag{15.10}$$

where $\lambda_0 = \lambda/3$ to take into account $T = \sum T_x/3$. (15.10) shows the typical Onsager behaviour: the variable $s_2$ is driven by the dual variable $\mu_2$; in this case, $s_2$ is driven back to zero by any non-zero $\mu_2$, taking it for granted that the signs of the two are always the same. In fact, in our model the relation between the intensive and extensive variables is determined, and in the linear approximation, can be found. This is how. First, in this approximation,

$$Z = 2e^{3\beta J} + 6e^{-\beta J};  \qquad \text{see Exercise 15.113.}$$

Then, to first order

$$s_1 = p_{+++} + p_{++-} + p_{+-+} + p_{+--} - p_{-++}$$
$$-p_{-+-} - p_{--+} - p_{---}$$
$$\approx 2Z^{-1}\{e^{3\beta J}(\mu_1 + \mu_2 + \mu_3) + e^{-\beta J}(3\mu_1 - \mu_3 - \mu_2)\}  \tag{15.11}$$

and similarly

$$s_2 \approx 2Z^{-1}\{e^{3\beta J}(\mu_1 + \mu_2 + \mu_3) + e^{-\beta J}(3\mu_2 - \mu_1 - \mu_3)\}  \tag{15.12}$$
$$s_3 \approx 2Z^{-1}\{e^{3\beta J}(\mu_1 + \mu_2 + \mu_3) + e^{-\beta J}(3\mu_3 - \mu_2 - \mu_1)\}.  \tag{15.13}$$

These can be solved for $\mu$, and the solution, for $\mu_2$ say, put in the form

$$\mu_2 = e^{\beta J}\kappa Z/8(-s_1 + 2s_2 - s_3) + \frac{Z}{2(3e^{3\beta J} + e^{-\beta J})}s_2,  \tag{15.14}$$

where by Exercise 15.114,

$$\kappa = \frac{e^{4\beta J} - 1}{3e^{4\beta J} + 1} > 0 \quad \text{if} \quad \beta > 0.$$

This gives us the equation of motion for $s_2$, from (15.10),

$$\delta s_2 = \lambda_0 \kappa \{(s_3 - s_2) - (s_2 - s_1)\} - \frac{4\lambda_0}{3e^{4\beta J} + 1} s_2. \qquad (15.15)$$

This dynamics has two parts, which are both determined by the single rate parameter $\lambda$ and the beta of the fixed point. The first represents diffusion of spins, with a beta-dependent rate determined by the theory in terms of the hopping parameter $\lambda_0$; the expression $s_1 - 2s_2 + s_3$ is the finite-difference approximation to the second spatial derivative of the spin field; in the continuum limit it gives the diffusion equation. This term conserves the total spin number. In addition, there is a decay term, with rate

$$4\lambda_0(3e^{4\beta J} + 1)^{-1}$$

converging to $\lambda$ as $\beta \to 0$, and vanishing as $\beta \to \infty$. Physically, at $\beta = \infty$ all spins become aligned, and spin-flips become rare. The equation of motion for the energy-density, $\delta \mathcal{E}_1$, can be written as an equation for the rate of change of the beta; up to first order, this gives (in the limit as the latttice size goes to zero) the heat equation for the temperature [Connaughton (1998)], in which the conductivity depends on temperature.

With more work, similar models in higher dimensions, and with more spin-sites, can be formulated.

## 15.4 Non-linear Heisenberg Model

We noted that in the Ising model we had to introduce the stochastic matrix $T$, and its parameter $\lambda$, by hand; the size of $\lambda$ governs all irreversible effects, such as decay and diffusion. In the analogous quantum model, the Heisenberg model, there is a Hamiltonian, which determines the conservative dynamics. Krylov argued that the rate of dissipation should be related to the dynamics, that is, how mixing it is. We do not completely agree; the dissipation might be related to the mixing properties of the dynamics of the fast variables, exactly the part omitted from the description. But it would be nice to keep the arbitrary addition of dissipation to a minimum. We now try to follow the dictum, that dissipation is loss of information, never of energy. The Hamiltonian should, therefore, be the full microscopic Hamiltonian of the system. It is often said that, in a reduced description, in which the fast variables are eliminated from the algebra, the entropy of the remaining, slow, variables does not include the entropy of the fast variables; it is therefore not equal to the von Neumann entropy [Lebowitz and Maes (2003)]. However, we saw in Chap. 14 that this is not true. In fact, von Neumann entropy is a property of the *state*, and is not one of the observables of the theory. In a large system, the information gets dispersed into inaccessible aspects of the state, such as high-level correlation functions between points that are far apart. To detect this information requires the simultaneous measurement of properties at two of these points, and this is denied to the observer. The true state and the approximate state give the same expectations of all the observables being measured, and in a thermodynamic system, this property is preserved under the approximate time evolution. The von Neumann entropy of the exact dynamics is conserved, whereas that of the approximate dynamics increases. It is the proof of the latter point that is very hard; we did it for a special model in Chap. 14, following the paper [De Roeck, Jacobs, Maes and Netočný (2003)].

We now show how dissipation can be introduced by the technique of higher-order temperatures, in the spirit of Ingarden. There is still some freedom, in the selection of which constraints are chosen to be preserved by the map maximising the entropy. That is, we are free to choose the level of description. In this way, a given microscopic theory can have many dissipative versions. In none of them is the Hamiltonian altered, or any degrees of freedom lost. Let us illustrate this idea with the Heisenberg ferromagnet. The space $\Lambda$ is taken to be finite in this section, and at each site $x \in \Lambda$ we have the spin algebra $\mathbf{M}_2$. Define $\mathcal{A} = \bigotimes_x \mathbf{M}_2(x)$, and choose the Hamiltonian $H = -\sum J\mathbf{S}_x.\mathbf{S}_y$, where the sum is over neighbours, and the vector $\mathbf{S} = \hbar\boldsymbol{\sigma}/2$, where $\boldsymbol{\sigma}$ is the triplet of Pauli matrices. The state will be denoted by the density matrix $\rho$. The energy on each bond will be regarded as within the level of description; let us also include the three components of the spin on each site within the level of description, but no other spin correlations. Clearly, this is our choice, and, as emphasised by Ingarden, is part of the model. The fewer variables that we include, the more randomness we introduce in one time-step. We have already shown that the Hamiltonian defines a one-parameter group of automorphisms $\tau_t$ of $\mathcal{A}$. Thus the reversible time-evolution is the map $A \mapsto e^{iHt}Ae^{-iHt} = \tau_t A$. We must decide on the size of the time-step $t_0$, the choice of which is part of the model; in fact, only the product $Jt_0$ enters, so we have not really introduced another parameter. Let us define the map $Q$ on the set of states of $\mathcal{A}$ by the Jaynes-Ingarden method: $\rho Q$ is the (unique) state determined by $\rho$ by the rule that its entropy is the largest among all states with the same expectation values of all the chosen operators: thus

$$\rho Q\left(\mathbf{S_x}.\mathbf{S_y}\right) = \rho\left(\mathbf{S_x}.\mathbf{S_y}\right)$$
$$\rho Q\left(\mathbf{S_x}\right) = \rho\left(\mathbf{S}_x\right) \qquad \text{for all } x \in \Lambda.$$

We call $Q$ the thermalising map. Then we define one step in the dynamics by $\rho \mapsto \rho\tau$, where

$$\rho\tau(A) = \rho Q(\tau_{t_0} A).$$

By construction, this obeys both laws of thermodynamics. The states $\rho$ on any orbit are all defined by density matrices of the form

$$\rho = Z^{-1} \exp\left\{ -\sum_{x,y} \beta(x,y) J \mathbf{S}_x.\mathbf{S}_y + \sum_x \boldsymbol{\mu}(x).\mathbf{S}_x \right\}$$

where the double sum is over neighbours. This has the form of a canonical state with space-dependent coupling and an external magnetic field, but it is not interpreted as an equilibrium state. The equations of motion of the local intensive variables are very implicit. We take the limit as $t_0 \to 0$ to simplify the formulae. Let $X$ be an operator in the set defining the level of description; for example, $X = S_x^{(i)}$ or $X = \mathbf{S}_x.\mathbf{S}_y$, where $x$, $y$ are nearest neighbours. The time-evolution of the mean of $X$ then obeys the equation of motion

$$\hbar\frac{d}{dt}\rho_{\beta,\boldsymbol{\mu}} \cdot X = i\rho_{\beta,\boldsymbol{\mu}} \cdot [H, X]. \qquad (15.16)$$

In this equation, $\beta$ and $\boldsymbol{\mu}$ stand for the collection of local intensive variables, which are functions of time. We get one such equation for each $X$, exactly the same as the number of variables $\beta$, $\boldsymbol{\mu}$. The left-hand side of the equation can be written in terms of the betas and mus and their time derivatives by writing

$$\rho_{\beta,\boldsymbol{\mu}} \cdot X = Tr\left( Z^{-1} \exp\left\{ -J\sum_{x,y} \beta(x,y)\mathbf{S}_x.\mathbf{S}_y + \sum_x \boldsymbol{\mu}(x).\mathbf{S}_x \right\} X \right).$$
$$(15.17)$$

The right-hand side can similarly be written in terms of $\beta$ and $\mu$, and no time derivatives of these variables appear on the right. In this way, we get a family of first-order differential equations for the local intensive variables. The left-hand side is linear in the time-derivatives. This defines the model we shall call the dissipative Heisenberg spin model. We note that the means of the three components of the total spin are not changed by the dynamics, because of the rotation-invariance of the Hamiltonian. If we impose periodic conditions, then the total momentum is conserved in mean too. Thus the dynamics does not lead to the canonical state at large times, but to a grand canonical state; there is not enough mixing by the (very symmetric) Hamiltonian to reach the canonical state.

Let us look at the general features of the model. The thermalising map $Q$ does not alter the mean values of any of the relevant variables, and the changes in one time-step, which constitute the visible movement towards equilibrium, are caused by the microscopic dynamics $\tau(t)$. Krylov would have liked that. The dynamics is given by a smooth but very non-linear flow through the states of the quantum system, and no energy or entropy of microscopic variables are omitted by the thermalising map. The von Neumann entropy can therefore be safely interpreted as the physical entropy, if we multiply by Boltzmann's constant. There is a reduced description in that the number of parameters needed to describe the dynamics is much less than that needed to describe a general density matrix. For example, if there are $N$ sites, and $N$ bonds, as in one dimension with periodic boundary conditions, then we need $3N$ components for $\mu$ and $N$ components for $\beta$, compared with the number of parameters, $2^{2N} - 1$, needed to describe a density matrix on $\mathbf{M}_{2^N}$. Both laws of thermodynamics hold, and the final destination of the dynamics depends on the initial state and also on the mixing properties of the Hamiltonian. The rate of dissipation is partly determined by the Hamiltonian, but can be changed by altering the level of description. For example, if only the Hamiltonian $H$ is chosen for the set defining the level of description, then the system reaches equilibrium in one time-step, however short. If every operator is chosen, then no dissipation can occur, and the dynamics is reversible, given by the Hamiltonian.

The theory is easily generalised to other choices of $\mathcal{A}$ and $H$. For example, in chemical kinetics with $n - 1$ hard-core atoms, we choose $\mathcal{A}_x = \mathbf{M}_n$. We can cause reactions to occur by choosing a Hamiltonian to include polynomials in the creators and annihilators which allow the desired transitions between occupied and unoccupied neighbouring sites. The level of description should then be chosen to include the local energy of the bonds, as well as the occupation numbers of the atoms at each site.

We can now return to the question, what are the many-time correlations in a theory based on the use of Jaynes states? Suppose that one time-step is the map $\rho(t) \mapsto \rho(t)TQ$, where $T$ is bistochastic, and $Q$ replaces a state by the appropriate Jaynes state. We have a flow through the states, but it is non-linear and cannot be thrown onto the algebra simply by duality. It gives us an increasing von Neumann entropy, and the correct change of the means of the 'slow' variables.

In his appendix to [Krylov (1979)] Sinai remarks that Krylov's call for a model with dynamical mixing led Kolmogorov (and later Sinai himself) to ask for tests that distinguish the concept of random as opposed to deterministic sequences. There are deterministic sequences that for practical purposes cannot be shown not to be random. This led eventually to the theory of chaos, which studies the case in which the dependence of the state $\rho(t)$ at time $t$ on the initial state $\rho(0)$ at time 0 is so sensitive to the detail of $\rho(0)$ that we can do just as well by replacing $\rho(t)$ with a mixed state, such as that of the Jaynes-Ingarden theory. The ergodic programme becomes justified if there are two time-scales in the problem; the variables very sensitive to the initial conditions must be treated statistically, and no detectable error is introduced thereby if the dynamics were at this scale not distinguishable from random. Further, when we have a theory with local structure, as in the Heisenberg model, then the reversible time-evolution of the correlations between remote parts becomes a very intricate problem, whose *complexity* increases rapidly as the distance between the points increases. These correlations cease to lie within the possible level of description. We are therefore led to Ingarden's formulation, because these correlations cannot be *calculated*. They can be *measured* and, according to statistical dynamics, will be found to be well described by the state $\rho Q$, at least after a long enough time and if they are far enough apart. To justify these claims takes us into the traditional (hard) subjects of ergodic theory and the thermodynamic limit, a small part of which was given in Chap. 14.

## 15.5   Estimation; the Cramér-Rao Inequality

Given two random variables $X$ and $Y$, we may ask whether they are related by a linear relation such as $X = aY + Z$, where $a$ is a constant and $Z$ is a small random error. We thus measure $N$ samples of $\omega$, and plot the found values $X(\omega_i), Y(\omega_i)$ on a graph. If the points lie nearly on a straight line, then we may assume that the answer to our question was yes. We are then left with finding the best value of the parameter $a$. One answer is to choose $a$ to minimise the variance of $Z = X - aY$. Now,

$$\mathcal{V}(X - aY) = \mathcal{V}(X) - 2a\mathcal{C}(X, Y) + a^2\mathcal{V}(Y),$$

and the 'best' value of $a$ minimises this, and so is equal to $\mathcal{C}(X,Y)/\mathcal{V}(Y)$. We can estimate this from the data, and this gives us the method of least squares, used since Gauss. The method has the following problem: the answer is dependent on the coordinates used to evaluate it. For, it is clear that we would not get the same answer by finding the best value of $a^{-1}$ such that $Y - a^{-1}X$ has the smallest variance, since the answer to this version is that $a^{-1} = \mathcal{C}(X, Y)/\mathcal{V}(X)$, which gives a different 'best value' for $a$. The first is called the regression of $Y$ on $X$, and the second is called the regression of $X$ on $Y$. One again gets different answers for $a$ by minimising the variance of other choices such as the variance of $a^{-\frac{1}{2}}X - a^{\frac{1}{2}}Y$. We need a measure of the difference between two random variables that does not depend on the choice of coordinates used to evaluate it. This was solved by Fisher, and is called the Fisher metric [Fisher (1925)].

Let $\{\rho_\eta(x)\}$ be a family $\mathcal{M}$ of probability densities, depending on a parameters $\eta \in \mathbf{R}$. We associate with $\rho \in \mathcal{M}$ the random variable (the 'score') $Y = \partial \log \rho_\eta / \partial \eta$, which has mean zero relative to $\rho$. Let $X$ be a random variable whose distribution $\rho$ is believed or hoped to lie in the manifold $\mathcal{M} = \{\rho_\eta\}$. We estimate the value of $\eta$ by measuring $X$ independently $m$ times, getting the data $x_1, \ldots, x_m$. An *estimator* $f$ is a function of $(x_1, \ldots, x_m)$ that is used for this estimate. So $f$ is a function of $m$ independent copies of $X$, and so is a random variable on the product sample space. To be useful, the estimator must be independent of $\eta$, which we do not (yet) know. We say that an estimator is *unbiased* if its mean is the desired parameter; it is usual to take $f$ as a function of $X$ and to regard $f(x_i)$, $i = 1, \ldots, m$ as samples of $f$. Then the condition that $f$ is unbiased becomes

$$\rho_\eta \cdot f := \int \rho_\eta(x) f_i(x) dx = \eta. \tag{15.18}$$

We use the notation $\rho \cdot f$ for the expectation of $f$ in the state $\rho$. A good estimator should also have only a small chance of being far from the correct value, which is its mean if it is unbiased. This chance is measured by the variance, say $V = \mathcal{V}(f)$. The *Fisher information* [Fisher (1925)] is defined as follows.

$$G := \int \rho_\eta(x) \left( \frac{\partial \log \rho_\eta(x)}{\partial \eta} \right)^2 dx. \tag{15.19}$$

This expression is of the same form whether we use the variable $x$, or another variable $y = y(x)$ related by a smooth invertible mapping, $y$. For, the density in the variable $y$ is $\rho_1(y) = \rho(x(y))\frac{dx}{dy}$, and so $\log \rho_1(y) = \log(\rho(x(y))) + \log \frac{dx}{dy}$. Also, as $\frac{\partial y}{\partial x}$ never vanishes, keeping $y$ constant is the same as keeping $x$ constant. Also, $\log \frac{dx}{dy}$ is independent of $\eta$, so its derivative relative to $\eta$ is zero. It follows that $\frac{\partial \log \rho_1(y)}{d\eta}|_{y=\text{const}} = \frac{\partial \log \rho(x)}{\partial \eta}|_{x=\text{const}}$.

Then we get

$$\int \rho_1(x(y)) \left( \frac{\partial \log \rho_1(y)}{\partial \eta} \Big|_{y=c} \right)^2 dy$$

$$= \int \rho(x(y)) \frac{dx}{dy} \left( \frac{\partial \log \rho(x)}{\partial \eta} \Big|_{x=c} \right)^2 dy$$

$$= G, \tag{15.20}$$

which proves the result.

We note that Eq. (15.19) gives the variance of the random variable $Y = \partial \log \rho_\eta / \partial \eta$, which has mean zero. $G$ is associated with the family $\mathcal{M} = \{\rho_\eta\}$; then Fisher showed that $V \geq G^{-1}$. For the proof, differentiate Eq. (15.18) with respect to $\eta$, to get

$$\int \frac{\partial \rho_\eta(x)}{\partial \eta} f(x) dx = 1, \tag{15.21}$$

which can be written

$$\int \left( \frac{\partial \log \rho}{\partial \eta} \right) (f(x) - \eta) \rho_\eta(x) dx = 1. \tag{15.22}$$

We note that since $Y := \frac{\partial \log \rho}{\partial \eta}$ has mean zero, Eq. (15.22) states that the correlation between $Y$ and $f$ is 1, so their covariance matrix becomes

$$\begin{pmatrix} G & 1 \\ 1 & V \end{pmatrix}.$$

This is positive definite, giving us Fisher's result, $V \geq G^{-1}$. This puts a limit on the reliability with which we can estimate $\eta$. Fisher termed $V/G^{-1}$ the *efficiency* of the estimator $f$. Equality in the Schwarz inequality occurs if and only if the two functions are proportional. Let $-\partial \xi / \partial \eta$ denote this factor when we have equality, $V = G^{-1}$. Then the optimal estimator occurs when $Y = -\partial \xi / \partial \eta (f - \eta)$, which gives

$$\log \rho_\eta(x) = - \int \frac{\partial \xi}{\partial \eta} (f(x) - \eta) d\eta. \tag{15.23}$$

Doing the integral, and adjusting the constant of integration to normalise the state, leads to

$$\rho_\eta(x) = Z^{-1} \exp\{-\xi f(x)\} \tag{15.24}$$

which is the 'exponential family'.

To estimate a finite number of parameters, say $\{\eta_i\}$, Fisher introduced the *information matrix*

$$G^{ij} = \int \rho_\eta(x) \frac{\partial \log \rho_\eta(x)}{\partial \eta_i} \frac{\partial \log \rho_\eta(x)}{\partial \eta_j} dx. \tag{15.25}$$

We note that $Y_i := \partial \log \rho / \partial \eta_j$ is a vector of random variables of zero mean, and that $G^{ij}$ is its covariance matrix. Again, this expression is unchanged if we used any smooth invertible function $y(x)$ instead of $x$ in the formula. This is also true if the probability space is also multi-dimensional, so that $dx = dx_1 \ldots dx_m$. Here $m$ need not be equal to $n$. We can then change the coordinates to $y_1, \ldots, y_m$, smooth invertible coordinates as functions of $x_1, \ldots, x_m$, and the expression for the Fisher matrix is the same. The proof is the same as in the one-dimensional case, except that we need to use the Jacobian condition

$$dy := dy_1 \ldots dy_m = \det\left(\frac{dy_i}{dx_i}\right) dx_1 \ldots dx_m. \qquad (15.26)$$

Fisher stated in [Fisher (1925)], and Rao [Rao (1945)] and Cramér proved [Cramér (1999)], that the variance $V$ of an unbiased estimator $f$ obeys the matrix inequality $V \geq G^{-1}$. This inequality expresses that, given the family $\rho_\eta$, there is a limit to the reliability with which we can estimate $\eta$. Again, using the above method, the best estimate is given by the exponential family,

$$\rho_\eta(x) = Z^{-1} \exp\left\{-\sum_j \xi^j f_j(x)\right\}. \qquad (15.27)$$

We now prove the Cramér-Rao inequality. Put $V_{ij} = \rho_\eta.[(f_i - \eta_i)(f_j - \eta_j)]$, the covariance matrix of $\{f_i\}$. Differentiate the condition for being unbiased, (15.18) with respect to $\eta_j$, and rearrange, to get

$$\int \rho_\eta(x) Y^i(x)(f_j(x) - \eta_j)\, dx = \delta_{ij}. \qquad (15.28)$$

This is the correlation between $Y^i$ and $f_j$. The covariance matrix of the $2n$ random variables $Y^i, f_j$ therefore is

$$\begin{pmatrix} G & I \\ I & V \end{pmatrix}. \qquad (15.29)$$

This is therefore a positive semi-definite matrix. If it is not definite, it has zero as an eigenvalue, which leads to $GV = I$, and the manifold must be the exponential family, as before. If it is definite, so is its inverse, which is found to be

$$\begin{pmatrix} (G - V^{-1})^{-1} & -G^{-1}(V - G^{-1})^{-1} \\ -V^{-1}(G - V^{-1})^{-1} & (V - G^{-1})^{-1} \end{pmatrix}. \qquad (15.30)$$

Thus the leading submatrices $(G - V^{-1})^{-1}$ and $(V - G^{-1})^{-1}$ are positive definite, and so are their inverses, giving the matrix inequality $V \geq G^{-1}$.

We can justify Jaynes's method as follows. We optimise the Cramér-Rao bound by choosing the two operators $V$ and $G$ to be mutual inverses; this holds when the state lies in the exponential family; this is the only class of states with this property; the means of the variables $f_j$ are used to find values of the variables $\eta_j$, so the Jaynes states are the ones that give us estimators of 100% efficiency.

## 15.6 Efron, Dawid and Amari

Recall that an affine map, $U$ (acting on the right) from one vector space $T_1$ to another, $T_2$, is one that obeys

$$(\lambda X U + (1 - \lambda)Y U) = \lambda X U + (1 - \lambda)Y U, \qquad (15.31)$$

for all $X, Y \in T_1$ and all $\lambda \in [0, 1]$.

The same definition works on an *affine space*, that is, a convex subset of a vector space. This leads to the concept of an affine connection.

Let $\mathcal{M}$ be a manifold and denote by $T_\rho$ the tangent space at $\rho \in \mathcal{M}$. Consider an affine map $U_\gamma(\rho, \sigma) : T_\rho \to T_\sigma$ defined for each pair of points $\rho, \sigma$ and each continuous path $\gamma$ in the manifold starting at $\rho$ and ending at $\sigma$. Let $\rho, \sigma$ and $\tau$ be any three points and $\gamma_1$ a path from $\rho$ to $\sigma$, and $\gamma_2$ any path from $\sigma$ to $\tau$.

**Definition 15.1.** *We say that $U$ is an* affine connection, *if $U_\emptyset = Id$ and*

$$U_{\gamma_1 \cup \gamma_2} = U_{\gamma_1} \circ U_{\gamma_2}. \qquad (15.32)$$

Let $X$ be a tangent vector at $\rho$; we call $X U_{\gamma_1}$ the parallel transport of $X$ to $\sigma$, along the path $\gamma_1$.

The differential of $U$ along a specified direction is called the covariant derivative.

$$\nabla_Y X := d/dt \, X U_\gamma(\rho, \gamma(t))|_{t=0} \qquad (15.33)$$

where $\{\gamma(t)\}$, $0 \le t \le 1$ is any path from $\rho$ to $\sigma$, which starts at $\rho$ in the direction $Y \in T_\rho$. Conversely, a covariant derivative defines a connection. This concept allows us to specify that two tangent vectors to the manifold at points $\rho$ and $\sigma$ are *parallel* if the parallel transport (along a specified curve) of one from $\rho$ to $\sigma$ is proportional to the other. A *geodesic* is a self-parallel curve on $\mathcal{M}$. A given metric $g$ defines a special connection (that of

Levi-Civita), and its geodesics are lines of minimal length, as measured by
the metric.

Estimation theory might be considered geometrically as follows. For
some reason, we expect the distribution of a random variable to lie on a
submanifold $\mathcal{M}_0 \subseteq \mathcal{M}$ of states. The data give us a histogram, which is
a distribution, but not a pretty one. We seek the point on $\mathcal{M}_0$ that is
'closest' to the data. Suppose that the sample space is $\Omega$, with $|\Omega| < \infty$.
Let us place all positive distributions, including the experimental one, in a
common manifold, $\mathcal{M}$. This manifold will have the Riemannian structure,
$G$, provided by the Fisher metric. We then draw the geodesic curve through
the data point that has shortest distance to the sub-manifold $\mathcal{M}_0$; it cuts
$\mathcal{M}_0$ at a point in $\mathcal{M}$, which is our estimate for the state. This procedure,
however, does not always lead to unbiased estimators. Efron [Efron (1975)]
and Dawid [Dawid (1975)] noticed that the Levi-Civita connection is not
the only useful one. First, the ordinary mixtures of densities $\rho_1, \rho_2$ leads to

$$\rho = \lambda \rho_1 + (1 - \lambda)\rho_2, \qquad 0 < \lambda < 1. \tag{15.34}$$

Done locally, this leads to a connection on the manifold, now called the
$(-1)$-Amari connection: two tangents are parallel if they are proportional
as functions on the sample space. This differs from the parallelism given by
the Levi-Civita connection. We need to use $(-1)$-geodesics to give unbiased
estimates for $f$.

There is another obvious convex structure, that obtained from the linear
structure of the space of centred random variables, (the scores). Take $\rho_0 \in \mathcal{M}$ and write $f_0 = -\log \rho_0$. Consider a perturbation $\rho_X$ of $\rho_0$, which we
write as

$$\rho_X = Z_X^{-1} e^{-f_0 - X}. \tag{15.35}$$

The random variable $X$ is not uniquely defined by $\rho_X$, since by adding
a constant to $X$, we can adjust the partition function to give the same
$\rho_X$. Among all these equivalent $X$ we can choose the *score* which has zero
expectation in the state $\rho_0$: $\rho_0 \cdot X = 0$. We can define a sort of mixture of
two such perturbed states, $\rho_X$ and $\rho_Y$ by

$$`\lambda \rho_X + (1 - \lambda)\rho_Y` := \rho_{\lambda X + (1-\lambda)Y}. \tag{15.36}$$

This is a convex structure on the space of states, and differs from that given
in Eq. (15.34). It leads to an affine connection, now called the $(+1)$-Amari
connection. How do these connections relate to the metric?

**Definition 15.2.** Let $G$ be a Riemannian metric on the manifold $\mathcal{M}$. A connection $\gamma \mapsto U_\gamma$ is called a *metric connection* if

$$G_\sigma(XU_\gamma, YU_\gamma) = G_\rho(X, Y) \tag{15.37}$$

for all tangent vectors $X, Y$ and all paths $\gamma$ from $\rho$ to $\sigma$.

The Levi-Civita connection is a metric connection, but the ($\pm$) Amari connections are not; they are, however, dual relative to the Rao-Fisher metric; let $\gamma$ be a path connecting $\rho$ with $\sigma$; then for all $X, Y$:

$$G_\sigma(XU^+(\rho, \sigma), YU^-(\rho, \sigma)) = G_\rho(X, Y). \tag{15.38}$$

Let $\nabla^\pm$ be the two covariant derivatives obtained from the connections $U^\pm$. Amari defines intermediate covariant derivatives

$$\nabla^\alpha = \frac{1}{2}(1 + \alpha)\nabla^+ + \frac{1}{2}(1 - \alpha)\nabla^-. \tag{15.39}$$

These uniquely define connections, $U^{(\alpha)}$, whose dual relative to $G$ is $U^{(-\alpha)}$. The Levi-Civita covariant derivative is the case $\alpha = 0$, which is self-dual and therefore metric, as is known. Amari shows that $\nabla^{(\pm)}$ define flat connections without torsion. Flat means that the transport is independent of the path, and 'no torsion' means that $U$ takes the origin of $T_\rho$ to the origin of $T_\rho$ around any loop: it is linear. In that case there are affine coordinates, that is, global coordinates in which the respective convex structure is obtained by simply mixing coordinates linearly. Amari shows that for $\alpha \neq \pm 1$, $\nabla^\alpha$ is not flat, but that the manifold is a sphere in the Banach space $\ell^p$, $1/p = -\alpha/2 + 1/2$. In particular, the case $\alpha = 0$ leads to the unit sphere in the Hilbert space $L^2$, and the Levi-Civita parallel transport is vector translation in this space, followed by projection back onto the sphere. The resulting affine connection is not flat, because the sphere is not flat.

The metric distance between measures is the Hellinger distance, and the natural coordinates are the square-roots of the densities, imitating the wave-functions of quantum mechanics. Similar results were obtained in infinite dimensions in [Gibilisco and Pistone (1998)].

In estimation theory, the method of maximum entropy for unbiased estimators makes use of the $\nabla^-$ connection. This is true also in the dynamics of neural nets, dense liquids, Onsager theory, Brownian particles in a potential and the Soret and Dufour effects; the micro-state after a small time is replaced by a macrostate, which is the same as the max-entropy estimation of the state by one on the manifold generated by exponentials of the macrovariables (or, slow variables). The (intractible) microdynamics is continuously projected in a rolling construction onto the (easier) manifold of exponential states. This idea was proposed by Kossakowski, and Ingarden. The resulting non-linear dynamics can be described thus: after each time-step of the linear dynamics of the system, Nature makes the best estimate of the state among those lying on the manifold.

## 15.7   Entropy Methods, Exponential Families

Gibbs knew that the state of maximum entropy, given the mean energy, is the canonical state. More generally, let $\Omega$ be a countable sample space, and let $\Sigma$ denote the set of probabilities (or *states*) $\rho$ on $\Omega$. Let $1, f_1, \ldots, f_n$ be linearly independent random variables, whose means we can measure. We want to find the 'best' choice for the state, given these means. According to Jaynes, the least prejudiced choice of $\rho$ is to maximise the entropy $S$ of the state subject to the $n+1$ constraints given by the normalisation and the given values of the means, $\eta_j := \rho \cdot f_j$ of $f_j, j = 1, \ldots, n$. We assume that the given values are possible values of the means, strictly inside the set of possible means.

**Theorem 15.3.** *The state of maximum entropy, given the values of the means, is unique and is given by an expression of the form*

$$\rho(\omega) = Z^{-1} \exp\left\{ -\sum_j \xi^j f_j(\omega) \right\} \tag{15.40}$$

*where*

$$Z = \sum_\omega \exp\left\{ -\sum_{j=1}^n \xi^j f_j(\omega) \right\}.$$

Note: These make up the *exponential manifold* $\mathcal{M}$ determined by $\mathcal{F} := \text{Span}\{f_1, \ldots, f_n\}$ and is parametrised by the $n$ Lagrange multipliers $\xi^1, \ldots, \xi^n$. The parameters $\xi^j$ are called the canonical coordinates on $\mathcal{M}$.

**Proof.**   The constrained maximum of $S$ is the same as the maximum of

$$S(\rho) - \xi^0 \sum_\omega \rho(\omega) - \sum_j \xi^j \sum_\omega \rho(\omega) f_j(\omega)$$

subject varying $\rho(\omega)$ with no constraints, by Lagrange's theorem. Since $S$ is strictly concave, this maximum is unique. The $\xi^j, j = 0, \ldots, n$ are to be determined by normalisation and the means.

We vary the variables $\rho(\omega)$:

$$0 = -\frac{\partial \sum_{\omega'} \rho(\omega') \log \rho(\omega')}{\partial \rho(\omega)} - \xi^0 - \sum_{j=1}^{n} \xi^j f_j(\omega)$$

$$= -\log \rho(\omega) - 1 - \xi^0 + \sum_{j} - \sum_{j=1}^{n} \xi^j f_j(\omega).$$

Thus $\rho_\xi(\omega) = \text{const.} \exp\left\{-\sum_{j=1}^{n} \xi^j f_j(\omega)\right\}$, which is in the exponential family. $\qquad\square$

The $\xi^j$ are determined by the given expectation values by the conditions $\rho_\xi \cdot f_j = \eta_j, j = 1, \ldots, n$. Thus the $\eta_j$ are also coordinates for the exponential manifold (the *mixture* coordinates). It is easy to show (see Exercise 15.115) that

$$\eta_j = -\frac{\partial \Psi}{\partial \xi^j}, \quad j = 1, \ldots, n; \quad V_{jk} = -\frac{\partial \eta_j}{\partial \xi^k}, \quad j, k = 1, \ldots, n, \qquad (15.41)$$

where $\Psi := \log Z$, and that $\Psi$ is a convex function of $\xi^j$. The Legendre dual to $\Psi$ is

$$\sup_{\xi} \left\{ \sum \xi^i \eta_i - \Psi(\xi) \right\},$$

and this is the entropy $S = -\rho \cdot \log \rho$. The dual relations are

$$\xi^j = \frac{\partial S}{\partial \eta_j} \qquad G^{jk} = -\frac{\partial \xi^j}{\partial \eta_k}. \qquad (15.42)$$

By the rule for Jacobians, $V$ and $G$ are mutual inverses, which shows that the Jaynes method does give the estimators their minimum possible variance: they are the estimators of 100% efficiency. We shall obtain a similar result in the case of quantum mechanics as well.

## 15.8 The Work of Pistone and Sempi

The notable work in [Pistone and Sempi (1995)] arises as a generalisation to infinitely many parameters of the theory of the best estimation of parameters of a probability distribution, using the data obtained by sampling. It is also sometimes called 'non-parametric estimation'. In 1995, Pistone and Sempi obtained a useful formalism, making use of an Orlicz space [Krasnoselski and Ruticki (1961); Rao and Ren (1992)]. From the point of view of quantum mechanics, the classical case corresponds to the special case where all observables generate an abelian algebra. We start with a brief review of the classical case.

Pistone and Sempi developed a theory of best estimators (of minimum variance) among all locally unbiased estimators, in classical statistical theory. Thus, there is a sample space, $\mathcal{X}$, and a given $\sigma$-ring $\mathcal{B}$ of subsets of $\mathcal{X}$, the measurable sets, representing the possible events. On $\mathcal{X}$ is given a positive measure $\mu$, which is used to specify the sets of zero measure, that is, the impossible events. It may not be true that $\mu$ is normalised; it is not required to be a probability. The probabilities on $\mathcal{X}$, which represent the possible states of the system, are positive, normalised measures $\nu$ on $\mathcal{X}$ that are equivalent to $\mu$. By the Radon-Nikodym theorem, we may write

$$d\nu = f d\mu$$

where $f(x) > 0$ $\mu$-almost everywhere and

$$\mathbf{E}_{d\mu}[f] := \int_{\mathcal{X}} f(x)\mu(dx) = 1.$$

Let $f_0$ be such a density. Pistone and Sempi sought a family of sets $N$ containing $f_0$, and which can be taken to define the neighbourhoods of the state defined by $f_0$. They then did the same for each point of $N$, and so on, thus constructing a topological space which they showed had the structure of a Banach manifold. Their construction was the following. Let $u$ be a random variable on $(\mathcal{X}, \mathcal{B})$, and consider the class of measures whose $f$ has the form

$$f = f_0 \exp\{u - \psi_{f_0}(u)\} \tag{15.43}$$

in which $\psi$, called the free energy, is finite for all states of a one-parameter exponential family:

$$\psi_{f_0}(\lambda u) := \log \mathbf{E}_{f_0 d\mu}[e^{-\lambda u}] < \infty \text{ for all } \lambda \in [-\epsilon, \epsilon]. \tag{15.44}$$

Here, $\epsilon > 0$. This implies that all moments of $u$ exist in the probability measure $d\nu = f_0 d\mu$ and the moment-generating function is analytic in a neighbourhood of $\lambda = 0$. The random variables $u$ satisfying (15.44) for some $\epsilon > 0$ are said to lie in the Cramér class. This class was shown [Pistone and Sempi (1995)] to be a Banach space, and so to be complete, when furnished with the norm

$$\|u\|_L := \inf \left\{ r > 0 : \mathbf{E}_{d\nu} \left[ f_0 \left( \cosh \frac{u}{r} - 1 \right) \right] < 1 \right\}. \tag{15.45}$$

The function

$$\Phi(x) = \cosh x - 1 \tag{15.46}$$

used in the definition (15.45) of the norm, is a Young function. That is, $\Phi$ is convex, and obeys

(1) $\Phi(x) = \Phi(-x)$
(2) $\Phi(0) = 0$
(3) $\lim_{x \to \infty} \Phi(x) = +\infty$.

The theory of Orlicz spaces [Krasnoselski and Ruticki (1961); Rao and Ren (1992)] shows that, given a Young function $\Phi$, one can define a norm by

$$\|u\|_\Phi := \sup_v \left\{ \int |uv| d\nu : v \in L^{\Phi^*}, \int \Phi^*(v(x)) d\nu \leq 1 \right\}, \qquad (15.47)$$

or with the equivalent *gauge norm*, also known as a Luxemburg norm, for any $a > 0$:

$$\|u\|_{L,a} := \inf \left\{ r > 0 : \int \Phi(r^{-1} u(x)) \nu(dx) \right\} < a. \qquad (15.48)$$

By the Luxemburg norm, denoted $\|u\|_L$ we shall mean the case when $a = 1$ [Krasnoselski and Ruticki (1961); Rao and Ren (1992)].

The epigraph of $\Phi$ is the set of points $\{(x,y) : y \geq \Phi(x)\}$. The epigraph is convex, and is closed if and only if $\Phi$ is lower semicontinuous. If so, the map $\lambda \mapsto \Phi(\lambda x)$ is continuous on any open set on which it is finite [Hiriart-Urruty and Lermaréchal (1993)]. Examples of Young functions are

$$\Phi_1(x) := \cosh x - 1 \qquad (15.49)$$

$$\Phi_2(x) := e^{|x|} - |x| - 1 \qquad (15.50)$$

$$\Phi_3(x) := (1 + |x|) \log(1 + |x|) - |x| \qquad (15.51)$$

$$\Phi^p(x) := |x|^p \qquad \text{defined for } 1 \leq p < \infty. \qquad (15.52)$$

Let $\Phi$ be a Young function. Then its Legendre-Fenchel dual,

$$\Phi^*(y) := \sup\{xy - \Phi(x)\} \qquad (15.53)$$

is also a Young function. It is lower semicontinuous, being the supremum of linear functions over a convex set. So $\Phi^{**}$ is lower semicontinuous; its epigraph is the closure of the epigraph of $\Phi$ (which is always the epigraph of a Young function, known as the lower semicontinuous version of $\Phi$ [Ekeland and Temam (1976)]). For example, $\Phi_2 = \Phi_3^*$ and $\Phi^p = \Phi^{q*}$ if $p^{-1} + q^{-1} = 1$.

*Equivalence.* We say that two Young functions $\Phi$ and $\Psi$ are equivalent if there exist $0 < c < C < \infty$ and $x_0 > 0$ such that

$$\Phi(cx) \leq \Psi(x) \leq \Phi(Cx) \qquad (15.54)$$

holds for all $x \geq x_0$. We then write $\Phi \equiv \Psi$; the scale of $x$ is then not relevant. For example, $\Phi_1 \equiv \Phi_2$. Duality is an operation on the equivalence classes:

$$\Phi \equiv \Psi \Longrightarrow \Phi^* \equiv \Psi^*. \tag{15.55}$$

*The $\Delta_2$-class.* We say that a Young function satisfies the $\Delta_2$-condition if and only if there exist $\kappa > 0$ and $x_0 > 0$ such that

$$\Phi(2x) \leq \kappa \Phi(x) \qquad \text{for all } x \geq x_0. \tag{15.56}$$

For example, $\Phi^p$ and $\Phi_3$ satisfy $\Delta_2$, but $\Phi_1$ and $\Phi_2$ do not.

*The Orlicz space and the Orlicz class.*

Let $(\Omega, \mathcal{B}, \nu)$ be a measure space obeying some mild conditions, and let $\Phi$ be a Young function. The *Orlicz class* defined by $(\Omega, \mathcal{B}, \nu)$ and $\Phi$ is the set $\hat{L}^{\Phi}(\nu)$ of real-valued measurable functions $u$ on $\Omega$ obeying

$$\int_{\Omega} \Phi(u(x))\nu(dx) < \infty. \tag{15.57}$$

It is a convex space of random variables, and is a vector space iff $\Phi \in \Delta_2$. The *Orlicz space* $L^{\Phi}$ associated with $\Phi$ and $\nu$ consists of measurable functions $u$ such that

$$\int_{\Omega} \Phi(\alpha u(x))\nu(dx) < \infty \text{ for some } \alpha > 0. \tag{15.58}$$

It is a vector space of random variables, and is the span of the Orlicz class. Up to sets of measure zero, $L^{\Phi}$ is a Banach space when furnished with the *Orlicz norm* or its equivalent, a Luxemburg norm.

Equivalent Young functions give equivalent norms, and $L^{\Phi}$ is separable iff $\Phi \in \Delta_2$.

Analogue of Hölder's inequality
We have the inequality

$$\int |uv|\nu(dx) \leq 2\|u\|_L \|v\|_{L^*}, \tag{15.59}$$

where $\|v\|_{L^*}$ uses $\Phi^*$ in (15.48). This leads to

$$L^{\Phi} \subseteq \left(L^{\Phi^*}\right)^*.$$

Examples. For $\Phi^p(x) := |x|^p$, the Orlicz space is the Lebesgue space $L^p$, and the dual Orlicz space is $L^q$, where $p^{-1} + q^{-1} = 1$. For these Young functions, the Orlicz spaces and the Orlicz classes coincide, and give us separable Banach spaces. For $\Phi_1$ we get a non-separable space, sometimes called the Zygmund space when $\Omega = \mathbf{R}$. It is the dual of $L^{\Phi_3}$, also known as the $L \log L$ space of distributions of finite differential entropy. For this space, the Orlicz class is properly smaller than the Orlicz space it spans.

See [Rao and Ren (1992); Krasnoselski and Ruticki (1961)] for classical Orlicz theory.

*Squeezing in logarithms*

When $\nu$ is discrete with countable support, the Orlicz spaces associated with $\Phi^p$ are the $p$-summable sequences $\ell^p$, $1 \le p \le \infty$. These form a nested family of Banach spaces, with $\ell^1$ the smallest and $\ell^\infty$ the largest. However, this is not the best way to look at Orlicz spaces. Legendre transforms come into their own in the context of a manifold, as a transform between the tangent space and the cotangent space at each point. There is only one manifold, but many coordinatizations. For the information manifold $\mathcal{M}$ of Pistone and Sempi, the points of the manifold are the probability measures $\nu$ equivalent to $\mu$, and these form a positive cone inside $L^1(\Omega, \mu)$. This cone can be coordinatized by the Radon-Nikodym derivatives $f = d\nu/d\mu$. The linear structure of $L^1(\Omega, d\mu)$ provides the tangent space with an affine structure, which is called the (-1)-affine structure in Amari's notation. Amari has suggested that we may also use the coordinates

$$\ell_\alpha(f) := f^{(1-\alpha)/2} \qquad -1 < \alpha < 1, \tag{15.60}$$

known as the Amari embeddings of the manifold into $L^p$, where $p = 2(1 - \alpha)^{-1}$, (since $f \in L^1$, we have $u = f^{(1-\alpha)/2} \in L^p$). Thus, the Orlicz spaces of all the Young functions $|u|^p$ give the same topology on the manifold, namely, that of $L^1$. So they do not help in eliminating states of infinite relative entropy. These coordinates do provide us with an interesting family of connections, $\nabla_\alpha := \partial/\partial\ell_\alpha$, which define the Amari affine structures.

We do better with the formal limit as $p \to \infty$. In the discrete case, the relative entropy is the limit as $\alpha \to 1$ of the Hasegawa-Petz $\alpha$-entropies

$$S(g|f) := \sum_x f(x)(\log f(x) - \log g(x)) \tag{15.61}$$

$$= \sum_x \lim_{\alpha \to 1} (1-\alpha)^{-1} \left( f(x) - f(x)^\alpha g(x)^{1-\alpha} \right). \tag{15.62}$$

It turns out that $S_\alpha(f|g)$ is the 'divergence' of the Fisher metric along the $\alpha$-geodesics. The relative entropy $S(g|f)$ arises as the divergence along the geodesics provided by the embedding

$$\ell_1(f) := \log f.$$

Thus the affine structure corresponds to the linear structure of the random variables $u$ where $f = f_0 e^u$, as in the theory of Pistone and Sempi. The topology given by the corresponding Young function $\Phi_3$ is not equivalent to that of $L^1$, but gives rise to the smaller space $L \log L$, as wanted.

The map

$$u \mapsto \exp \{u - \psi_{f_0}(u)\} f_0 := e_{f_0}(u) \qquad (15.63)$$

maps the unit ball in the Cramér class into the class of probability distributions that are absolutely continuous relative to $\mu$. We can identify $\psi$ as the free energy by writing $f_0 = \exp\{-h_0\}$. Then $f = \exp\{-h_0 - u - \psi_f(u)\}$ and $h_0$ appears as the 'free Hamiltonian', and $u$ as the perturbing potential, of the Gibbs state $f d\mu$. Random variables $u$ and $v$ that differ by a constant give rise to the same distribution. The map (15.63) becomes bijective if we adjust $u$ so that $\mathbf{E}_{d\mu}[f_0 u] = 0$; that is, $u$ has zero mean in the measure $f_0 d\mu$. Such a $u$ is called a *score* in statistics. The corresponding family of measures, $f_0(\lambda u) d\mu$, is called a one-parameter exponential family. In [Pistone and Sempi (1995)], a neighbourhood $N$ of $f_0$ consists of all distributions in some exponential family, as $u$ runs over the Cramér class at $f_0$. Similarly, Pistone and Sempi define the neighbourhood of any $f \in N$, and so on; consistency is shown by proving that the norms are equivalent on overlapping neighbourhoods. They thus construct the information manifold $\mathcal{M}$, modelled on the Banach space functions of Cramér class. This Banach space is identified with the tangent space at any $f \in \mathcal{M}$. The manifold $\mathcal{M}$ is furnished with a Riemannian metric, the Fisher metric, which at $f \in \mathcal{M}$ is the second Fréchet differential of $\psi_f(u)$. The Pistone-Sempi manifold is flat with respect to the $(+1)$-connection given by the addition of the potentials $u$. However, it is not a trivial manifold, as these coordinates are only valid on the Orlicz *class* of a point, not the whole manifold.

We shall construct a quantum analogue of this manifold, as in [Streater (2000, 2008a,b)]. We shall thus extend the construction of the quantum version of the theory, as introduced in [Streater (2004); Grasselli and Streater (2000); Gibilisco and Isola (1999)], from the cases with the potential limited to $H$-operator bounded, or $\epsilon$-$H$-form bounded, or in some $L^p$, to a suitable limit $p \to \infty$, to include the case analogous to the Zygmund space: namely, we shall include all cases where the potential $X$ is $H$-form bounded with a small enough bound.

## 15.9   The Finite-Dimensional Quantum Info-Manifold

In this section we study the quantum version of estimation theory, but limit the algebra of observables to $M_n$, the space of $n \times n$ matrices. States are then given by $n \times n$ density operators $\rho$. Suppose that we want to find the best estimate for a state $\rho$, unknown to us, but whose means $\rho \cdot X_1, \cdots, \rho \cdot X_N$ are measured. In quantum mechanics the operators $X_i, \ldots, X_N$ might not commute with each other. At this point, some physicists object: we cannot observe non-commuting operators simultaneously [Fujiwara and Nagaoka (1995)]. However, in estimation theory we postulate that we have a sequence of independent copies of the unknown state $\rho$. We may use the first $m$ of these, and observe the values taken by $X_1$; say that the values are $x_{1j}, j = 1, \ldots, x_{1m}$. Then would estimate the mean of $X_1$ in this state to be the average value $m^{-1} \sum_j x_{1j}$. We may use the samples from $m+1$ to $2m$ to observe the values taken by $X_2$, and their average gives the value we assign to $\rho \cdot X_2$; in this way, we get estimators for the mean of each $X_i, i = 1, \ldots, N$.

In 1957 Jaynes suggested that of all states with the given means, the 'most likely' one would be the state of maximum entropy. We showed that this idea is correct when we are in classical probability. In the quantum version of Jaynes's axiom, Jaynes uses the von Neumann formula for the entropy. We shall show that this too is correct, even when the $X_1, \ldots, X_N$ do not all commute with each other. Thus, we shall show that the best estimator for the state is the density operator in the exponential family generated by the $X_i$; thus, it is of the form $\rho = \exp\{-\sum_i \beta_i X_i\}$, and the $\beta_i$ are determined by the given means for $X_i, i = 1, \ldots, N$. We shall choose one of the observables to be the unit operator, with the mean equal to 1. This condition gives $\rho \cdot 1 = 1$, that is, $\mathrm{Tr}\rho = 1$, which ensures that the state is normalised.

In the classical case (and later in the quantum case too) Chentsov asked whether the Fisher-Rao metric was unique. Any manifold has a large number of different metrics on it; apart from those that differ just by a constant factor, one can multiply a metric by a positive space-dependent factor. There are many others. Chentsov therefore imposed conditions on the metric. He saw the metric (and the Fisher metric in particular) as a measure of the distinguishability of two states. He argued that if this is to be true, then the distance between two states must be *reduced* by any stochastic map; for, a stochastic map must 'muddy the waters', reducing our ability to distinguish states. He therefore considered the class of metrics $G$ that are reduced by any stochastic map on the random variables. Recall that in classical probability (3.6):

**Definition 15.4.** A stochastic map is a linear map on the algebra of random variables that preserves positivity and takes 1 to itself.

Chentsov was able to prove that the Fisher-Rao metric (15.19) is unique, among all metrics, being the only one (up to a constant multiple) that is reduced by any stochastic map.

In quantum mechanics, instead of random variables we use the algebra of matrices $M_n$. We limit discussion to the faithful states, which we take to form the manifold $\mathcal{M}$; it is a genuine manifold, and not one of the non-commutative manifolds without points that occur in Connes's theory. See [Holevo (1982); Helstrom (1976)] for early work on quantum estimation. In infinite dimensions, we add the requirement that the state have finite von Neumann entropy. The natural morphisms in the quantum case are the completely positive maps $T : M_n \to M_n$ that preserve the identity:

**Definition 15.5.** A linear map $T : M_n \to M_n$ is said to be a *quantum stochastic morphism* if

(1) $TI = I$
(2) $T \otimes I_j$ is positive on $M_n \otimes M_j$ for all integers $j = 1, 2, \dots$.

Here $I \in M_n$ is the unit observable and $I_j$ is the $j \times j$ unit matrix. Chentsov attempted to find out whether his uniqueness theorem of classical probabiltiy does or does not extend to quantum theory, but he died before completing this work. We say that a metric on the set of states has the Chentsov property if it is reduced by every quantum stochastic morphism. Petz [Petz (1996)] then constructed all metrics on $\mathbf{M}_n$ with the Chentsov property. He showed that the uniqueness of the metric (up to a multiple) does not follow from the requirement that the distance between any two states, defined by the metric, is reduced or left the same by every quantum stochastic morphism $T$; there are infinitely many inequivalent metrics with this property. The GNS (short for Gelfand-Naimark-Segal) and BKM (short for Bogoliubov-Kubo-Mori) metrics are in common use in quantum estimation, and both are decreased, or left the same, by every such $T$. However, these two metrics are not proportional.

As is the case for classical probability, there are several affine structures on the manifold of density matrices. The first comes from the mixing of the states, and is called the $-1$-affine structure. Coordinates for a state $\rho$ in a 'hood of $\rho_0$ provided by $\rho - \rho_0$, a traceless matrix which can be taken to be small. The whole tangent space at $\rho$ is thus identified with the set of traceless matrices, and this is a vector space with the usual rules for adding matrices. Obviously, the manifold is flat relative to this affine structure.

The +1-affine structure is constructed as follows. Since a state $\rho_0 \in \mathcal{M}$ is faithful we can write $H_0 := -\log \rho_0$ and any $\rho$ near $\rho_0 \in \mathcal{M}$ as

$$\rho = Z_X^{-1} \exp\{-H_0 - X\} \tag{15.64}$$

for some Hermitian matrix $X$; we see that $X$ is ambiguous up to the addition of a multiple of the identity. We choose to fix $X$ by requiring $\rho_0 \cdot X = 0$, and call $X$ the 'score' of $\rho$. Then the tangent space at $\rho$ can be identified with the set of scores; let us denote this tangent space by $\partial \mathcal{M}_\rho$. The +1-linear structure on $\partial \mathcal{M}_\rho$ is given by matrix addition of the scores. If the quantum Hilbert space is of infinite dimension, so that $\dim \mathcal{H} = \infty$, we shall require that $X$ be a small form-perturbation of $H_0$; we also require that the generalised mean of $X$ be zero. Corresponding to these two affine structures, there are two affine connections, whose covariant derivatives are denoted $\nabla^{(\pm)}$. Following [Hasegawa and Petz (1996)] one can also define interpolating affine structures, similar to the classical case as in Eq. (15.39).

As an example of a metric on $\mathcal{M}$, let $\rho \in \mathcal{M}$, and for $X, Y$ in $\partial \mathcal{M}_\rho$ define the *GNS* metric by

$$G_\rho(X, Y) = \operatorname{Re} \operatorname{Tr}[\rho X Y]. \tag{15.65}$$

We remarked above that this metric is reduced by all quantum stochastic morphisms $T$; that is, it obeys

$$G_{T^*\rho}(TX, TX) \leq G_\rho(X, X), \tag{15.66}$$

in accordance with Chentsov's idea. $G$ is positive definite since $\rho$ is faithful. This has been used in [Helstrom (1976)] in the theory of quantum estimation theory. The theory has been extended to include pure states by [Fujiwara and Nagaoka (1995)]. However, Nagaoka has remarked that if we take this metric, then the $(+1)$ and the $(-1)$ affine connections are not dual; the dual to the $(-1)$ affine connection, relative to this metric, is not flat and has torsion. This might lead one to choose a different metric, with respect to which these two connections are dual. It is known that the BKM-metric has this property, as well as being a Chentsov metric. It is the only Chentsov metric, up to a factor, for which this is true [Grasselli and Streater (2001)].

We seek a quantum analogue of the Cramér-Rao inequality. Given a family $\mathcal{M}$ of density operators, parametrised by a real parameter $\eta$, we seek an estimator $X$ whose mean we can measure in the true state $\rho_\eta$. To be unbiased, we require $\operatorname{Tr} \rho_\eta X = \eta$, which, as in the classical case gives

$$\operatorname{Tr}\left\{\rho_\eta \rho_\eta^{-1} \frac{\partial \rho_\eta}{\partial \eta}(X - \eta)\right\} = 1. \tag{15.67}$$

Here we used that $\eta \mathrm{Tr}\partial\rho/\partial\eta = 0$. A state obeying this condition, at $\eta = 0$, is said to be *locally unbiased*. It is tempting to regard $L_r = \rho^{-1}\partial\rho/\partial\eta$ as a quantum analogue of the Fisher info; it has zero mean, and the above equation says that its covariance with $X - \eta$ is equal to 1. However, $\rho$ and its derivative need not commute, so $Y$ is not Hermitian, and is not popular as a measure of quantum information. Helstrom, and Petz and Toth [Helstrom (1976); Petz and Toth (1993)] get round this by using the idea of a *logarithmic derivative*. Let $g$ be a real or complex scalar product on the space of matrices; we say that a matrix $L$ is the $g$-logarithmic derivative of the family $\rho_\eta$ if for any matrix $X$,

$$\frac{\partial\rho_\eta \cdot X}{\partial\eta} = g(L^*, X). \tag{15.68}$$

Writing $X - \eta I$ for $X$ in this equation gives

$$\frac{\partial\rho}{\partial\eta} \cdot (X - \eta) = g(L^*, X - \eta) = 1. \tag{15.69}$$

The symmetric logarithmic derivative uses the real part of the *GNS* metric for $g$, so that

$$\frac{\partial}{\partial\eta}\mathrm{Tr}(\rho_\eta X) = \frac{1}{2}\mathrm{Tr}[\rho_\eta(L_s X + X L_s)]. \tag{15.70}$$

As remarked above, another metric in Chentsov's allowed class is the BKM metric; let $X$ and $Y$ have zero mean in the state $\rho$. Then put

$$g_\rho(X, Y) = \int_0^1 \mathrm{Tr}\left[\rho^\alpha X \rho^{1-\alpha} Y\right] d\alpha. \tag{15.71}$$

This is a positive-definite scalar product on the space of self-adjoint matrices, known as the BKM metric. The corresponding logarithmic derivative, $L_B$, is defined such that

$$\frac{\partial}{\partial\eta}\rho_\eta \cdot X = \int_0^1 \mathrm{Tr}\left[\rho_\eta^\lambda L_B \rho_\eta^{1-\lambda} X\right] d\lambda \tag{15.72}$$

and is given explicitly by

$$L_B = \int_0^\infty (\lambda + \rho_\eta)^{-1}\frac{\partial\rho_\eta}{\partial\eta}(\lambda + \rho_\eta)^{-1}d\lambda. \tag{15.73}$$

Equation (15.67) then says that at $\eta = 0$,

$$1 = g_\rho(X, L_B) \leq g_\rho(X, X)^{\frac{1}{2}} g_\rho(L_B, L_B)^{\frac{1}{2}}, \tag{15.74}$$

by the Schwarz inequality. This is the quantum Cramér-Rao inequality. Now the Schwarz inequality for a vector space gives equality only if the two vectors are proportional. Thus, to find the best estimator for a self-adjoint operator $X$, we seek the $\rho_\eta$ such that $X$ is proportional to $L_B$. Now, note the equation

$$\frac{\partial}{\partial \eta}\left[e^{-H-\xi(\eta)X}\right] \cdot Y = \int_0^1 \rho_0^\alpha \left(-\frac{\partial \xi}{\partial \eta}X\right)\rho_0^{1-\alpha}d\alpha \cdot Y \qquad (15.75)$$

$$= g_{\rho_0}(L_B, Y) \qquad (15.76)$$

where $\rho_0 = e^{-H}$. This shows that the exponential state $\rho = e^{-H-\xi X}$ has $X$ and $L_B$ proportional, and the unbiased condition fixes the value of $\xi$. Since the metric is positive definite, this is the only state with this property. Thus, the Jaynes states give the best estimators, in the sense that their variance is the minimum possible.

For estimators for several parameters we get a version of the inequality in matrix form [Petz (2002)]. Thus, we want to estimate the parameters $\eta_1, \ldots, \eta_n$ as the 'means' of hermitian matrices, $X_1, \ldots, X_n$, which are to be locally unbiased estimators. We remarked that there is no need for the matrices to commute with the state, or among themselves. Thus, the state is one of a family $\rho_\eta$ which obeys the quantum analogue of Eq. (15.21):

$$\frac{\partial \rho_\eta \cdot (X_k - \eta_k)}{\partial \eta_j}\frac{\partial \rho_\eta}{\partial \eta_j} \cdot X_k = \delta_{jk}, \qquad j, k = 1, \cdots, n. \qquad (15.77)$$

Let us try a state

$$\rho_\eta = Z^{-1}\exp\{-H - \sum_\ell \xi_\ell X_\ell\}$$

in the exponential family. Then for each $k$ we get at $\eta = 0$ the equation

$$\frac{\partial \rho \cdot X_k}{\partial \eta_j} = \int_0^1 \rho^\alpha \sum_\ell \frac{\partial \xi_\ell}{\partial \eta_j}X_\ell \rho^{-(1-\alpha)} X_k \, d\alpha$$

$$= g_B(\sum_\ell \frac{\partial \xi_\ell}{\partial \eta_j}X_\ell, X_k). \qquad (15.78)$$

Thus, $L_B$ is proportional to a sum of $X_k$. Since the exponential function is convex, this is the unique possibility for a state in which $L_B$ is proportional to a sum of $X_k$. The values of $\xi_\ell$ are then determined by the requirement that the expectation values are those obtained as the ones found by experiment. Thus, Jaynes was correct also in the case of several parameters, when he postulated that the entropy should be a maximum, since this also leads to the exponential family.

The BKM metric $g$ has other desirable properties, apart from entering in Kubo's 'theory of linear response'. For the metric $g$, the connections with covariant derivatives $\nabla^{(\pm\alpha)}$ are dual, and there are coordinates for $\nabla^\alpha$, namely, it is the unit sphere in the (finite-dim.) Banach space $\mathcal{C}_p$, the Schatten class with norm $\|X\|_p = (\mathrm{Tr}|X|^p)^{1/p}$. The case $p = 2$, or $\alpha = 0$, leads to the Hilbert space of Hilbert-Schmidt operators. More, the Massieu function $\log Z$ is the generating function for all the connected Kubo functions, and in particular, the mean is the first derivative, and the metric is the second, as in Eq. (15.78). The entropy is again the Legendre transform of the Massieu function $\Psi(X) = \log Z$ provided that we use the BKM metric:

$$S(Y) = \sup\{g(X,Y) - \Psi(X)\}$$

and the reciprocal relations of Eq. (15.42) hold. It follows that the Cramér-Rao inequality for the BKM-metric is achieved exactly for the exponential family, among states in the whole neighbourhood of the exponential family, agreeing with the method of maximum entropy. Thus, we justify Jaynes, in that his method of choosing the state with the maximum entropy, given the means, is the estimate of the state with the smallest variance, among all states. We shall obtain a similar result in infinite dimensional Hilbert space.

## 15.10    Araki's Expansionals and the Analytic Manifold

Araki [Araki (1973)] has considered the case where $\rho$ is a *KMS* state on a $W^*$-algebra. He then perturbed the state by adding bounded operators to the Hamiltonian; for the perturbed KMS state has a finite relative entropy, which has a convergent Kubo perturbation expansion, defining an analytic function in the Banach space of bounded perturbations. We try to follow this for unbounded perturbations, in the case where the $W^*$-algebra is $\mathcal{B}(\mathcal{H})$. In that case, write $\rho = e^{-H}$ and $\sigma = Z_X e^{-H-X}$ where $X$ is bounded. Of course, $\mathrm{Tr}\, e^{-H} = 1$, $Z_X = \mathrm{Tr}\, e^{-H-X}$ hold; we can also assume that $\rho \cdot X = 0$ without loss of generality; thus, $X$ is a score. Then the relative entropy is

$$S(\sigma|\rho) := \mathrm{Tr}\,\rho(\log\rho - \log\sigma), \tag{15.79}$$

and we have

**Theorem 15.6.** $S(\sigma|\rho) = \log Z_X$.

**Proof.**    We have

$$S(\sigma|\rho) = \mathrm{Tr}\,\rho\left(-H + H + X + \log\mathrm{Tr}\, e^{-H-X}\right).$$

This gives the result, since $\rho \cdot X = 0$. $\qquad\qquad\qquad\qquad\square$

One can prove [Hiai (1995)] that this obeys

$$S(\sigma|\rho) \geq \frac{1}{2}\|\rho - \sigma\|_1^2,$$

improving the result of Lemma 9.6. We now extend the class of perturbations $X$ to forms that are small relative to $H$.

Let $\Sigma$ be the set of density operators on $\mathcal{H}$, and let int $\Sigma$ be its interior, the faithful states. We shall deal only with systems described by $\rho \in \text{int}\,\Sigma$; this means that for a free Schrödinger particle, or system of such, we are limited to systems inside a finite volume of real space. Then we would expect the entropy to be finite. The following class of states turns out to be tractable. Let $p \in (0, 1)$ and let $\mathcal{C}_p$, denote the set of operators $C$ such that $|C|^p$ is of trace class. This is like the Schatten class, except that we are in the less popular case, $0 < p < 1$, for which $C \mapsto (\text{Tr}[|C|^p])^{1/p}$ is only a quasi-norm. Let

$$\mathcal{C}_< = \bigcup_{0<p<1} \mathcal{C}_p. \qquad\qquad\qquad (15.80)$$

One can show that the entropy

$$S(\rho) := -\text{Tr}[\rho \log \rho] \qquad\qquad\qquad (15.81)$$

is finite for all states in $\mathcal{C}_<$. We take the *underlying set* of the quantum info manifold to be

$$\mathcal{M} = \mathcal{C}_< \cap \text{int}\Sigma. \qquad\qquad\qquad (15.82)$$

For example, this set contains the case $\rho = \exp\{-H_0 - \psi_0\}$, where $H_0$ is the Hamiltonian of the quantum harmonic oscillator, and $\psi_0 = \text{Tr}\exp\{-H_0\}$. In this example, we may take any $p \in (0,1)$. The set $\mathcal{M}$ includes most other examples of non-relativistic physics. It contains also the case where $H_0$ is the Hamiltonian of the free relativistic field, in a box with periodic boundary conditions. Furthermore, all these states have finite von Neumann entropy. In limiting the theory to faithful states, we are imitating the decision of Pistone and Sempi that the probability measures of the information manifold should be equivalent to the guiding measure $\mu$, rather than, say, merely absolutely continuous. Here, the trace is the quantum analogue of the measure $\mu$. Thus in general, an element $\rho$ of $\mathcal{M}$ has a self-adjoint logarithm, and can be written

$$\rho = \exp(-H) \qquad\qquad\qquad (15.83)$$

for some self-adjoint $H$, which is non-negative, since $\operatorname{Tr} \exp(-H) = 1$. Note that the set $\mathcal{M}$ is not complete relative to any quasi-norm $\| \cdot \|$.

Our aim is to cover $\mathcal{M}$ with balls with centre at a point $\rho \in \mathcal{M}$, each belonging to a Banach space; we have a Banach manifold when $\mathcal{M}$ is furnished with the topology induced by the norms; for this, the main problem is to ensure that various Banach norms, associated with points in $\mathcal{M}$, are equivalent at points in the overlaps of the balls.

Let $\rho_0 \in \mathcal{M}$ and write $H_0 = -\log \rho_0 + cI$. We choose $c$ so that $H_0 \geq I$, and we write $R_0 = H_0^{-1}$ for the resolvent at $0$. We define a 'hood of $\rho_0$ to be the set of states of the form

$$\rho_V = Z_V^{-1} \exp - (H_0 + V), \tag{15.84}$$

where $V$ is a sufficiently small $H_0$-bounded form perturbation of $H_0$. The necessary and sufficient condition to be Kato-bounded is that

$$\|V\|_0 := \|R_0^{1/2} V R_0^{1/2}\|_\infty < \infty. \tag{15.85}$$

The set of such $V$ make up a Banach space, $\mathcal{T}(0)$, with (15.85) as norm. The first result is that $\rho_V \in \mathcal{M}$ for $V$ inside a small ball in $\mathcal{T}(0)$.

We define the $(+1)$-affine connection by transporting the score $V - \operatorname{Tr} \rho V$ at the point $\rho$ to the score $V - \operatorname{Tr} \sigma V$ at $\sigma$. This connection is flat and torsion-free, since it patently does not depend on the path between $\rho$ and $\sigma$. The $(-1)$-connection can be defined in $\mathcal{M}$ since each $\mathcal{C}_p$ is a vector space.

Araki proved [Araki (1973)] that if $V$ is bounded, the Kubo expansion converges:

$$\log Z_V = \sum_{n=0}^\infty (n!)^{-1} \int_0^1 \prod d\alpha_i \delta(\sum \alpha_i - 1) \operatorname{Tr} (\rho^{\alpha_1} V \ldots \rho^{\alpha_n} V). \tag{15.86}$$

We prove in [Grasselli and Streater (2000)] that the series converges also for our class of perturbations, said to be $\epsilon$-bounded, and that there are equivalent norms on overlapping regions. However, this does not cover all cases of perturbation, since $\epsilon$-boundedness is a restriction to smoother directions in the tangent space. Because of this, the norm is unlikely to be equivalent to an Orlicz norm. So we now turn to the largest class of states known to make up a statistical manifold analogous to that of Pistone and Sempi: the set of all small-enough form-bounded perturbations of $H_0$.

## 15.11   The Quantum Young Function

The function

$$\Phi(x) = \cosh x - 1 \tag{15.87}$$

used in the definition (15.45) of the Orlicz norm, is a Young function; see Definition 15.8. The classical theory of Orlicz spaces can use any Young function in Definition 15.45, to give various Luxemburg norms, not all equivalent to each other; see [Krasnoselski and Ruticki (1961); Rao and Ren (1992)]. It would appear, then, that to define a quantum Orlicz space would require the definition of quantum Young functions. Possibly the first attempt to do this was done by W. Kunze [Kunze (1990)]. This author takes a classical Young function $\Phi$, and writes the corresponding quantum Young function $\Phi(X)$ as a function of the operator, $X$, but considers only functions of the form $\Phi(X) = \Phi(|\tilde{X}|)$, where the tilde denotes the 'reordered' value of the modulus. This is well defined for any classical Young function $\Phi$, as we can use the spectral theorem for the self-adjoint operator $|\tilde{X}|$ to define the function. This gives rise to a norm, but it would seem that it fails to take account of the quantum phase between operators, and so might not be the correct quantum version. However, some use of this idea has been used in [Al-Rashid and Zegarlinski (2008)]. The main problem is that it fails to make sense when $X$ is a quadratic form. What will work, however, is the Young function chosen in [Streater (2000, 2004)].

Let $\rho := e^{-H}$ be a density operator. For any quadratic form $X$, defined on $\operatorname{Dom} H^{1/2}$, and small-enough relative to the form of $H$, we define

$$\Phi_H(X) := \frac{1}{2}\operatorname{Tr}\left(e^{-(H+X)} + e^{-(H-X)}\right) - 1. \qquad (15.88)$$

We drop the subscript $H$ until we need to distinguish norms at different values of $H$. Then we claim

**Theorem 15.7.** $\Phi(X)$ *obeys*

*(1)* $\Phi(X)$ *is finite for all forms* $X$ *with small-enough* $H$*-form bound, say* $\|X\|_K < c$.
*(2)* $X \mapsto \Phi(X)$ *is convex on the set of* $H$*-small-enough forms.*
*(3)* $\Phi(X) = \Phi(-X)$.
*(4)* $\Phi(0) = 0$ *and if* $X \neq 0$, *then* $\Phi(X) > 0$ *or* $\Phi(X) = \infty$.

This is proved when $X$ is an $H$-small operator in [Streater (2000, 2004)]. The map $X \mapsto \operatorname{Tr} e^{H+X}$ remains convex when extended to $H$-form-bounded forms $X$ of small-enough $H$-form bound. We now give a proof of this. $H$ has discrete spectrum, say $\lambda_1 \leq \lambda_2 \leq \ldots$; since $\exp(-H)$ is of trace class, we see that $\lambda_n \to \infty$ as $n \to \infty$. Let $H = \sum_j \lambda_j P_j$ be the spectral resolution of $H$, and put

$$Q_n := \sum_{j=1}^{n} P_j, n = 1, 2, \ldots.$$

Each normalised eigen-vector, say $\psi_j, j = 1, 2, \ldots$, of $H$ lies in $\text{Dom } H^{1/2}$, so we have that $X_n := Q_n X Q_n$ is a bounded operator. Moreover, for any sufficiently small $H$-form bounded $X$, we have

$$\langle \psi_j, e^{(H+X)} \psi_j \rangle = \lim_{n \to \infty} \langle \psi_j, e^{-(H+X_n)} \psi_j \rangle. \tag{15.89}$$

Since exponentials are positive, the traces are absolutely convergent, and we have for any $H$-form small-enough $X$ and $Y$, and any $\lambda \in (0,1)$,

$$\text{Tr}\, e^{-(H+\lambda X + (1-\lambda)Y)} = \sum_j \langle \psi_j, e^{-(H+\lambda X + (1-\lambda)Y)} \psi_j \rangle$$

$$= \sum_{j=1}^{\infty} \langle \psi_j, e^{-\lambda(H+X) + (1-\lambda)(H+Y)} \psi_j \rangle$$

$$= \sum_{j=1}^{\infty} \lim_{n \to \infty} \langle \psi_j, e^{-(\lambda(H+X_n) + (1-\lambda)(H+Y_n))} \psi_j \rangle$$

$$\leq \sum_{j=1}^{\infty} \lim_{n \to \infty} \left\{ \langle \psi_j, \left( \lambda e^{-(H+X_n)} + (1-\lambda)e^{-(H+Y_n)} \right) \psi_j \rangle \right\}$$

by convexity with bounded perturbations

$$= \sum_{j=1}^{\infty} \left\{ \lambda \langle \psi_j, e^{-(H+X)} \psi_j \rangle + (1-\lambda) \langle \psi_j, e^{-(H+Y)} \psi_j \rangle \right\}$$

$$= \lambda \text{Tr}\, e^{-(H+X)} + (1-\lambda) \text{Tr}\, e^{-(H+Y)}.$$

Thus, $\text{Tr}\, e^{-(H+X)}$ is convex in $X$ in a 'hood of $X = 0$, and so $\Phi(X)$ is too.

The closure of the epigraph of $\lambda \mapsto \Phi(\lambda X)$ defines a lower-semicontinuous quantum Young function, which we now again call $\Phi$, and which we take to define the norm

$$\|X\|_L := \inf \left\{ r > 0 : \Phi\left( \frac{X}{r} \right) < 1 \right\}. \tag{15.90}$$

The function $\lambda \mapsto \Phi(\lambda X)$ is continuous in $\lambda \in (-\epsilon, \epsilon)$ provided that it is finite in the interval $(-\epsilon, \epsilon)$. For, we proved [Streater (2004)] that the value of $\text{Tr}\, e^{-(H+\lambda X)}$ lies between $\text{Tr}\, e^{-(1+\lambda a)H}$ and $\text{Tr}\, e^{-(1-\lambda a)H}$, and so is continuous at $\lambda = 0$. For a general $\lambda$, we replace $H$ by $H + \lambda' X + \lambda X$ and show that this is continuous at $\lambda = 0$. Another proof of these facts follows from the Golden-Symanzik-Thompson inequality, in the form:

**Theorem 15.8.** *Let $A, B$ be positive self-adjoint operators on the separable Hilbert space $\mathcal{H}$, such that $\text{Dom}\, A^{1/2} \cap \text{Dom}\, B^{1/2} := \mathcal{D}$ is dense in $\mathcal{H}$. Let $A \stackrel{\circ}{+} B$ be the form sum of $A$ and $B$, defined on $\mathcal{D}$. Then $A \stackrel{\circ}{+} B$ is positive, and so has a positive operator extension, the Friedrichs extension, which we again call $A \stackrel{\circ}{+} B$. Then*

$$\text{Tr}\, e^{A \stackrel{\circ}{+} B} \leq \text{Tr}\left(e^A e^B\right). \tag{15.91}$$

Simon gives a sketch of the proof of this, in his book [Simon (2005)], Theorem 8.5. He used a result of Kato [Kato (1967)], and showed for any symmetric norm $\|\,.\,\|$, (that is, the norm obeys $\|X\| = \|X^*\|$) defined on the semigroups, that if

$$e^{A/2} e^B e^{A/2}$$

has finite $\|\,.\,\|$, then so has $e^{A+B}$; and then we have

$$\left\|e^{A+B}\right\| \leq \left\|e^{A/2} e^B e^{A/2}\right\|.$$

As an alternative, there is the proof of [Hiai (1995)].

We apply this form of the inequality to states in our choice, $\mathcal{M}$ as the underlying set of the manifold, as in Eq. (15.82). So, let $H > 0$ be self-adjoint on $\mathcal{H}$ be such that $e^{-\beta_1 H}$ is of trace-class; let $X$ be an $H$-form-bounded quadratic form, and put $\beta(H \stackrel{\circ}{+} X) = A$ and $(1-\beta)H = B$. Then $\text{Dom}\, A, B = \text{Dom}\, H^{1/2}$ is dense in $\mathcal{H}$, and $H \stackrel{\circ}{+} \beta X = A \stackrel{\circ}{+} B$. Moreover, we have

$$\text{Tr}\, e^{-(H \stackrel{\circ}{+} \beta X)} = \text{Tr}\, e^{A \stackrel{\circ}{+} B} \leq \text{Tr}(e^A e^B) = \text{Tr}\left(e^{-\beta(H \stackrel{\circ}{+} X)} e^{-(1-\beta)H)}\right)$$

$$\leq \|e^{-\beta(H \stackrel{\circ}{+} X)}\|_\infty \, \|e^{-(1-\beta)H}\|_1 \tag{15.92}$$

by Hölder's inequality. It follows from our assumption that $e^{-(1-\beta)H}$ is of trace-class for all $1 - \beta > \beta_1$, so the perturbed state is of trace-class for all $0 < \beta < 1 - \beta_1$. Similarly, one shows that $e^{-\alpha(H \stackrel{\circ}{+} \beta X)}$ is of trace class for all $\alpha$ close enough to 1, and so it then lies in $\mathcal{M}$, (15.82).

The first patch of the manifold around the point $\rho = e^{-H}$ is defined in a way following [Pistone and Sempi (1995)]; it consists of spheres of all forms $X$ such that the norm $\|X\|_L$ is sufficiently small. The topology of this set is that determined by this Orlicz norm. At a point $X$ with $\rho_X := e^{-(H+X+\psi(X))} \in \mathcal{M}$, we could use the norm

$$\|Y\|_X = \inf\left\{r > 0 : \Phi_{H+X+\psi(X)}\left(\frac{Y}{r}\right) < 1\right\} \tag{15.93}$$

and proceed as do Pistone and Sempi; to make sense, on points lying in the overlap of these two spheres, we need to prove that the topologies given by the norms $\|Y\|_L$ and $\|Y\|_X$ are equivalent. This is done below. So we here construct a quantum analogue of the Pistone-Sempi manifold, following [Streater (2008a,b)]. Our norm is smaller than any of the $L^p$ norms in [Gibilisco and Isola (1999)], and our space of states is larger. Our quantum Young function does not obey the $\Delta_2$ condition, and the space of states $\mathcal{M}$ is not separable.

We note that if $\Phi$ is a quantum Young function, satisfying (1)-(4) in Theorem 15.7, then its Legendre transform is a quantum Young function for the space of density operators with their convex structure. That is,

$$\Psi(Y) := \sup\{g_B(X,Y) - \Phi(X)\} \qquad (15.94)$$

is a quantum Young function on the space of $Y$, being a set of density operators of finite entropy. Thus when $\Phi$ is our choice of quantum Young function, acting on $X$, which is a sufficiently small form perturbation of $H$, then $\Psi$ is the quantum analogue of $\Phi_3$, Eq. (15.49). The Hölder inequality is then

$$\mathrm{Tr}(XY) \leq 2(\|X\|_\Phi \|Y\|_\Psi), \qquad (15.95)$$

the analogue of Eq. (15.59).

The author's proposal [Streater (2004)] in Eq. (15.88) for a quantum Young function, is one of the non-commutative versions of the classical Young function $\cosh x - 1$. Jenčová [Jenčová (2003)] first proposed a different function, closer to that in [Al-Rashid and Zegarlinski (2008)]; she has now chosen one built out of Araki's relative entropy, [Araki (1976)] which is limited to bounded scores $X$, defined for any $W^*$-algebra; for the case of $\mathcal{B}(\mathcal{H})$ it reduces to the usual relative entropy $\log \mathrm{Tr} e^{-H-X}$ of the states $e^{-H}$ and $e^{-H-X-\psi(X)}$, and gives the same Young function as we use here [Jenčová (2006)]. Her theory is worth studying more closely. As seen above, our own version allows perturbations of the Hamiltonian by *forms*. In particular, it allows the perturbation of the state $\rho = e^{-\beta H}$ by a small multiple of $H$; this shows that our 'hood of $\rho$, a canonical state, contains canonical states with nearly the same temperature.

Let us add to the remarks in [Streater (2000, 2004)]. First, we may write $\beta H = H - (1-\beta)H$; then we have that the operator $(1-\beta)H$ is $H$-small. Thus the perturbation theory of [Streater (2008a)] shows that the free energy $\log \mathrm{Tr} \exp\{\beta H\}$ is indeed analytic in $\beta$ lying in a 'hood of $\beta = 1$. We conclude that the function $\mathrm{Tr} \exp\{-\beta H\}$ is analytic if it is finite in a 'hood of $\beta = 1$. Note that in this theory, $H$ is not a given Hamiltonian of some dynamics in the theory; rather, $H$ is a positive self-adjoint operator that determines the state $\rho$ of interest.

## 15.12 The Quantum Cramér Class

We perturb a given state $\rho \in \mathcal{M}$ by adding a potential $X$ say, to $H$, in analogy with the classical theory where the potential is $u$ as in (15.43). Suppose that $X$ is a quadratic form on $\mathcal{H}$ such that $\text{Dom}X \supseteq \text{Dom}H^{1/2}$ and there exist positive $a, b$ such that

$$|X(\phi, \phi)| \leq a \left\langle H^{1/2}\phi, H^{1/2}\phi \right\rangle + b\|\phi\|^2 \qquad (15.96)$$

for all $\phi \in \text{Dom } H^{1/2}$. Then we say that $X$ is form-bounded relative to $H$. The infimum of all $a$ satisfying (15.96) for some $b > 0$ is called the $H$-form bound of $X$; we shall denote the form bound by $\|X\|_K$, in honour of T. Kato. It is a semi-norm on the linear set of forms bounded relative to $H$. It is well known that if $\|X\|_K < 1$, then $H + X$ defines a semi-bounded self-adjoint operator. Furthermore, if $\|X\|_K$ is small enough, less than $a < 1 - \beta_0$, then by Lemma 4 of [Streater (2000)], we have

$$e^{b\beta}\text{Tr}\left(e^{-(1-a)H\beta}\right) \geq \text{Tr}\left(e^{-(H+X)\beta}\right) \geq e^{-b\beta}\text{Tr}\left(e^{-(1+a)H\beta}\right). \qquad (15.97)$$

It follows that $\exp(-\beta(H+X))$ is of trace class for all $\beta > \beta_X := \beta_0/(1-a)$, which is less than 1. Thus $\rho_X := \exp -(H + X + \psi(X)) \in \mathcal{M}$ for all forms $X$ with form-bound less than $1 - \beta_0$. Forms obeying this condition will be said to be of 'small-enough Kato seminorm'. Here, $\psi(X) := \text{Tr}[\exp - (H + X)]$. Two perturbations that differ by a multiple of the identity operator give rise to the same perturbed state; as in the classical case, we define the 'scores' to be $H$-form-bounded forms $X$ with zero expectation;

$$\text{Tr} \exp\{-H\}X = 0. \qquad (15.98)$$

Where $X$ is an H-small-enough form, the defining inequality for this shows that $X$ has finite mean in the state $\rho$; subtracting this mean from $X$ gives us a score.

We define the Cramér class (for the state $\rho = \exp\{-H\}$) to be the set of all $H$-form-bounded scores $X$ of small enough semi-norm $\|X\|_K$. If $X$ is a bounded operator, it has zero Kato semi-norm, and so (when the algebra is $\mathcal{B}(\mathcal{H})$) our theory contains that of Jenčová.

Now, the spectrum of $H$ is positive; for if it contains zero, then the trace of $\rho$ would be greater than 1. Thus $H^{-1}$ is bounded, and we see that the Cramér class (of the state $\rho = \exp(-H)$) contains all $H$-small forms $X$ with

$$\|X\|_1 := \|H^{-1/2}XH^{-1/2}\| < 1.$$

For then we see that

$$X(\phi, \phi) = \langle H^{1/2}\phi, H^{-1/2}XH^{-1/2}H^{1/2}\phi \rangle \le \|X\|_1 \langle H^{1/2}\phi, H^{1/2}\phi \rangle.$$

This tells us that $a = \|X\|_1$ and $b = 0$; and see from (15.97) that $\psi(\lambda X)$ is finite if $|\lambda| < 1$ and continuous at $\lambda = 0$, since its value is sandwiched between

$$\text{Tr}\left(e^{(1-\lambda)\beta H}\right) \text{ and } \text{Tr}\left(e^{(1+\lambda)\beta H}\right).$$

## 15.13 The Parameter-Free Quantum Manifold

In Theorem 15.7 the map $X \mapsto \Phi_H(X)$, where

$$\Phi_H(X) := \frac{1}{2}\text{Tr}\left[(\exp\{-H + X\} + \exp\{-H - X\})\right] - 1 \qquad (15.99)$$

is shown to obey the axioms of a quantum Young function. Thus [Streater (2008a)], the Luxemburg definition

$$\|Y\|_L := \inf\{r > 0 : \Phi_H(Y/r) < a\} \qquad (15.100)$$

defines a norm on the space of $H$-bounded forms $Y$. We now prove [Streater (2008b)] that the two norms of a form in the overlap of the 'hoods of two states are equivalent, and this is main purpose of the present section.

**Theorem 15.9.** *Let $\rho := \exp -H \in \mathcal{M}$ and let $X$ be a form which is small-enough relative to $H$. Then the Luxembury norms, denoted by $\|Y\|_L$ and $\|Y\|_X$, relative to both $H$ and $H + X + \psi(X)$, are equivalent: there exists a constant $C$ such that*

$$C^{-1}\|Y\|_L \le \|Y\|_X \le C\|Y\|_L$$

*holds for all forms $Y$ that are bounded and small enough relative to both $H$ and $H + X$.*

**Proof.** It is known that two norms are equivalent if and only if they define the same topology on the vector space. Moreover it is enough to prove this at the origin, since the space is a vector space. So it is enough to prove that any convergent net $\{Y_n\}n \in \mathcal{N}$, going to zero relative to one norm, goes to zero relative to the other.

(i) Suppose that $\|Y_n\|_L \to 0$ as $n \to \infty$; then $\|Y_n\|_X \to 0$ as $n \to \infty$. Suppose not. Then there exists a net $Y_n$ such that $\|Y_n\|_L \to 0$ but $\|Y_n\|_X$ does not go to zero. Then there exists $\delta > 0$ and a subnet $Y_{n'}$ such that $\|Y_{n'}\|_L \to 0$ but for all $n'$ we have

$$\|Y_{n'}\|_X \geq \delta.$$

Then either $\|Y_{n'}\|_X$ has a bounded subnet, or it has only unbounded subnets; but the latter is impossible, since it lies in a bounded set, the small ball in $\|\cdot\|_X$. Then, without loss in generality we may assume that there is a net $\{\|Y_n\|_X\}$ that is convergent to a positive limit, $\delta$. By renormalising the terms of the net and considering large enough $n'$ we get a net, say $Z_n$, with $\|Z_n\|_L \to 0$, $\|Z_n\|_X = \delta$ and where $Z_n$ is such that both $H \pm Z_n$ and $H + X \pm Z_n$ are self-adjoint for all $n$. In terms of the Young function, this gives

$$\delta = \|Z_n\|_X = \inf\left\{ r : \Phi_{H+X}\left(\frac{Z_n}{r}\right) < 1 \right\} \qquad (15.101)$$

$$\|Z_n\|_L = \inf\left\{ s : \Phi_H\left(\frac{Z_n}{s}\right) < 1 \right\}. \qquad (15.102)$$

We may choose our $\Phi$ to be lower semi-continuous, which is continuous where it is finite. So for each $n$ the inf is achieved at $r = \delta = \|Z_n\|_X$ and $s = \|Z_n\|_L$. Let us introduce the notation, for any $H$-bounded form $X$:

$$\Psi(X) = \mathrm{Tr}\, \exp\{-H - X\} \qquad (15.103)$$

to give the equations

$$1 = \frac{1}{2}\Psi\left(X + \psi(X) + \frac{Z_n}{\delta}\right) + \frac{1}{2}\Psi\left(X + \psi(X) - \frac{Z_n}{\delta}\right) - 1$$

$$1 = \frac{1}{2}\Psi\left(\frac{Z_n}{\|Z_n\|_L}\right) + \frac{1}{2}\Psi\left(\frac{Z_n}{\|Z_n\|_L}\right) - 1.$$

Therefore for all $n$,

$$4 = \mathrm{Tr}\, \exp\left\{-\left(H + X + \psi(X) + \frac{Z_n}{\delta}\right)\right\} +$$
$$\mathrm{Tr}\, \exp\left\{-\left(H + X + \psi(X) - \frac{Z_n}{\delta}\right)\right\} \qquad (15.104)$$

$$4 = \mathrm{Tr}\, \exp\left\{-\left(H + \frac{Z_n}{\|Z_n\|_L}\right)\right\} +$$
$$\mathrm{Tr}\, \exp - \left\{-\left(H - \frac{Z_n}{\|Z_n\|_L}\right)\right\}. \qquad (15.105)$$

We now show that it is not possible for these equations to hold for any sequence $\{Z_n\}$ with $\|Z_n\|_L \to 0$.

For large $n$ we can ensure that $\frac{\|Z_n\|_L}{\delta} X$ is small enough relative to $\left(1 - \frac{\|Z_n\|_L}{\delta}\right)(H + X)$. We can use the Golden-Thompson inequality,

$$\operatorname{Tr} e^{-A-B} \leq \operatorname{Tr} e^{-A} e^{-B}, \tag{15.106}$$

with

$$A = \frac{\|Z_n\|_L}{\delta}\left(H + \frac{Z_n}{\|Z_n\|_L}\right)$$

$$B = \left(1 - \frac{\|Z_n\|_L}{\delta}\right)\left(H + X + \psi(X) - \frac{\|Z_n\|_L}{\delta - \|Z_n\|_L}(X + \psi(X))\right).$$

Then

$$A + B = H + X + \psi(X) + \frac{Z_n}{\delta}$$

and we get

$$\operatorname{Tr} \exp - \left\{H + X + \psi(X) + \frac{Z_n}{\delta}\right\}$$

$$\leq \operatorname{Tr} \exp\left\{-\frac{\|Z_n\|_L}{\delta}\left(H + \frac{Z_n}{\|Z_n\|_L}\right)\right\} \times$$

$$\exp\left\{-\left(1 - \frac{\|Z_n\|_L}{\delta}\right)\right.$$

$$\left.\left(H + X + \psi(X) - \frac{\|Z_n\|_L}{\delta - \|Z_n\|_L}(X + \psi(X))\right)\right\}.$$

The Hölder inequality for traces states that for positive operators $R, S$, and numbers $p, q \geq 1$ such that $p^{-1} + q^{-1} = 1$, we have

$$\operatorname{Tr} RS \leq \|R\|_p \|S\|_q,$$

where the norm, say, $\|R\|_p = (\operatorname{Tr} R^p)^{1/p}$. Replacing $R$ with $R^{1/p}$ and $S$ with $S^{1/q}$, the Hölder inequality says

$$\operatorname{Tr} R^{1/p} S^{1/q} \leq (\operatorname{Tr} R)^{1/p} (\operatorname{Tr} S)^{1/q}. \tag{15.107}$$

Use this with $p = \delta/\|Z_n\|_L > 1$ and $q = \frac{\delta}{\delta - \|Z_n\|_L}$; we get

$$\operatorname{Tr} \exp - \left\{H + X + \psi(X) + \frac{Z_n}{\delta}\right\} \leq$$

$$\epsilon_n \left(\operatorname{Tr} \exp - \left\{H + \frac{Z_n}{\|Z_n\|_L}\right\}\right)^{\|Z_n\|_L/\delta} \tag{15.108}$$

where
$$\epsilon_n := \left[ \operatorname{Tr} \exp\{-H - X - \psi(X) - \frac{\|Z_n\|_L}{\delta - \|Z_n\|_L}(X + \psi(X))\} \right]^\gamma,$$
where $\gamma = \frac{\delta}{\delta - \|Z_n\|_L}$. Since $\epsilon_n \to \operatorname{Tr} \exp\{-H - X - \psi(X)\} = 1$ as $\|Z_n\|_L \to 0$ and since $\operatorname{Tr} \exp\{-H - Z_n/\|Z_n\|_L\} < 4$, we get, by letting $n \to \infty$, that the left-hand side obeys
$$\operatorname{Tr} \exp\{-H - X - \psi(X) - Z_n/\delta\} \leq 4^{\|Z_n\|_L/\delta} \epsilon_n \to 1 \text{ as } n \to \infty.$$
Similarly,
$$\operatorname{Tr} \exp\{-H - X - \psi(X) + Z_n/\delta\} \to 1 \text{ as } n \to \infty,$$
so the sum converges to 2 at most. This contradicts (15.104). Therefore our assumption, (i), that there exists a net $Y_n$ with $\|Y_n\|_L \to 0$, but $\|Y_n\|_X \nrightarrow 0$, is false, and we have proved that the topology given by $\|Y\|_L$ is stronger than (or equivalent to) the topology given by $\|Y\|_X$.

(ii) We may replace $H$ by $H + X + \psi(X)$ in the above argument, since by Lemma (4) of [Streater (2000)], the state $\exp\{-H + X - \psi(-X)\}$ lies in $\mathcal{M}$. Then the state $\exp -H$ gets replaced by $\exp\{-H - X + \psi(X)\}$ and the same argument with a possibly larger $\beta_0$, but still less than 1, then shows that the topology given by $\|Y\|_X$ is stronger than (or equivalent to) that given by $\|Y\|_L$. Hence the topologies are equivalent, by combining (i) and (ii), and so the norms are equivalent. $\qquad \square$

We can now return to the method of Jaynes; in the 'hood of a state $\rho \in \mathcal{M}$, those states with infinite BKM-norm will not give good estimators, since their variance is infinite. For the states of finite BKM-norm, we can use the proof used in the case of finite dimension to show that the best esimator lies in the exponential manifold generated by the random variables $X_1, \ldots, X_n$ which have been measured.

To summarise: we have proved that the state space is a Banach manifold $\mathcal{M}$ modelled on a Banach space, our Orlicz space, which is flat under the $(+1)$-connection. It is not trivial, however, since there are no coordinates around $\rho = e^{-H}$ that cover the whole space $\mathcal{M}$, or even the part connected to $\rho$. Thus, the set of states with coordinates $X$ near $\rho$ is the Orlicz class, not the whole Banach space; the latter consists of all $X$ with finite norm. The manifold $\mathcal{M}$ can nevertheless be proved to be convex [Streater (2000)]. We note that $\Phi$ does not satisfy the $\Delta_2$-condition, and so leads to a non-separable Banach manifold. Pistone has advocated a smaller, separable manifold in the classical case. This is the completion of the local algebra of bounded measurable functons, $L^\infty$, in the Orlicz norm, instead of allowing all small-enough $H_0$-bounded form perturbations. Then, Grasselli [Grasselli (1999)] has shown the equivalence of the norms based on two nearby points of the manifold. In the quantum case, this idea of Pistone has been worked out by Jenčová [Jenčová (2006)] for a general $W^*$-algebra $\mathcal{A}$ using the relative entropy of Araki [Araki (1976)] in place of $\operatorname{Tr}\exp\{-H-X\}$ in the Young function. She gets the completion of $\mathcal{A}$ as the Orlicz space modelling the tangent space. When the algebra is $\mathcal{B}(\mathcal{H})$, the relative entropy and the logarithm of our $\operatorname{Tr}\exp\{-H-X\}$ are equal when $X$ is bounded (15.79). We would expect, however, that in quantum mechanics, the smaller manifold of Pistone, Grasselli and Jenčová would not be enough even for non-relativistic estimation theory. It might contain, or be equal to, the space of states that are obtained by perturbing $H$ with the set of $H$-infinitesimal forms; this means that the Kato seminorm of the perturbation is zero. This is clearly a small subset of our choice, and we prefer to extend the space from $\mathcal{A}$ to all $H$-small-enough forms, as far as is practicable at present.

## 15.14 Exercises

Exercise 15.109. Show that the state of maximum entropy, subject to having mean energy $\overline{E}$ and a given variance $V$ ot the energy, is of the form

$$\rho = Z^{-1}\exp\{-\beta H - \beta'(H-\overline{E})^2\} \qquad (15.110)$$

where $\beta'$ and $Z$ are determined by $\operatorname{Tr}\rho_{\beta,\beta'} = 1$ and $\rho(H-\overline{E})^2 = V$ .

Exercise 15.111. Find the general form of the state of three spin $1/2$ atoms which maximises the entropy subject to fixed values of the three mean spins $S_1$, $S_2$, $S_3$, and of the three Ising bond energies $-JS_1S_2$, $-JS_2S_3$ and $-JS_3S_1$.

Exercise 15.112. Derive (15.10) from (15.9) in the approximation where $\mu_x$ and $\beta - \beta_x$ are small.

Exercise 15.113. Show that

$$Z = 2e^{3\beta J} + 6e^{-\beta J}$$

for the dynamic Ising model at equilibrium.

Exercise 15.114. Solve (15.11), (15.12) and (15.13) for $\mu_x$ to obtain (15.14), and write (15.10) in the form of (15.15).

Exercise 15.115. Let

$$\rho_\xi(\omega) = Z^{-1} \exp\left\{ -\sum_j \xi^j f_j(\omega) \right\}$$

be an exponential state. Show that if $\Psi := \log Z$ then

$$\eta_j = -\frac{\partial \Psi}{\partial \xi^j}, \qquad j = 1, \ldots, n$$

and

$$V_{jk} = -\frac{\partial \eta_j}{\partial \xi^k}, \qquad j, k = 1, \ldots, n.$$

Show that the Legendre dual to $\Psi$, namely, $\sup_x i \left\{ \sum_i \xi^i \eta_i - \Psi \right\}$, is equal to the entropy of $\rho$, and obtain the dual relations Eq. (15.42).

# Bibliography

Alberti, P. and Uhlmann, A., **Stochasticity and Partial Order**, VEG Deutscher Verlag, Berlin, 1981

Alicki R. and Messer, J., *Nonlinear Quantum Dynamical Semigroups for Many-body Open Systems*, Journal of Statistical Physics, **32**, 299-312, 1983. See also Alicki, R., and Lendl, K. **Quantum Dynamical Semigroups and Applications**, Lecture Notes in Physics, **286**, Springer, 1987

Al-Rashid, M. H. A., and Zegarlinski, B., Non-commutative Orlicz spaces associated with a state, *to appear in Studia Mathematica*

Amari, S.-I. **Differential Geometric Methods in Statistics**, Lecture Notes in Statistics, **28**, Springer-Verlag, New York, 1985

Amit, D., **Modelling Brain Function**, Cambridge University Press, 1989

Amit, D. and Verbin, Y., **Statistical Physics: an introductory course**, World Scientific, 1999

Araki, H., *Hamiltonian formalism in quantum field theory*, Journal of Mathematical Physics, **1**, 492-, 1960

Araki, H., *Expansional in Banach Algebra*, Ann. sci. école normale sup. **46**, 67-84, 1973

Araki, H., *Relative entropy for states of von Neumann algebras* Publ. RIMS Kyoto Univ. **11**, 809-833, 1976

Araki H. and Woods, E. J., *Representations of the Canonical Commutation Relations Describing a Nonrelativistic Free Bose Gas*, Journal of Mathematical Physics, **4**, 637-662, 1963

Araki, H. and Wyss, W., *Representations of Canonical Anticommutation Relations*, Helvetica Physica Acta, **37**, 136-159, 1964

Arrhenius, S., Z. Phys. Chem. (Leipzig), **4**, 226-, 1889

Aspect, A., Grangier, P. and Roger, G., *Experimental realization of EPR-Bohm gedankenexperiment. A new violation of Bell's inequalities* Phys. Rev. Letters, **49**, 91-94, 1982

Atkins, P. W.,**The Elements of Physical Chemistry**, Oxford University Press, third ed., 2001

Atkins, P. W. and de Paula, J., **Physical Chemistry**, 7$^{\text{th}}$ edition, Oxford University Press, 2006.

Bailyn, M., **A Survey of Thermodynamics**, Amer. Institute of Physics, 1995

Balescu, R., **Equilibrium and Nonequilibrium Statistical Mechanics**, John Wiley, 1975

Balescu, R., **Statistical Dynamics**, World Scientific, 1997

Balian, R., **From Microphysics to Macrophysics: Methods and Applications of Statistical Physics** vol. 1, study ed., Springer-Verlag, 2006

Balian, R., **From Microphysics to Macrophysics: Methods and Applications**, vol. 2, study ed., Springer-Verlag, 2006

Balslev, E., Manuceau, J. and Verbeure, A., *Representations of Anti-commutation Relations and Bogoliubov Transformations*, Communications in Mathematical Physics, **8**, 313-326, 1968

Balslev, E. and Verbeure, A., *States on Clifford Algebras*, Communications in Mathematical Physics, **7**, 55-76, 1968

Bapat, R. B. and Raghavan, T. E. S., **Nonnegative Matrices and Applications**, Encyclopedia of Mathematics and its Applications, University of Cambridge Press, 1997

Birkhoff, G.D., Univ. Nac. Tucuman Rev. Ser. **A5**, 147, 1946. See: Ando, T. *Majorization, Doubly Stochastic Matrices and Comparison of Eigenvalues*, Sopporo, 1982; Linear Algebra and its Applications, **118**, 163-248, 1983

Bratteli, O. and Robinson, D. W., **Operator Algebras and Quantum Statistical Mechanics**, Vols. I and II, Springer-Verlag, 1979, 1981

Broadwell, J. E., *Study of rarified shear flow by discrete velocity method*, Journal of Fluid Mechanics, **7**, 401-414, 1964

Chapman, S. and Cowling, T. G., **The Mathematical Theory of Nonuniform Gases**, 3$^{rd}$ edition, Cambridge University Press, 1970

Chentsov, N. N., **Statistical Decision Rules and Optimal Inference**, Translations of Mathematical Monographs, Amer. Math. Soc., Providence, R. I., **58**, 1982

Chentsov, N. N., and Morozova, E. A., *Markov maps in non-commutative probability theory and mathematical statistics*, **Probability Theory and Mathematical Statistics**, Vol. II, Vilnius, 1985, 287-310, VNU Sci Press, Utrecht, 1987

Choi, M. D., *Positive Linear Maps on $C^*$-algebras*, Canadian Journal of Mathematics, **24**, 520-529, 1972; *Completely Positive Linear Maps on Complex Matrices*, Linear Algebra and Applications, **10**, 285-290, 1975

Clarke, B. L., *Stability of Complex Reaction Networks*, Advances in Chemical Physics, **43**, 1-216, 1980

Collie, C. H., **Kinetic Theory and Entropy**, Longman Higher Education, 1982

Connaughton, C., Ph. D. thesis, King's College London, 1998

Cramér, H., **Mathematical Methods of Statistics**, nineteenth printing, Princeton University Press, 1999

Davies, E. B., **Quantum Theory of Open Systems**, Academic Press, 1976

Davies, E. B., *A Model of Heat Conduction*, Journal of Statistical Physics, **18**, 161-170, 1978

Davies, E.B., **Science in the Looking Glass**, Oxford University Press, 2003

Dawid, A., *Discussion of a paper by Bradley Efron*, Annals of Statistics, **3**, 1231-

1234, 1975; *Further comments on a paper by Bradley Efron*, ibid, **5**, 1249, 1977

De Roeck, W., Jacobs, T., Maes, C. and Netočný, K., *An extension of the Kac ring model*, J. Phys. A, **36**, 11547-11559, 2003

Duffield, N. G., *The Continuum Limit of Dissipative Dynamics in H. Fröhlich's Pumped Phonon System*, Helvetica Physica Acta, **61**, 363-378, 1988

Efron, B., *Defining the curvature of a statistical problem*, Annals of Statistics, **3**, 1189-1242, 1975. See also *The geometry of exponential families, ibid*, **5**, 457-458, 1977

Einstein, A., *Zur Quantentheorie der Strahlung*, Physikalische Zeitschrift, **18**, 121-, 1917. Translated into English in: van der Waerden, B. L., **Sources of Quantum Mechanics**, Dover, New York, 28-32, 1969

Einstein, A., Podolsky, P. and Rosen, N., *Can quantum-mechanical description of reality be considered complete?* Physical Review, **47**, 777-780, 1935

Ekeland, I. and Temam, R., **Convex Analysis and Variational Problems**, North Holland, 1976

Eyring, H., Lin, S. H. and Lin, S. M., **Basic Chemical Kinetics**, Wiley, New York, 1980

Fisher, R. A., *Theory of Statistical Estimation*, Proceedings of the Cambridge Philosophical Society, **22**, 700-725, 1925

Fowler, R. H., **Statistical Mechanics**, 2nd Ed., Cambridge, 1936.

Freidlin, M. J. and Wentzell, A. D., **Random Perturbations of a Dynamical System**, Springer-Verlag, 1984

Fröhlich, H., *Long-range Correlations and Energy Storage in Biological Systems*, International Journal of Quantum Chemistry, **2**, 641-649, 1968

Fujita, H. and Kato, T., *on the Navier-Stokes initial value problem I*, Arch. Rational Mech. and Analysis, **16**, 269-315, 1964

Fujiwara, A., and Nagaoka, H., *Quantum Fisher metric and estimation for pure state models*, Physics Letters A, **201**, 119-124, 1995

Gibilisco, P., and Isola, T., *Connections on statistical manifolds of density operators by geometry of non-commutative $L^p$-spaces*, Infinite Dim. Anal., Quantum Prob., and Related Topics, **2**, 169-, 1999

Gibilisco, P. and Pistone, G., *Connections on Nonparametric Statistical Manifolds by Orlicz Space Geometry*, Infinite Dimensional Analysis, Quantum Probability and Related Topics, **1**, 325-347, 1998

Glauber, R. J., *Time-dependent Statistics for the Ising Model*, Journal of Mathematical Physics, **4**, 294 (1963)

Gorini, V., Kossakowski, A. and Sudarshan, E. C. G., *Completely Positive Dynamical Semigroups of N-level Systems*, Journal of Mathematical Physics, **17**, 821-825, 1976

Grandy, W. T., *The origins of entropy and irreversibility*, Open Systems and Information Dynamics, **1**, 183-196, 1992

Grasselli, M. R., *Dual Connections in Nonparametric Classical Information Geometry*, Ann. Inst. Statist. Math., in press; ArXiv. math-ph/9910031

Grasselli, M., and Streater, R. F., *Quantum information manifolds for epsilon-bounded forms*, Reports on Mathematical Physics, **46**, 325-335, 2000

Grasselli, M. R., and Streater, R. F., *Uniqueness of the Chentsov metric in quantum information theory*, Infinite-dim. Analysis Quantum Prob. and Related Topics, **4**, 173-182, 2001

Griffiths, R. B., *Peierls' Proof of Spontaneous Magnetization in a Two-dimensional Ising Ferromagnet*, Physical Review A, **136**, 437-439, 1964. See also: Griffiths, R. B., *Rigorous Results and Theorems*, in: **Phase Transitions and Critical Phenomena**, Editors: Domb, C. and Green, M. S., Vol. 1, Academic Press, 1972

Guldberg, C. M., and Waage, P., Forh. Vid. Selsk. Christiania, **35**, 92-, 111-, 1864. See also: **Etudes sur les affinité chimiques**, Brøgger and Christie, Christiania, Oslo, 1867; J. Prakt. Chem., **19**, 69-, 1879

Gzyl, H., **The method of maximum entropy**, World Scientific, 1995

Haag, R., *On quantum field theory*, Mat.-Fys. Medd. Danske Videns. Selskapet, **29**, No.12, 1-37, 1955

Haag, R., **Local Quantum Physics**, Springer-Verlag, 1993?

Haag, R., Hugenholtz, N. M., and Winnink, M., *On the Equilibrium States in Quantum Statistical Mechanics*, Communications in Mathematical Physics **5**, 215-236, 1967

Hasegawa, H. and Petz, D., *On the Riemannian metric of $\alpha$-entropies of density matrices*, Letters in Mathematical Physics, **38**, 221-225, 1996

Helstrom, C. W., **Quantum Detection and Estimation Theory**, Academic Press, 1976

Hiai, F., *Trace-norm convergence of exponential product formula*, Letters in Mathematical Physics, **33**, 147-158, 1995

Hiriart-Urruty, J.-P. and Lermaréchal, **Convex Analysis and Minimisation Algorithms I**, Springer-Verlag, 1993

Ho, T. G., Landau, L. J., and Wilkins, A. J., *On the Weak Coupling Limit of a Fermi Gas in a Random Potential*, Reviews in Mathematical Physics, **5**, 209-298, 1993

Holevo, A. S., **Probabilistic and Statistical Aspects of Quantum Theory**, North Holland, Amsterdam, 1982

Horn, F., *Necessary and Sufficient Conditions for Complex Balancing in Chemical Kinetics*, Archive for Rational Mechanics and Analysis **49**, 172-186, 1972-73

Horn, F. and Jackson, R., *General Mass Action Kinetics*, Archive for Rational Mechanics and Analysis, **49**, 81, 1972

Ingarden, R. S., *Information theory and variational principles in statistical theories*, Ser. Math-Astro-Phys. **11**, 541-, 1963

Ingarden, R. S., *Towards mesoscopic thermodynamics: small systems in higher-order states*, Open Systems and Information Dynamics, **1**, 75-102, 1992

Ingarden, R. S., Sato, Y., Sugawa, K., and Kawaguchi, M., *Information thermo-dynamics and differential geometry*, Tensor **33**, 347-, 1979

Ingarden, R. S., Janysek, H., Kossakowski, A., and Kawaguchi, T., *Information geometry of quantum mechanical systems*, Tensor, **37**, 105-111, 1982

Jaynes, E. T., *Prior Probabilities*, IEEE Transactions on Systems Science and Cybernetics, SSC-4, 227-241, 1968; *Information Theory and Statistical Mechanics I, II*, Physical Review **106**, 620-; **108**, 171-190, 1957

Jenčová, A., Affine connections, duality and divergences for a von Neumann algebra, arXiv/math-ph/0311004, 2003

Jenčová, A., *A construction of a non-parametric quantum information manifold*, Journal of Functional Analysis, **239**, 1-20, 2006

Jordan, P. and Wigner, E. P., *Über das Paulische Äquivalenzverbot*, Zur Physik, **47**, 631-651, 1928

Kac, M., in **Proceedings of the Third Berkeley Symposium on Mathematical Statistics and Probability**, Berkeley, 1955

Kac, M., Ch 14, **Probability and Related Topics in Phys Sci**, Interscience, 1959

Kato, T., **Perturbation Theory for Linear Operators**, Springer-Verlag, 1966

Kato, T., *Trotter's product-formula for an arbitrary pair of self-adjoint contraction semi-groups*, Advances in Math. Supplement Stud., **3**, 185-195, Academic Press, New York, 1978

Katz, A., **Principles of Statistical Mechanics; The Information Theory Approach**, W. H. Freeman and Company, 1967

Keizer, J., **Statistical Thermodynamics of Nonequilibrium Processes**, Springer, New York, 1987

Kendal, D. G., *On infinite doubly-stochastic matrices and Birkhoff's problem*, Journal London Math. Society **35**, 81-84, 1960

Kiegerl, G. and Schürrer, F., *Energy Conservation and H-Theorem in the Scalar Non-linear Equation and its Multigroup Representation*, Physics Letters A, **148**, 158-163, 1990

Kifer, Y., **Random Perturbations of Dynamical Systems**, Birkhaüser, Boston, 1988

Kingman, J. F. C., Journal of Applied Probability, **6**, 1-18, 1969

Klauder, J. R., Ph. D. thesis, Princeton, 1959

Koseki, S., *Isothermal Statistical Dynamics*, Ph. D. Thesis, King's College, 1993

Koseki S. and Streater, R. F. *The statistical dynamics of activity-led reactions*, Open Systems and Information Dynamics, **2**, 77-94, 1993

Kossakowski, *On the quantum informational thermodynamics*, Bulletin de l'Academie Polonaise des Sciences, **17**, 263-267, 1969

Krasnoselski M. A. and Ruticki, Ya. B., **Convex Functions and Orlicz Spaces**, P. Noordhoff, 1961

Kraus, K., *General State Changes in Quantum Theory*, Annals of Physics, **64**, 311-335, 1971. See also: Kraus, K., **States, Effects and Operations**, Lecture Notes in Physics, **190**, Springer-Verlag, Berlin, 1983

Krylov, N. S., **Works on the Foundation of Statistical Physics**, Princeton University Press, 1979

Kubo, R., *Statistical Mechanical Theory of Irreversible Processes*, Journal of the Physical Society of Japan **12**, 570-586, 1957

Kümmerer B. and Maassen, H., *The Essentially Commutative Dilations of Dynamical Semigroups on* $\mathbf{M}_n$, Communications in Mathematical Physics, **109**, 1-, 1987

Kunze R. A., and Segal, I. E., **Integrals and Operators**, Springer, 1978

Kunze, W., Non-commutative Orlicz spaces and generalized Arens algebras, *Math. Nachr.*, **147**, 123-, 1990

Laidler, K. J., **Chemical Kinetics**, $3^{rd}$ ed., Harper-Collins, New York, 1987

Landau, L. J., *On the violation of Bell's inequality in quantum theory*, Physics Letters, **A120**, 54-56, 1987

Landau, L. J., *Observation of Quantum Particles on a Large Space-time Scale*, Journal of Statistical Physics, **77**, 259-310, 1994

Landau L. J., and Streater, R. F., *On Birkhoff's Theorem for Doubly Stochastic Completely Positive Maps of Matrix Algebras*, Linear Algebra and its Applications, **193**, 107-127, 1993

Lanford O. E. and Robinson, D. W., *Approach to Equilibrium of Free Quantum Systems*, Communications in Mathematical Physics, **24**, 193-219, 1972

Lebowitz, J. L., and Maes, C., *Entropy - a dialogue*, pp. 269-276, in **Entropy**, eds. Greven, A., Keller, G. and Warnecke, G., Princeton Univ. Press, 2003

Leppington, F., private communication.

Leray, J., *Essai sur le mouvement d'un liquide visceux emplissant l'espace*, Acta Math. **63**, 193-248, 1934 and references therein.

Lindblad, G., *Completely Positive Maps and Entropy Inequalities*, Communications in Mathematical Physics, **40**, 147-151, 1975

Lindblad, G., *On the Generators of Quantum Dynamical Semigroups*, Communications in Mathematical Physics, **48**, 119-130, 1976

Lions, P. L., **Mathematical Topics in Fluid Mechanics**, Vols. 1, 2; Clarendon Press, Oxford, 1996, 1998

Majewski, W. A., *The Detailed Balance Condition in Quantum Statistical Mechanics*, Journal of Mathematical Physics, **25**, 614-, 1984. See also: Majewski, W. A., *Dynamical Semigroups in the Algebraic Formulation of Statistical Mechanics*, Fortschritte der Physik, **32**, 89-, 1984

Majewski, W. A., and Streater, R. F., *Detailed Balance and Quantum Dynamical Maps*, Journal of Physics, **A31**, 7981-7995, 1998

Manuceau, J. and Verbeure, A., *Quasifree states of the CCR Algebra and Bogoliubov Transformations*, Communications in Mathematical Physics, **9**, 293-302, 1968

Marcinkiewicz, J., Mathematisches Zeitschrift, **44**, 612-618, 1939

Martin, P. C. and Schwinger, J., *Theory of many-particle systems: I*, Physical Review **115**, 1352-1373, 1959

McKean, H. P., *Propagation of Chaos for a Class of Nonlinear Parabolic Equations*, in: Lecture Series in Differential Equations II, A. K. Aziz (Ed.), 177, Van Nostrand Reinhold, 1969

Meyer, D. A. *Quantum strategies*, Physical Review Letters, **82**, 1052-1055, 1999

Monaco R. and Preziosi, L., **Fluid Dynamic Applications of the Discrete Boltzmann Equation**, World Scientific, 1991. See also Bellomo, N., Palczewski, A. and Toscani, G., **Mathematical Topics in Nonlinear Kinetic Theory**, World Scientific, 1988

Moshinsky, M. and Winternitz, P., *Quadratic Hamiltonians in Phase Space and their Eigenstates*, Journal of Mathematical Physics, **21**, 1667-1681, 1980

Neveu, J., **Mathematical Foundations of the Calculus of Probability**, Holden-Day, San Francisco, 1965

Ohya M. and Petz, D., **Quantum Entropy and its Use**, Springer-Verlag, 1993

Omnès, R., **Understanding Quantum Mechanics**, Princeton University Press, 1999.

Omnès, R. **Converging Realities: Towards a Common Philosophy of Physics and Mathematics**, Princeton University Press and Oxford, 2005

Ostwald, F. W., J. Prakt. Chem. **29**, 385-, 1884

Peierls, R. E., *On Ising's Model of Ferromagnetism*, Proceedings of the Cambridge Philosophical Society, **32**, 477-481, 1936

Penrose, O., **Foundations of Statistical Mechanics**, Pergamon, 1970

Petz, D., *Monotone metrics on matrix spaces*, Linear Algebra and Applications, **244**, 81-96, 1996

Petz, D., *Covariance and Fisher information in quantum mechanics*, Journal of Physics A, **35**, 929-939, 2002

Petz, D., and Toth, G., *The Bogoliubov inner product in quantum statistics*, Letters in Mathematical Physics, **27**, 205-216, 1993

Pistone, G., and Sempi, *An infinite-dimensional geometric structure on the space of all probability measures equivalent to a given one*, Annals of Statistics, **33**, 1543-1561, 1995

Platt, O., **Ordinary Differential Equations**, Holden-Day, San Francisco, 1971

Ponce, G., Racke, R., Sideris, T., and Titi, E. S., *Global stability of large solutions to the 3D Navier-Stokes equations* Communications in Mathematical Physics, **159**, 329-341, 1994.

Pusz, W. and Woronowicz, S. L., *Passive States and KMS States for General Quantum Systems*, Communications in Mathematical Physics, **58**, 273-290, 1978

Rao, C. R., *Information and accuracy attainable in the estimation of statistical parameters*, Bulletin Calcutta Mathematics Society, **37**, 81-91, 1945

Rao, M. M. and Ren, Z. D., **Theory of Orlicz Spaces**, Marcel Decker, 1992

Renyi, A., **The Theory of Probability**, North-Holland, 1970

Revuz, D., **Markov Chains**, North-Holland, 1975

Riesz, F. and St.-Nagy, B., **Functional Analysis**, Frederick Ungar Publishing, New York, 1955

Rivet, J.-P. and Boon, J. P., Lattice Gas Hydrodynamics, Cambridge University Press, 2001

Robinson, D. W., *A Theorem Concerning the Positive Metric*, Communications in Mathematical Physics, **1**, 89-94, 1965

Robinson, D. W., *The Ground State of the Bose Gas*, Communications in Mathematical Physics, **1**, 159-174, 1965

Robinson, D. W., *The Statistical Mechanics of Quantum Spin Systems*, Communications in Mathematical Physics **7**, 337-348, 1968

Rondoni, L., *A Stochastic Treatment of Reaction and Diffusion*, Ph.D. thesis, Virginia Tech., 1991

Rondoni, L., *Complex Chemical Reactions: a Probabilistic Approach*, Nuclear Science and Engineering, **112**, 392-405, 1992. See also: McQuarrie D. A.,

*A stochastic Approach to Chemical Kinetics*, Supplement Review Series in Applied Probability, **8**, Methuen, 1967; Oppenheim, I., Shuler, K. E. and Weiss, G. H., *Stochastic Processes in Chemical Physics: the Master Equation*, MIT Press, 1977.

Rondoni, L., *Long Time Behaviour of "X-led" Reactions*, Journal of Mathematical Physics, **35**, 1778-1795, 1994

Rondoni, L. and Streater, R. F., *The Brussellator with Thermal Noise*, Proceedings of the XXVI Symposium on Mathematical Physics, Torun, 1993. pp 59-64, Ed. by Jamiolkowski, A., Majewski, W. A., Michalski, M. and Mrugala, R., Nicolas Copernicus University Press, 1994

Rondoni, L. and Streater, R. F., *The Statistical Dynamics of the Brussellator*, Open Systems and Information Dynamics, **2**, 175-194, Torun, 1994

Ruelle, D., *Statistical Mechanics: Rigorous Results*, W. A. Benjamin, New York, 1969

Ruskai, M. B., Commun. Math. Physics, **26**, 280-289, 1972

Safarov, Y., *Birkhoff's theorem and multidimensional numerical range*, Journal of Functional Analysis **222**, 61-97, 2005

Segal, I. E., *Foundations of the Theory of Infinite-dimensional Systems, II*, Canadian Journal Of Mathematics, **13**, 1-18, 1961

Segal, I. E., *Mathematical Problems of Relativistic Physics*, American Mathematical Society, Providence, 1963

Sewell, G., *Quantum Theory of Collective Phenomena*, Oxford University Press, 1986.

Shale, D. and Stinespring, W., *States on the Clifford Algebra*, Annals of Mathematics, **80**, 365-, 1964

Shear, D. B., *Stability and uniqueness of the equilibrium point in chemical reaction systems*, J. Chemical Physics, **48**, 4144-4147, 1968

Simon, B. *The Statistical Mechanics of Lattice Gases I*, Princeton University Press, 1993

Simon, B., **Trace Ideals and their Applications**, second edition, Mathematical Surveys and Monographs, vol. 120, American Math Society, 2005

Spohn, H., **Large Scale Dynamics of Interacting Particle Systems**, Springer-Verlag, 1991

Spohn, H. and Lebowitz, J. L., *Irreversible Thermodynamics for Quantum Systems Weakly Coupled to Thermal Reservoirs*, Advances in Chemical Physics, **38**, 127-, 1977

Stinespring, W. F., *Positive Functions on $C^*$-algebras*, Proceedings of the American Mathematical Society, **6**, 211-216, 1955. If you can find it, see also the beautiful pamphlet, Evans, D. E. and Lewis, J. T., *Dilations of irreversible evolutions in algebraic quantum theory*, Communications of the Dublin Institute for Advanced Studies, Series A, No. 24, 1977

Stokes, G. G., *On the effect of the internal friction of fluids ...*, Trans. Cambridge Philosophical Soc., **9**, 24-62, 1856

Størmer, E., *Positive Linear Maps of Operator Algebras*, Acta Mathematica, **110**, 233-278, 1963. See also: Størmer, E., *Positive Linear Maps of $C^*$-algebras*, Lecture Notes in Physics, **29**, Springer-Verlag, 1974

Streater, R. F., *Spontaneous Breakdown of Symmetry in Axiomatic Theory*, Proc. Royal Society, **278 A**, 510-518, 1965

Streater, R. F., *The Heisenberg Ferromagnet as a Quantum Field Theory*, Communications in Mathematical Physics, **6**, 233-247, 1967

Streater, R. F., *Convergence of the Iterated Boltzmann Map*, Publications of the Research Institute in Mathematical Sciences, Kyoto, **20**, 913-927, 1984

Streater, R. F., *Convergence of the Quantum Boltzmann Map*, Communications in Mathematical Physics, **98**, 177-185, 1985

Streater, R. F., *Entropy and the Central Limit Theorem in Quantum Mechanics*, Journal of Physics **A 20**, 4321-4330, 1987

Streater, R. F., *Symmetry Groups and Non-abelian Cohomology*, Communications in Mathematical Physics, **132**, 201-215, 1990

Streater, R. F., *The F-Theorem for Stochastic Models,* Annals of Physics, **218**, 255-278,, 1992

Streater, R. F., *Heat Conduction in a Fermionic Crystal*, Journal of Physics A: Math. Gen. **26**, 1553-1557, 1993

Streater, R. F., *The Free-energy Theorem*, pp. 137-147 of **Quantum and Noncommutative Analysis**, Ed. Araki, H., Ito, K. R., Kishimoto, A. and Ojima, I., Kluwer, 1993

Streater, R. F.,*Detailed Balance and Free Energy,* Proceedings of the Conference on Quantum Information Theory and Open Systems, Kyoto; Ed. Ohya, M. et al., RIMS Kokyuroku Series, Kyoto, 1993

Streater, R. F., *Stochastic Models of Cotransport*, Transport Theory and Statistical Physics, **22**, 1-37, 1993

Streater, R. F., *Statistical Dynamics with Sources and Sinks*, pp 15-20 in: **Quantum Chemistry and Technology in the Mesoscopic Level**, Ed. Hasegawa, H., Physical Society of Japan, 1994

Streater, R. F., *Convection in a Gravitational Field,* Journal of Statistical Physics, **77**, 441-448, 1994

Streater, R. F., *Entropy and KMS Norms*, pp. 405-413, in **Stochastic Analysis and Applications in Physics**, Eds. Cardoso, A. I., de Faria, M., Potthoff, J., Sénéor, R. and Streit, L., Kluwer, 1994

Streater, R. F., *Statistical Dynamics with Thermal Noise*, Proc. Edinburgh Conference on Stochastic Processes, Edited by Etheridge, A., Cambridge University Press, 1995

Streater, R. F., The information manifold for relatively bounded potentials, in **Problems of the Modern Mathematical Physics**, Eds. Vladimirov, V. S., Slavnov, A. A., and Katanaev, M. O., *Proc. V. A. Steklov Institute Math.*, **228**, 205-223, 2000

Streater, R. F., *Classical and quantum probability*, Jour. of Mathematical Physics, **41**, 2000

Streater, R. F., *Corrections to fluid dynamics*, Open Systems and Information Dynamics, **10**, 3-30, 2003

Streater, R. F., *Quantum Orlicz spaces in information geometry*, Open Systems and Information Dynamics, **11**, 350-375, 2004

Streater, R. F., **Lost Causes in and beyond Physics**, Springer-Verlag, 2007

Streater R. F., The set of density operators modelled on a quantum Orlicz space, Lecture given at Nottingham, July 2006. Pages 99-109 of **Quantum Stochastics and Information**, Statistics, Filtering and Control. Eds. Belavkin, V. P. and Gutǎ, M., World Scientific, 2008

Streater, R. F., *The Banach manifold of quantum states*, in **Algebraic and Geometric Methods in Statistics**, Eds. Gibilisco, P., Riccomagno, E., Rogantin, M.-P., and Wynn, H. P., Cambridge University Press, 2008

Streater, R. F., and Wightman, A. S., **PCT, Spin and Statistics and All That**, Benjamin-Cummings, 1964. Latest ed: Princeton University Press, 2000

Thirring, W., **Quantum Mechanics of Large Systems**, Springer-Verlag, 1980

Tolman, R. C., **Foundations of Statistical Mechanics**, Oxford, 1938.

Tregub, S. L., Soviet Math. **30**, 105-, 1986

Tsirelson B. S., *Quantum analogues of Bell's inequalities* (Russian), Zap. Nauchn. Sem. Leningrad Odtel. Mat. Inst. Steklov (LOMI), **142**, 174-194, 1985

Uehling, E. A. and Uhlenbeck, G. E., *Transport Phenomena in Bose-Einstein and Fermi-Dirac Gas*, Physical Review, **43**, 552-561, 1933. For a more modern version, see Balescu, R., above.

van Hove, L., *Quantum Mechanical Perturbations giving rise to a Statistical Transport Equation*, Physica, **21**, 517-540, 1955

van t'Hoff, J. H., **Etudes de dynamique chimique**, Muller, Amsterdam, 1884

von Neumann, J., **Mathematical Foundations of Quantum Mechanics**, Princeton University Press, 1955, translated by Robert T. Beyer from **Mathematische Grundlagen der Quantenmechanik**, 1932

Wehrl, A., Reports on Mathematical Physics, **16**, 353-, 1979

Wehrl, A., Reviews of Modern Physics, **50**, 221-, 1987

Wichmann, E. H., *Density Matrices arising from Incomplete Measurements*, Journal of Mathematical Physics, **4**, 884-896, 1963

Wightman, A. S., *Quantum field theory in terms of vacuum expectation values*, Physical Review, **101**, 860-, 1956

Wightman, A. S., Appendix, in Israel, R., **Convexity in the Theory of Lattice Gases**, Princeton University Press, 1979

Wigner, E. P. and Yanase, M. M., *Information content of distributions*, Proc. Nat. Acad. Sci. USA, **49**, 910-918, 1963

Wilhelmy, L., Pogg. Ann., **81**, 413-, 499-, 1850

Woyszynski, W. A. and Saichev, A. I., **Distributions in the Physical and Engineering Sciences**, Vol. 1, Birkhauser Verlag, 1996

Yau, H. T., Lecture at Conference of the International Assoc. of Mathematical Physics, Imperial College, 2000 (unpublished)

# Index